Historical Geology

ROBERT J. FOSTER

Macmillan Publishing Company
New York
Collier Macmillan Canada, Inc.
Toronto
Maxwell Macmillan International Publishing Group
New York Oxford Singapore Sydney

Senior Editor: Robert McConnin
Production Editor: Rex Davidson
Art Coordinator: Lorry Woost
Cover Designer: Brian Deep

Cover photograph © Frank Balthis.

This book was set in Meridien.

Printed in the United States of America

Macmillan Publishing Company
866 Third Avenue, New York, New York 10022

Collier Macmillan Canada, Inc.

Library of Congress Cataloging-in-Publication Data

Foster, Robert J. (Robert John), 1929 Apr. 19–
Historical geology / Robert J. Foster.
p. cm.
"Rewritten version of part of the 5th edition of General geology"–
–Galley pref.
Includes bibliographical references and index.
ISBN 0-675-21355-X
1. Historical geology. I. Foster, Robert J. (Robert John), 1929
Apr. 19– General geology. II. Title.
QE28.3.F67 1991
551.7—dc20 90-43649
CIP

Printing: 1 2 3 4 5 6 7 8 9 Year: 1 2 3 4

Preface

Physical geology is the only geology course that most students ever take. They miss getting a glimpse into the immensity of geologic time, the development of our planet, and the evolution of its life. Currently there is great concern for the future of the earth because of pollution and other environmental issues, ranging from the extinction of organisms to the heating caused by the greenhouse effect. It is difficult to put such issues into perspective without understanding how the earth works and what has happened to it in the past. This text accomplishes that goal.

The reason few general education students take historical geology is obvious to every student and professor—there is little extra time in the crowded curriculum. Perhaps a short manageable historical text will help. The precedent is clear; for more than 20 years the short but comprehensive books of the *Earth Science Series* have been used in a variety of courses never dreamed of when they were first published.

This text can be used in a traditional historical geology course. It can be used with any physical geology text to help cover the essential elements of historical geology. It can stand alone as a first course in geology. It can add depth to an earth science course, a physical science course, or a biology course.

LEARNING AIDS

As new terms are introduced, they appear in boldface type and their definitions are given in italics. At the end of each chapter a summary recaps the main points. There are two types of questions: some test the students' understanding of the chapter and are factual and straightforward; others encourage the students to apply what they have learned and lead to ideas that go beyond the content of the text. The supplementary readings for each chapter are mainly articles in current periodicals, and it is hoped they will stimulate outside reading. A comprehensive glossary and appendices on rocks, geologic and topographic maps, and plates of fossil-forming organisms should aid the student as well as add flexibility in using the text.

ACKNOWLEDGMENTS

This book is an extensively rewritten version of part of the fifth edition of *General Geology,* and my debt to the many people who aided me over the years with that book is immense. My wife, Joan, helped in many ways, including editing, proofing, and indexing. The experienced professionals of the Macmillan staff were enormously helpful. I am also grateful to the following colleagues who reviewed the entire text or various chapters and offered many helpful suggestions: Richard L. Bowen, University of Southern Mississippi; Douglas H. Erwin, Michigan State University; Bryan Gregor, Wright State University, OH; Dennis Hibbert, North Seattle Community College; William S. McLoda, Mountain View College, TX; William N. Mode, University of Wisconsin–Oshkosh; Michael E. Nelson, Fort Hays State University, KS; Darlene S. Richardson, Indiana University of Pennsylvania; and Richard E. Thoms, Portland State University.

Robert J. Foster

Contents

1
How the Earth Works

WHY STUDY EARTH HISTORY?

Geology has contributed a great deal to civilization both intellectually and economically. **Historical geology** *is concerned with the origin and development of all parts of the earth, including the atmosphere, the oceans, and life, as well as the solid rock body earth.* To understand where the earth is in the universe, how it formed, and its subsequent history helps us to know who we are and what our future may be. What has happened in the past is the best guide we have to what can happen in the future. Taking good care of our earth—it's the only one we have—is very important considering population increases and our destructive abilities.

GEOLOGY IS SIMPLE—THE BASIC LAWS OF GEOLOGY

Geologic Time—The Age of the Earth

Among the great concepts gained from geological studies are an understanding of the great age of the earth and the development of an absolute time scale. *Geology differs from most other sciences in that it is concerned with absolute time.* Time appears in the equations of physics and chemistry, but these sciences are generally concerned with rates of change, and the time is relative, not absolute. Geologic time extends back almost 5000 million years to when the earth formed. Thus geology is concerned with immense lengths of time when measured against human

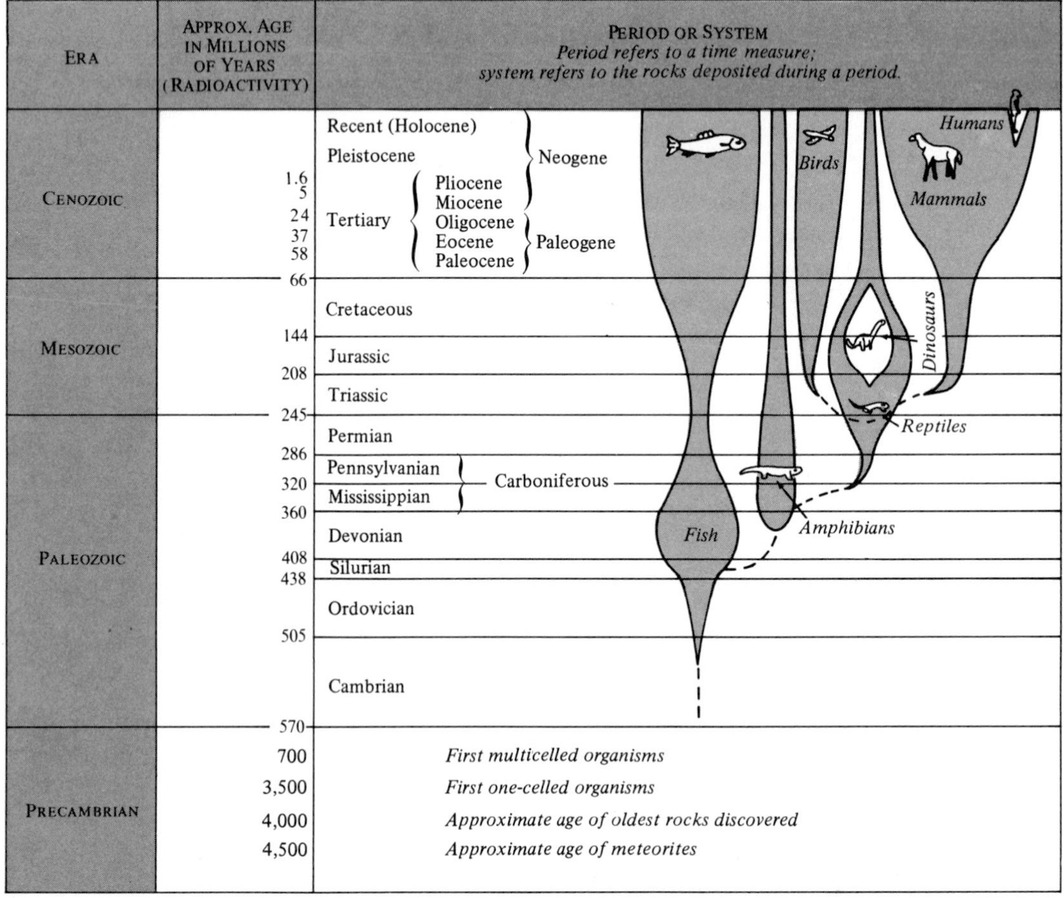

FIGURE 1.1
The geologic time scale. Shown to the right above is a very simplified diagram showing the development of life. Not included on the diagram are many types of invertebrate fossils such as clams, brachiopods, corals, sponges, snails, and so forth, which first appeared in the Cambrian or Ordovician and have continued to the present.

experience. It is difficult to comprehend the lengths of time involved in geologic processes, but this must be done to appreciate geology fully. (See Fig. 1.1.)

Uniformitarianism

Uniformitarianism (also called actualism) *is the fundamental principle that underlies most of geology and is simply that the present processes occurring on the earth have occurred throughout geologic time.* Thus ancient rocks can be interpreted in terms of present processes. In other words, any structure in old rocks must have been formed by processes now going on upon the earth. A simple statement of uniformitarianism is *the present is the key to the past.* Although this simple statement of uniformitarianism has limitations, it was very important to the development of geology. Uniformitarianism was not always accepted. We see erosion going on around us, note the material carried by muddy rivers, and conclude that rivers eroded the valleys in which they flow. Not too many years ago some geologists believed that valleys, especially deep valleys such as Yosemite, were formed by great earthquakes that split open the earth and that the rivers then flowed into them simply because they were the lower areas. This theory of sudden changes is called catastrophism.

Uniformitarianism implies slow processes going on over great lengths of time. Rapid processes, such as earthquakes, landslides, and volcanic eruptions, do occur, but, in general, do so at widely separated times, so their average rate is slow.

Original Horizontality

Layers of sediments deposited in a lake or the ocean are originally deposited in more or less horizontal layers. This is termed **original horizontality;** if the beds or layers are not horizontal, postdepositional deformation is implied.

Superposition

The third principle is called **superposition** and states that if a series of beds or layers of sedimentary rocks have not been disturbed, the oldest beds are at the bottom and the youngest at the top.

HOW THE EARTH WORKS

Internal Structure of the Earth

Earthquake waves have revealed the layered structure of the earth. The details of the analysis are best seen in Figure 1.2. Three main layers were soon distinguished: the **crust,** a very thin shell between 4.8 and 56 km (3 and 35 mi) thick; the **mantle,** the next layer, about 2898 km (1800 mi) thick; and the **core,** with a radius of about 3418 km (2123 mi). The velocity of earthquake waves changes abruptly at the contacts between these layers, but, as we will soon see, other boundaries more difficult to detect are also important.

Other smaller differences in seismic travel times (velocity) reveal a layering within the crust itself, as shown in Figure 1.3. *The continental crust is believed to be granitic* because intrusive and metamorphic rocks of this composition underlie the sedimentary rocks that cover much of the surface and because the seismic velocity of granitic rocks is similar to the observed seismic velocities. In the same way, *the oceans are believed to be underlain by basalt,* and the lower layer under the continents is believed to be basaltic also. Although this layer has the composition of basalt,

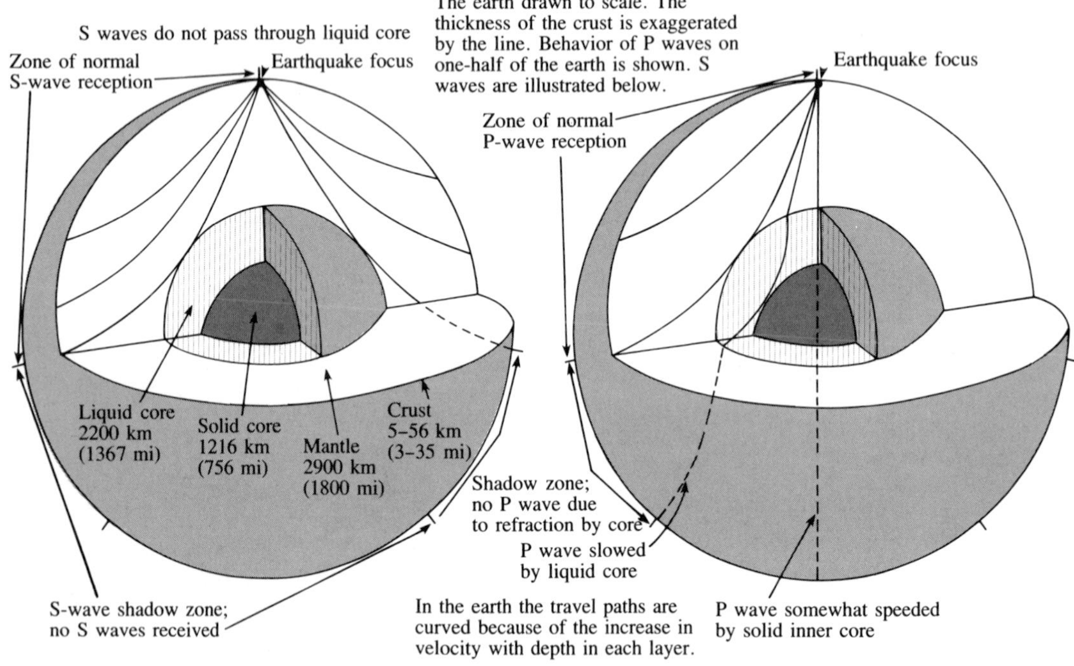

FIGURE 1.2
Behavior of P and S waves in the mantle and the core reveals the layered structure of the earth.

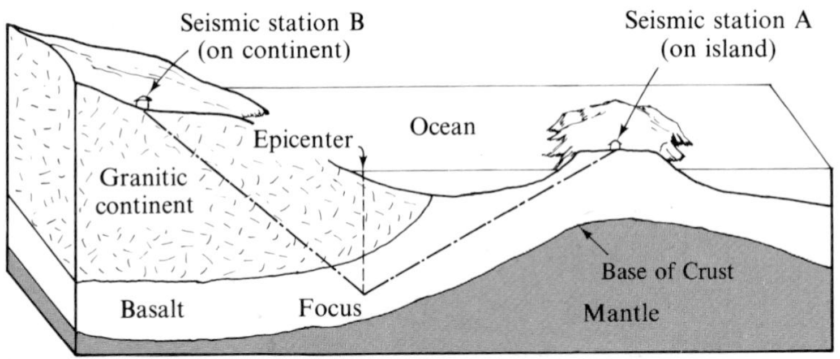

FIGURE 1.3
Block diagram showing paths of seismic waves in oceanic basalt and continental granite. Stations A and B are equidistant from the focus of the earthquake, but station A receives waves first because they travel faster in basalt than in granitic rock.

seismic data suggest that it is in the form of gabbro or amphibolite, both of which have the same chemical composition as basalt.

The upper mantle is also layered, but the layers here could not be detected until sensitive seismographs were available. At depths between about 70 km (40 mi) and 250 km (150 mi), seismic shear waves (S waves) are both slowed and attenuated. This is called the **low-velocity zone.** The top of this zone varies in depth between 50 km (30 mi) and 125 km (75 mi). The low-velocity zone is in the upper mantle. The slowing has not been detected everywhere under the continents, but it is generally present under the oceans. *The rocks above the low-velocity zone are brittle and are called the* **lithosphere.** (See Fig. 1.4.) The lithosphere is the crust and the upper mantle. *The plates that move on the earth's surface are composed of lithosphere.*

The layer below the lithosphere is called the **asthenosphere;** this is the layer in which the lithosphere's plates move. *The asthenosphere begins at the low-velocity zone and is a weak, or soft, zone in the upper mantle.* At the temperature and pressure that exist at the depth of the asthenosphere, the upper mantle rock is very close to its melting point. It is estimated that between 1 and 10 percent of the asthenosphere is melted. This partial melting would cause the slowing and attenuation of shear waves observed in the low-velocity zone. Partial melting of the upper mantle rocks has also been suggested as the origin of the basalt that rises at the mid-ocean ridges.

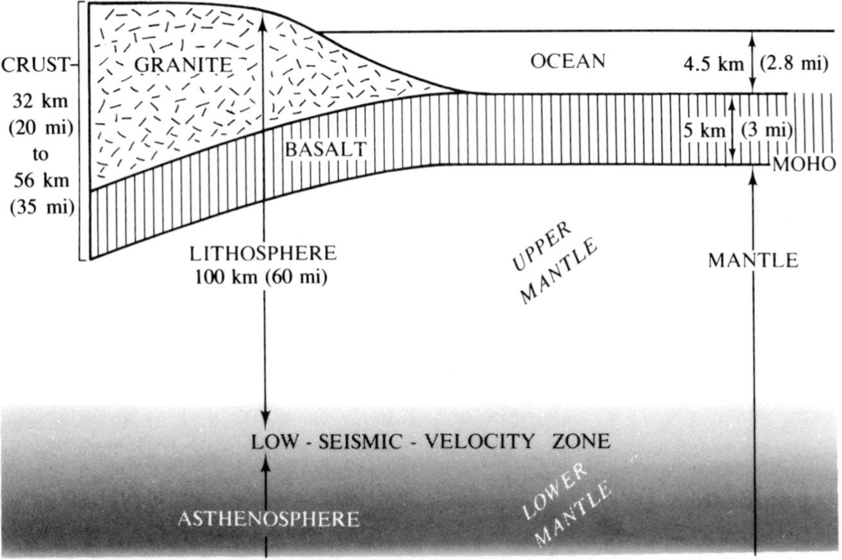

FIGURE 1.4
The crust and upper mantle form the lithosphere, in which the rocks behave as solids. In the asthenosphere, below the low-velocity zone, the rocks flow.

Internal vs. External Processes

The earth is a dynamic planet whose surface is constantly undergoing change. Given the length of geologic time, one can comprehend that the hills and mountains are not eternal features, but that eventually they will be worn flat by erosion. Erosion is the external process, and it is powered by gravity and the sun's energy. *If it were not for the earth's internal energy, erosion would complete its work and the earth would be as smooth as a billiard ball.* Earthquakes and volcanoes reveal a restless interior whose activities tend to increase the size and elevation of the continents. Deformed rocks that have been bent into folds and broken also bear mute testimony to the power of the earth's internal energy.

The Mobile Earth—Internal Energy

Continental Drift The idea of continental drift has been with us since the first accurate maps showed that Africa and South America fit together. During the last part of the nineteenth century, European geologists were deciphering the extremely deformed rocks in the Alps and American geologists were mapping the West, which was just opening up. As a result, they took little notice of some startling discoveries made in India, Australia, and South Africa. At each of these places a peculiar flora, called the *Glossopteris* flora after its most abundant genus, was found in beds overlying glacial deposits. Since then, the same sequence has been found in South America and Antarctica. The glacial deposits are from continental, not mountain, glaciers and rest on glacially polished and striated (scratched) basement rocks. (See Fig. 1.5.) The striations suggest that in many cases the ice moved from areas that are now occupied by oceans. The moraines contain many boulders of rock types not found on that continent, and in some cases the foreign boulders match well with outcrops on the adjoining continent. The questions immediately arise: How could extensive glaciers form in these tropical areas, and how did the *Glossopteris* flora spread across the oceans now separating these areas? (See Fig. 1.6.) These areas are so widespread that shifting the rotational axis of the earth does not help because even in the most favorable location, some of the glacial areas are still within ten degrees of the equator. If all the continents of the southern hemisphere had been in contact at one time, these occurrences would be explained; but that large continent would have to break up

---

FIGURE 1.5
Evidence for continental drift. A. *Glossopteris.* This leaf has been found in beds overlying ancient glacial deposits in India, Australia, South Africa, and Antarctica. B. Glacial polish in Australia giving clear evidence of ancient glaciers there. Ice moved toward upper right. C. Striated boulder from ancient glacial moraine in Australia. D. Glacial polish from glaciers of the same age in Africa. Etching by Bushmen on left. Parts B, C, and D, photos by Warren Hamilton.

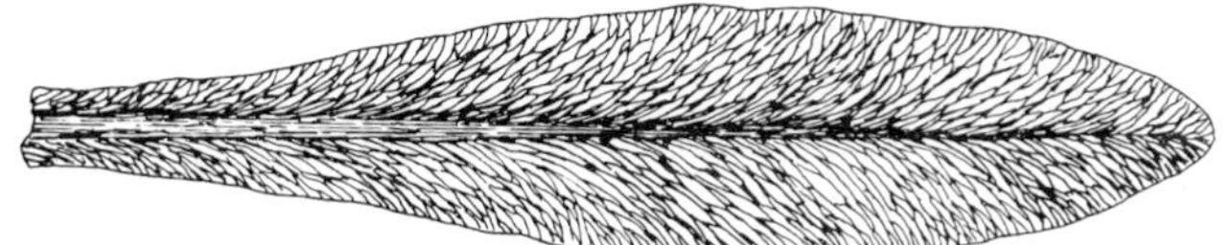

A

B

C

D

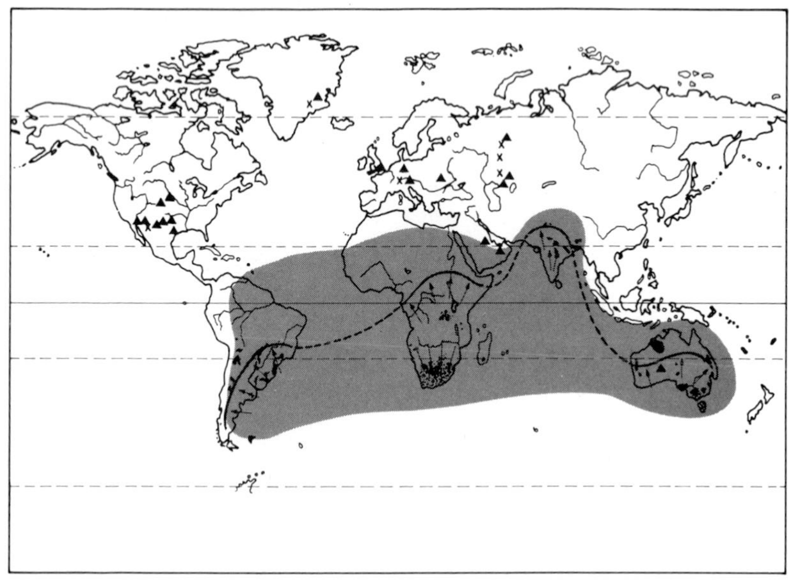

FIGURE 1.6
Map showing the present distribution of the ancient glaciation of the southern hemisphere. The area involved in the hypothetical southern continent existing before drifting is shaded. The areas of glacial deposits are indicated and extend from north of the equator to near the south pole. Reefs (crosses) and evaporites (triangles) far to the north indicate much warmer climates. Reefs and evaporites from Opdyke, 1962.

and pieces of it would have to drift (or move) to explain the present geography. (See Fig. 1.7.) Thus began the idea of continental drift. Today the evidence is even more compelling because of the discovery in Antarctica of *Glossopteris* and other fossils, indicating a humid, temperate climate at one time.

The *Glossopteris* flora found just above the glacial deposits is believed to be evidence of previous connection among the southern continents. It is possible that seeds (spores) could have floated across the oceans to distribute these plants, but it

FIGURE 1.7
Reconstruction of the ancient continent of the southern hemisphere. The glaciers and their directions of movement are indicated.

seems rather unlikely. Today, a newly formed volcanic island soon develops plants, but the seeds are carried by birds. The first birds did not appear until tens of millions of years after the supposed separation and drift. Land bridges and island chains have been suggested in the past, but the known oceanic structure precludes this possibility.

The swimming reptile *Mesosaurus,* found as fossils in both South America and southern Africa, is also strong evidence favoring connection of these two continents. Although a swimmer, this aquatic reptile lived in fresh water, and it is very unlikely that it could swim across an ocean. Other reptiles that lived during the time of the alleged connection have been found on the southern continents. The distribution of some of these reptiles requires a single connected continent.

Piecing together the continents is like putting together a jigsaw puzzle whose pieces do not quite match. Fitting together the various layers of sedimentary rocks is like putting together a multilayer jigsaw puzzle with many of the pieces missing or distorted. Radiometric dates and deformation of the rocks both show a good match between the basement rocks of western Africa and those of northern Brazil. (See Fig. 1.8.)

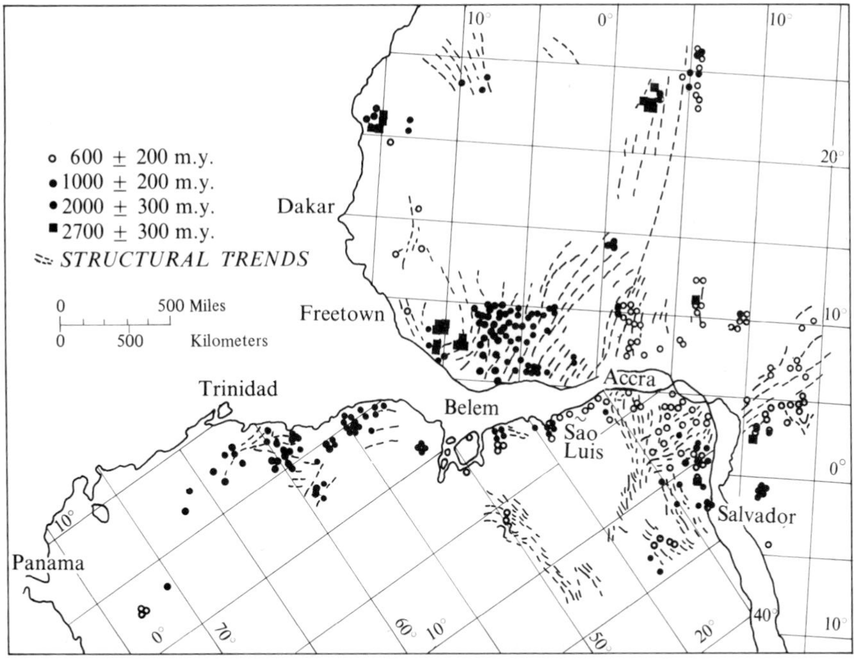

FIGURE 1.8
West Africa and South America fitted together, showing the coincidence in both structural trends and radiometric ages of the rocks (m.y. is million years). From Hurley and others, 1967. Copyright 1967 by the American Association for the Advancement of Science.

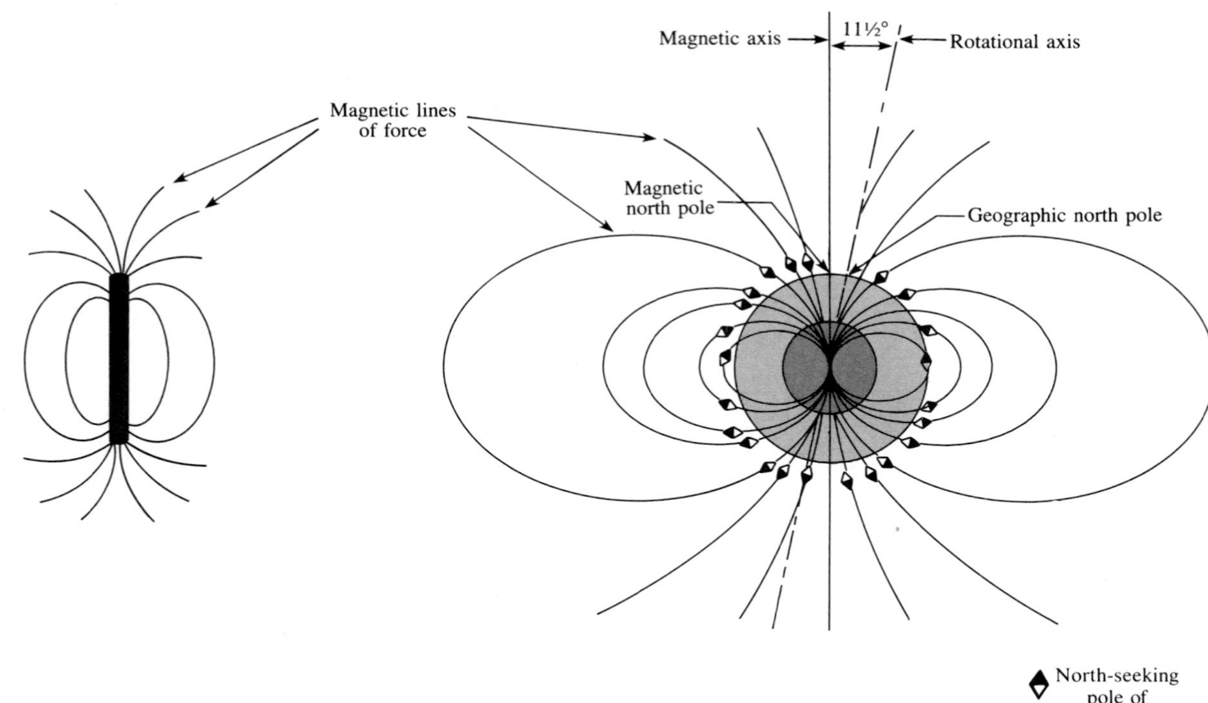

FIGURE 1.9
The earth's magnetic field is much like that of a bar magnet. The magnetic poles do not coincide with the geographic poles that are determined by the rotation axis. A compass needle that is free to move in the vertical plane will point in the directions shown.

The fit of the continents on a map seems to be too good to be accidental. The fit is even better if, instead of the shoreline, the true edge of the continent, the continental shelf, is used. Later deformation or erosion does spoil the match at some places. There are, however, problems. In most reconstructions, Spain must be rotated, and there is no room for much of Central America; however, these problems seem small in the overall view. Most reconstructions are based on magnetic data, so they will be discussed next.

Earth's Magnetic Field

It is the earth's magnetic field that aligns a compass needle and makes it point toward magnetic north. The earth behaves much as if it had a giant bar magnet in its core. (See Fig. 1.9.) However, even though the core is made of iron and nickel, no such magnet can exist at the temperatures there. Iron loses its magnetism when heated above the **Curie point** temperature — well below its melting point. Also, the earth's magnetic field is not that of a simple **dipole,** *which is the field of a bar magnet.* (See Fig. 1.10A.) The earth's field consists of a dipole field and a nondipole field. One effect of the nondipole field is that the south magnetic pole, at the edge of

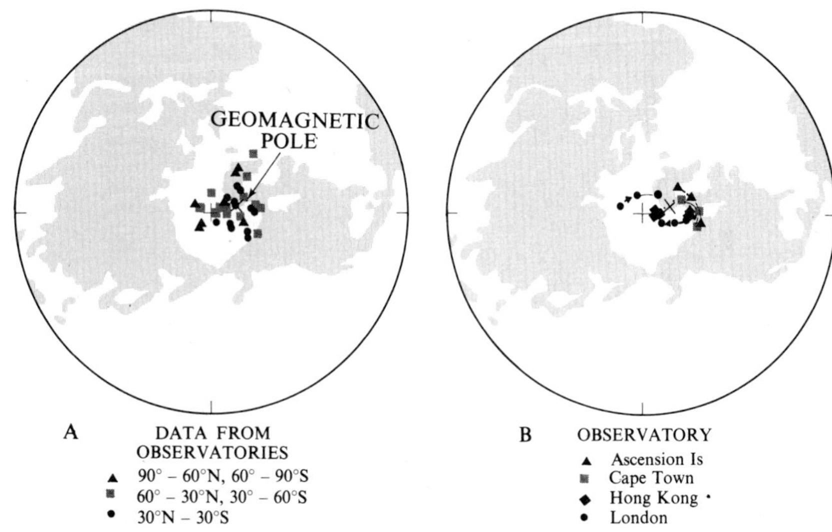

FIGURE 1.10
Variations in the earth's magnetic field. A. Variations in position of magnetic north pole computed in 1945 from direction and inclination of the earth's magnetic field at various places. These calculations were made assuming a dipole field and show the effect of the nondipole part of the earth's field. B. Changes in the position of the magnetic north pole with time. Time between points is 40–50 years. In both figures the cross shows the position of the north geographic pole, and the X, the average 1945 position of the north geomagnetic pole. From Cox and Doell, 1960.

Antarctica, is 2093 km (1300 mi) from the point that is antipodal to the north magnetic pole, which lies in northern Canada. However, for the most part, *the earth's field can be considered nearly a dipole field.*

The earth's magnetic field causes a compass to point toward the magnetic pole, but if the compass is pivoted to move in the vertical direction also, it will indicate inclination as well. (See Fig. 1.9.) At the magnetic equator, the inclination of the field is horizontal; at the poles, vertical; and in between, it varies regularly. The intensity of the field at the poles is about twice that at the equator. At some places records of the direction and inclination of the earth's field have been kept for hundreds of years, and they reveal changes in direction and inclination. (See Fig. 1.10.) More recent studies also show variation in intensity. Study of the *remanent magnetism* of rocks (to be discussed) indicates that numerous times in the geologic past the direction of the earth's field has reversed, i.e., the north pole of a compass needle would point toward the present location of Antarctica. The timing of these reversals in the last 150 million years is well enough known that they can be used

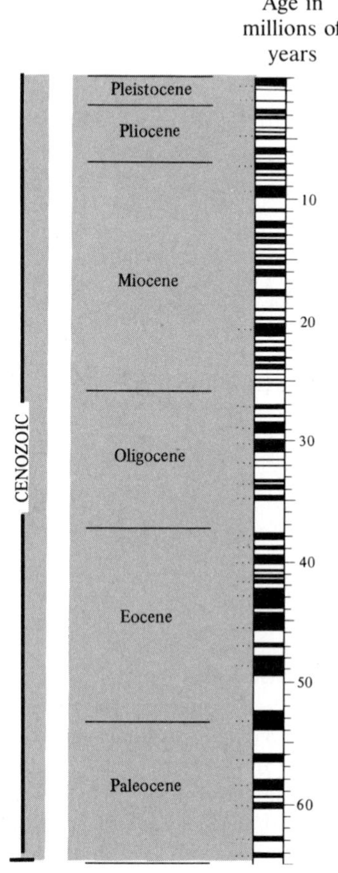

FIGURE 1.11
Reversals of the earth's magnetic field. From data by Cox.

to date rocks. (See Fig. 1.11.) Any theory for the origin of the earth's magnetic field must explain these observations.

The earth's magnetic field is believed to originate from electric currents in the core. It is a well-known law in physics that a moving current creates a magnetic field, and conversely, a current is induced in a conductor moving in a magnetic field. Electrical currents in the core could start in a number of ways, such as through a weak battery action caused by compositional differences. Once started, these currents could be amplified by a dynamo action in the earth's core. A dynamo produces electric current by moving a conductor in a magnetic field. In the earth the core is the conductor, and because it is fluid, it can move. The mechanism envisioned is similar to that of a dynamo, in which an electric current is produced and fed to an electric motor that drives the dynamo. Friction and electrical resistance would prevent such a perpetual motion machine from continuing to operate unless more energy is added. In the earth this additional energy can come from movement of the fluid core. Calculation shows that very little additional energy needs to be added in the case of the earth; the necessary energy for core motion, probably in the form of convection, could come from many sources.

Thus, a current is somehow started in the earth's core, and this current produces a magnetic field. The conducting core moves through the magnetic field, and a current is induced in this part of the core, starting the whole process over. (See Fig. 1.12.) In a stationary earth, the movements of the core would probably be more or less random and the magnetic fields produced would probably cancel each other. The rotation of the earth tends to orient the motions of the core so that the

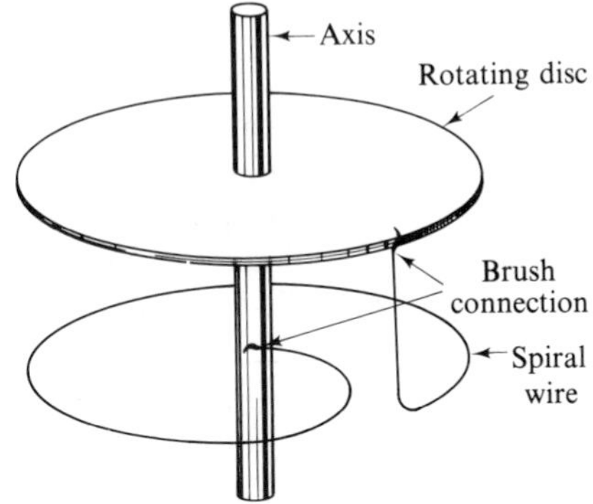

FIGURE 1.12
The disc dynamo illustrates in a general way how the earth's magnetic field is believed to originate. The disc is attached to the axis and rotates with the axis. The spiral wire is in electrical contact with the axis and the disc by means of brushes. To start this self-sustaining dynamo, suppose that a current is flowing in the spiral wire. This current produces a magnetic field; the disc is a conductor moving through the field, and so a current is induced in the disc. This current flows to the spiral wire by means of the brush and so continues the processes.

motions and, therefore, the currents are in planes perpendicular to the rotational axis. Viewed from outside the earth, the overall current produced would be parallel to latitudes. Such currents produce a dipole magnetic field centered on the rotational axis.

A field produced in this way would have all of the characteristics of the earth's magnetic field. The main field would be a dipole field; superimposed on this would be other fields caused by variations in the core's motion. This can account for the variations in intensity and location of the earth's field. The earth's dipole field does not coincide with its rotational axis, however, as this theory predicts, although if averaged over a long enough period, the dipole field would perhaps appear centered on the rotational axis. The positions of the magnetic pole deduced from remanent magnetism of Pleistocene and younger rocks cluster around the rotational axis, suggesting that this is true.

Remanent Magnetism

Many of the recent data on continental drift have come from measuring the remanent magnetism of rocks. **Remanent magnetism** *is the magnetism frozen into a rock at the time the rock forms.* For instance, in the cooling of basalt, when the temperature reaches a certain point, which is well below the crystallization temperature, the magnetic minerals, mainly magnetite, are magnetized by the earth's field. As rocks cool below this temperature, their magnetism is not greatly affected except by very strong fields, many times stronger than any in the earth. Thus it is possible to measure the direction and inclination of the earth's field at the time the basalt cooled. The inclination can be used to calculate the distance to the magnetic pole.

Remanent magnetic data for rocks of different ages have been collected on each of the continents. The study of magnetic data from old rocks is called *paleomagnetism* (old magnetism). The magnetic poles for each age (geologic period) on each continent group in a fairly small area. However, in general, for any given continent, each geologic period has a different location for the magnetic pole, and the poles for any period fall in different locations for each continent. (See Fig. 1.13.) Here, then, is evidence of continental drift. If the magnetic pole moved, it should be at the same place during each period for each continent; instead, each continent gives a different location for each period. Thus, if the interpretation of the paleomagnetic data is correct, the continents must have moved relative to each other. (See Fig. 1.14.) The magnetic pole may also have moved. Note in Figure 1.13 that the magnetic data indicate no drift of North America relative to Europe since Eocene time.

Paleomagnetic data have certain limitations. They tell the direction and distance to the magnetic pole. Thus they show the rotation of a continent and its latitude. (See Fig. 1.15.) They do not indicate the longitude—the continent may be anywhere on the indicated latitude, either north or south of the equator. This gives the imagination rather free rein in proposing reconstructions.

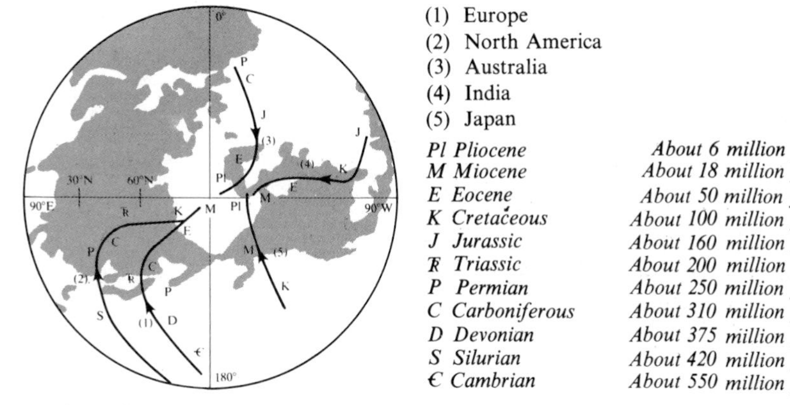

FIGURE 1.13
On this diagram, for each continent the position of the magnetic pole relative to the present position of each continent for each geologic period is indicated by the letters. The lines drawn through these locations show the apparent drift of the magnetic pole for each continent. The diagram indicates that movement of the continents relative to each other must have occurred but does not show the route of drifting because of the limitations of paleomagnetic data, shown in Figure 1.15. From Cox and Doell, 1960.

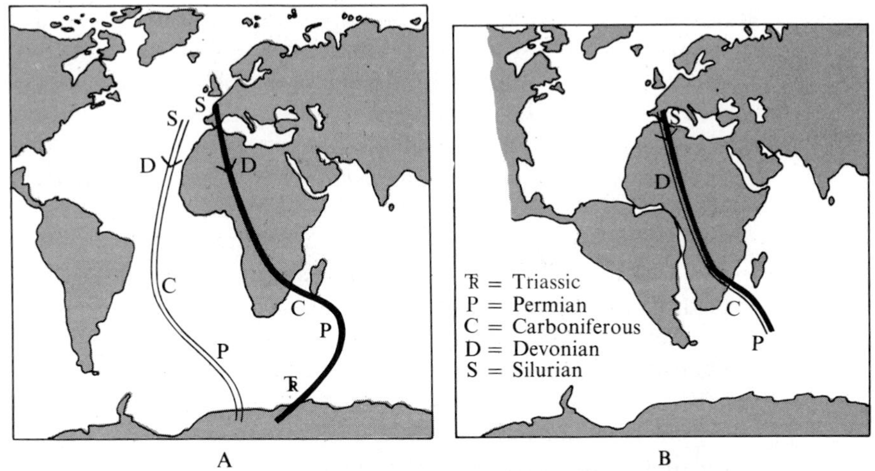

FIGURE 1.14
A. The curves show the position of the magnetic pole for each geologic period relative to the present positions of the continents. B. Shows that if South America and its polar curve are moved so that the two polar curves match, the two continents join. The position at which they join is geographically and geologically the most likely.

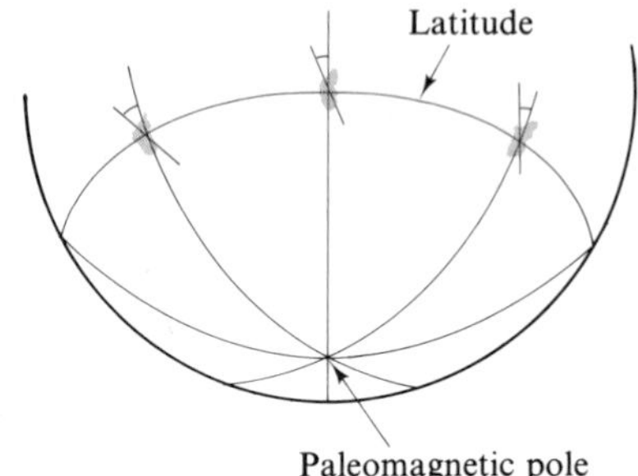

FIGURE 1.15
Paleomagnetic data can show only the latitude and orientation of a continent. The horizontal component of the paleomagnetism points the direction toward the pole, and the vertical inclination gives the latitude. The continent could have been in that orientation anywhere on that latitude.

Magnetic Stripes

The earth's magnetic field is impressed on a rock at the time the rock is formed. Ocean-floor basalt, because it contains small amounts of magnetite, an especially magnetic mineral, faithfully records the earth's magnetic field. Magnetic surveys near the mid-ocean ridge show bands of alternating high and low magnetic intensity (magnetic stripes) that parallel the ridge, as shown in Figure 1.16. The meaning of these magnetic bands was not clear until magnetic studies were made on the continents. It was soon discovered that many rocks recorded a magnetic field opposite (or reversed) to the present magnetic field. Magnetic studies of each flow in areas of thick accumulations of basalt flows showed that these reversed magnetic field rocks did not occur in a random way but, rather, that if radiometric dates of the rocks were also obtained, the magnetic reversals occurred simultaneously all over the earth. Starting with the youngest rocks, it has been possible to determine the times of reversals accurately enough to use the magnetic direction to date some young rocks. (See Fig. 1.11.) *These reversals suggest that the bands of alternating high and low magnetic intensities paralleling the mid-ocean ridge could be caused by bands of ocean-floor basalt that have normal and reversed magnetic fields.* The normally magnetized basalt would reinforce the present magnetic field, giving a high magnetic intensity; the reverse-magnetized rocks would partially cancel the present field, giving a lower than normal magnetic intensity. Careful comparison of the magnetic intensity on each side of the mid-ocean ridge shows that the magnetic intensity on one side is nearly the mirror image of that on the other side. (See Fig. 1.17.) The final argument is to construct a model of the ocean floor by calculating the positions of bands of normal and reversed magnetic basalt and then to calculate the theoretical magnetic field produced by such a model. As shown in Figure 1.18, this model produces a field almost identical to that found in nature.

Sea-Floor Spreading Sea-floor spreading is a much newer idea than continental drift. Although it had been proposed much earlier, it was finally accepted in the 1960s when the evidence for it became overwhelming. All of the

FIGURE 1.16
Magnetic anomalies in part of the ocean along the west coast of North America. Positive magnetic anomalies are indicated. They are believed to be formed where the remanent magnetism of the underlying rocks reinforces the earth's field. The intervening white areas have lower than normal magnetism and may be places where the underlying rocks have reversed magnetism. The straight lines are faults that displace the pattern. The arrows indicate short lengths of oceanic ridge, and the pattern is parallel and symmetrical to the ridges. After Vine, 1966. Copyright 1966 by the American Association for the Advancement of Science.

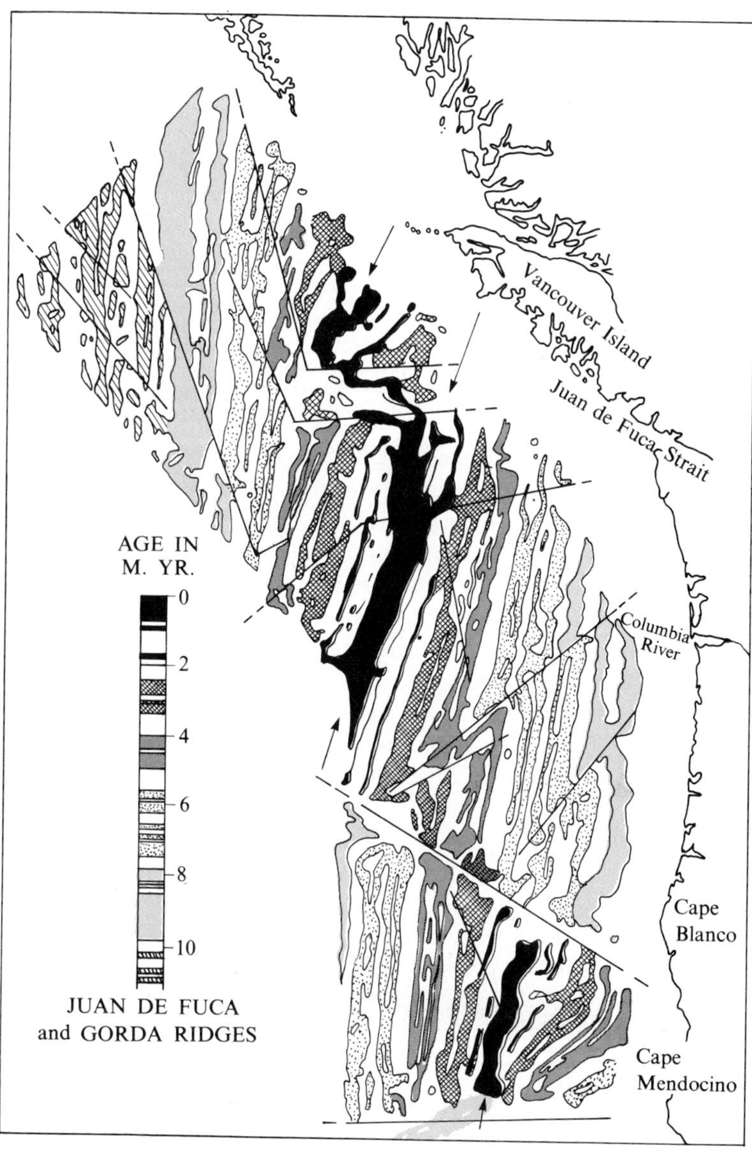

ocean bottoms have ridges or rises in them, called **mid-ocean ridges** (also called **spreading centers** and **divergent plate boundaries**). These are areas of volcanic activity and shallow earthquakes. *New oceanic lithosphere is created at the ridges by the volcanic activity.* (See Fig. 1.19.) This is shown by the age and thickness of the rocks on the ocean floors, which have been sampled by drilling. The age of the volcanic bedrock of the oceans is progressively older as one moves away from the ridges in both directions. In the same way, the sediments above the bedrock become thicker,

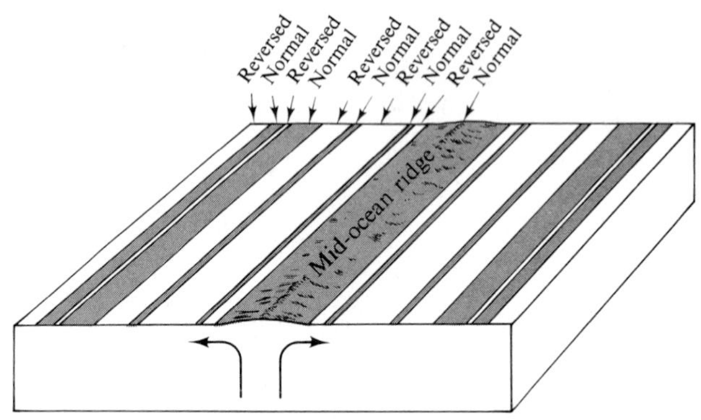

FIGURE 1.17
Formation of magnetic anomalies at a mid-ocean ridge. In this hypothesis the lavas that form the ocean floor cool at the ridge and are magnetized by the earth's field at the time they cool. Convection currents carry the lavas away from the ridge and so create magnetic anomaly patterns parallel and symmetrical to the ridge.

and the deepest sediments older, with distance from the mid-ocean ridges. The conclusion seems inescapable that new oceanic lithosphere is being created at the mid-ocean ridges and that the ocean floors are moving away from the ridges. The rate of movement is only a few centimeters per year, but given the earth's great age, the total movement is very large.

If new lithosphere is created at the mid-ocean ridges, then lithosphere must be consumed somewhere; otherwise, the earth must expand. **Volcanic arcs** *are places where the oceanic lithosphere apparently returns to the interior of the earth.* (See Fig. 1.20.) At many places the volcanic arcs form volcanic island arcs—islands formed by volcanoes that appear as curves or arcs on a map. The volcanic arcs are geologically active areas with deep earthquakes as well as the volcanic activity.

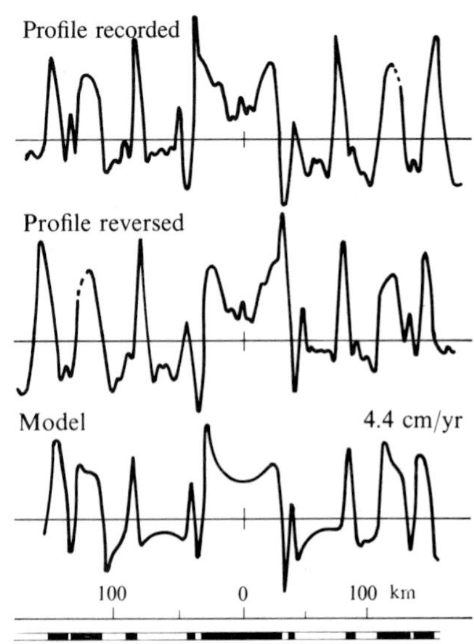

FIGURE 1.18
Magnetic profiles across the East Pacific Rise. The two upper curves are an actual magnetic profile, shown both as recorded and reversed. The lower curve was computed from a model, assuming a spreading rate of 4.4 cm per year. After Vine, 1966. Copyright 1966 by the American Association for the Advancement of Science.

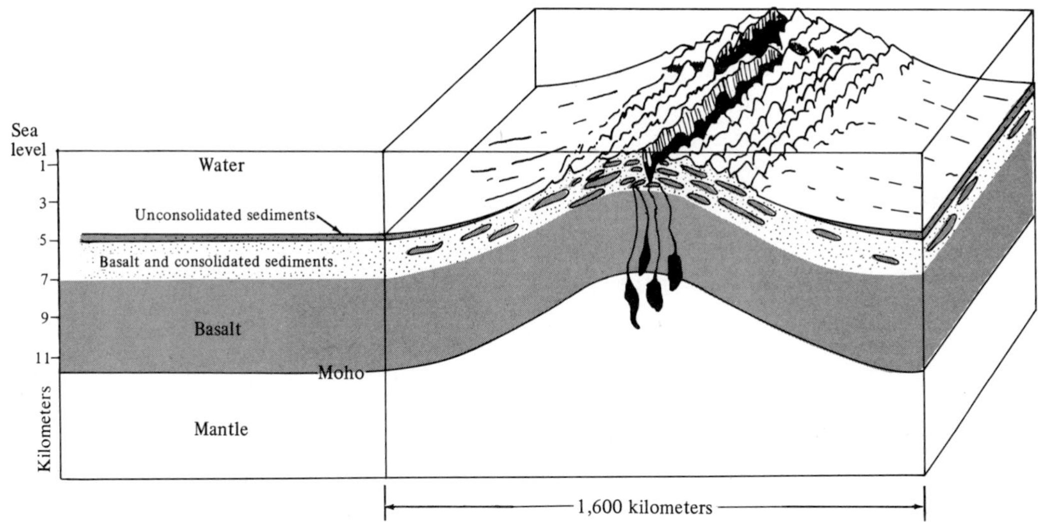

FIGURE 1.19
Mid-ocean ridge (spreading center, divergent plate boundary). A cross-section of typical ocean floor is shown on the left. The ocean crust thins at the ridge. Note the vertical exaggeration. See Glossary for definition of Moho.

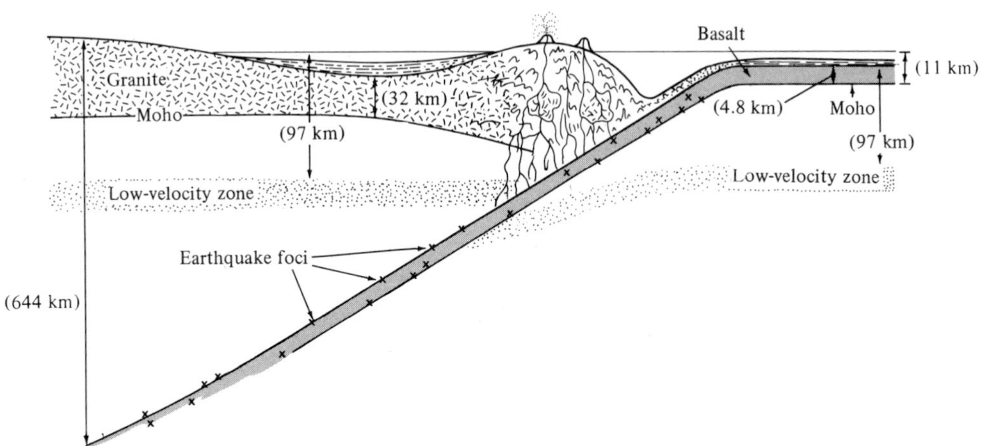

FIGURE 1.20
Volcanic island arc and subduction zone, showing location of earthquake foci. Because of the distances involved, this figure is not drawn to scale. See Glossary for definition of Moho and other terms.

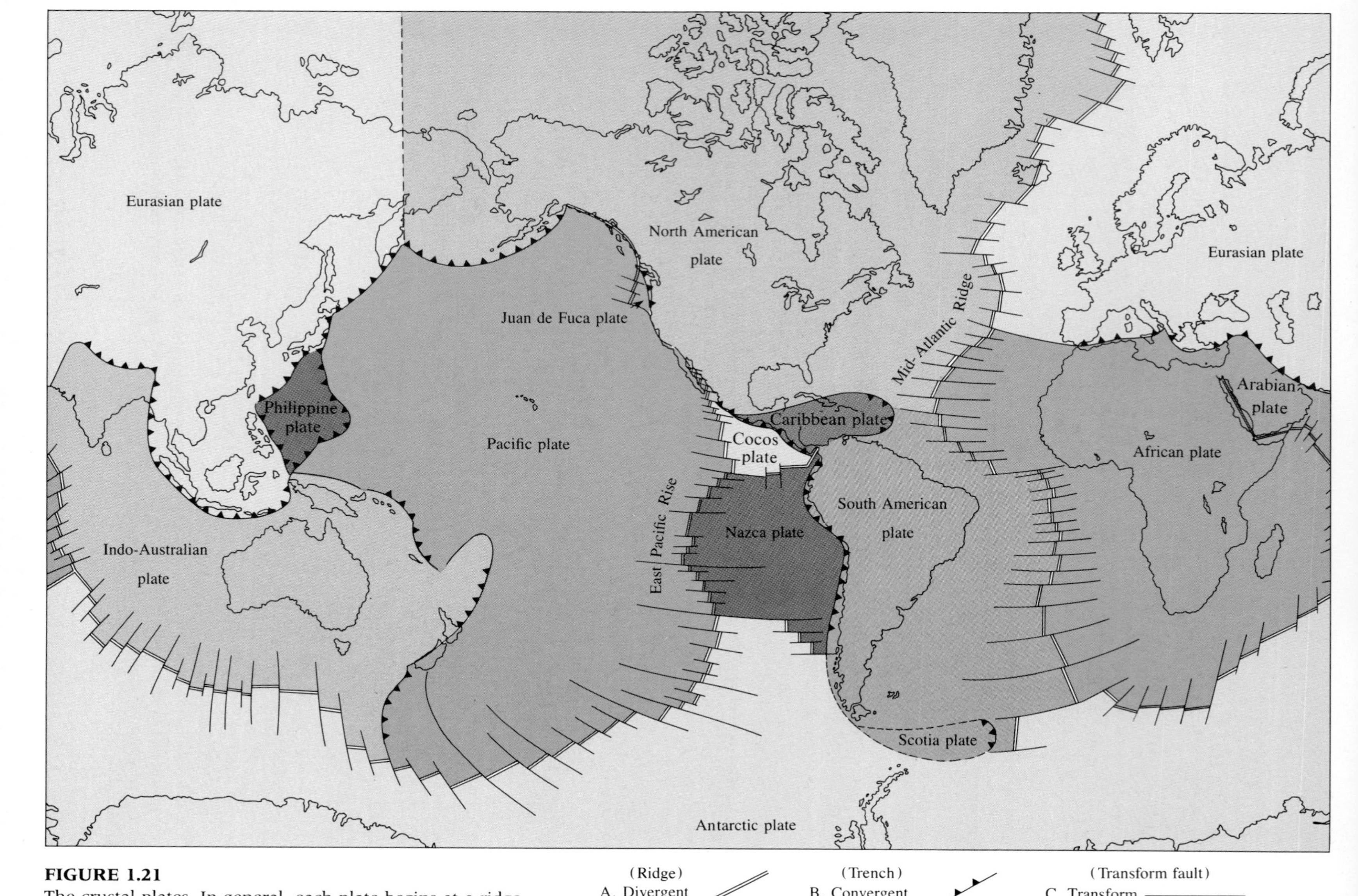

FIGURE 1.21
The crustal plates. In general, each plate begins at a ridge (double lines) and moves toward a trench (saw-toothed lines). Transform faults are shown by thin lines. Courtesy

It is believed that the oceanic lithosphere returns to the interior of the earth at the volcanic arcs because the earthquakes occur on a plane sloping under the volcanic arc. (See Fig. 1.20.) Movement of a slab of lithosphere would cause the earthquakes. The volcanoes could be the result of partial melting of the slab of lithosphere. Downward movement of the ocean floor is also suggested by the deep submarine trenches in front of volcanic arcs. (See Fig. 1.20.) *Thus, in sea-floor spreading, new oceanic lithosphere is created at the mid-ocean ridges and moves slowly toward the volcanic arcs, where it is consumed.*

Plate Tectonics Continental drift and sea-floor spreading come together as parts of plate tectonics. In **plate tectonics,** *the surface of the earth is believed to be composed of a number of thin slabs, or plates, that move across the surface.* (See Fig. 1.21.) The plates shown in Figure 1.21 are outlined by the locations of earthquakes. *The plates are formed at the mid-ocean ridges, and they return to the earth's interior at* **convergent plate boundaries** *such as the volcanic island arcs. Where the plates pass each other, earthquakes are also caused by the motion, and these plate boundaries are called* **transform faults.**

The continents are underlain by granitic rock—rocks less dense than the ocean-floor rocks. *The continents ride atop the ocean-floor basalt, and so as the plates move, they carry the continents with them.* (See Fig. 1.22.) In this way, the same mechanism that causes sea-floor spreading is the cause of continental drift. Movements of the continents through geologic time are shown in Figure 1.23.

As the plates move across the earth's surface, they collide from time to time. The type of collision is determined by the composition of the plates at the point of collision. *If ocean floor collides with continent, the ocean floor is consumed at the resulting convergent plate boundary and a volcanic island arc forms.* (See Fig. 1.24.) *If ocean floor meets ocean floor, one plate is consumed and a volcanic island arc forms.* (See Fig. 1.25.) *If continent collides with continent, a mountain range is formed.* (See Fig. 1.26.) In all of these collisions, the rocks are deformed by the collision. The rocks, especially the layers of sedimentary rocks, are bent and folded and at places are broken and faulted as well. Igneous rocks are also formed, by partial melting of the descending plate, where plates are consumed. The rising of the resulting magma causes

FIGURE 1.22
Possible role of sea-floor spreading or convection currents in continental drift. Formation of ridge beneath continent could result in splitting of the continent and movement to a trench.

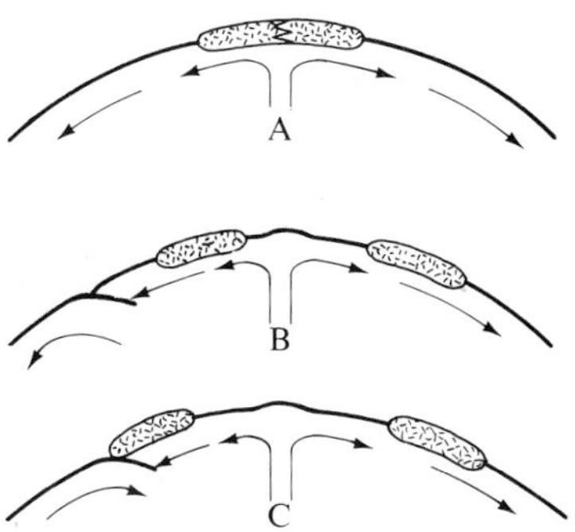

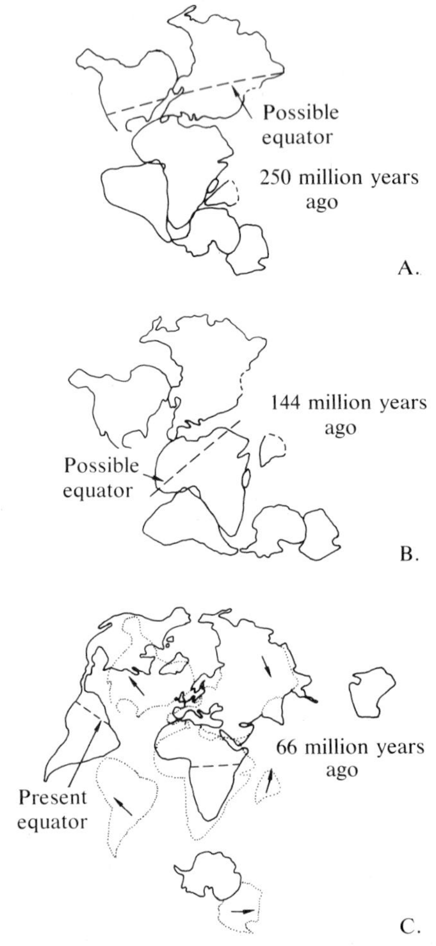

FIGURE 1.23
Movement of continents. A. Late Paleozoic time, 250 million years ago. B. End of Jurassic Period, 144 million years ago. C. End of Cretaceous Period, 66 million years ago, is shown by dotted lines. Present positions are shown by the solid lines. Based on Deitz and Holden, 1970.

metamorphism and deformation. This, then, is the origin of the folded and faulted rocks that form most of the earth's mountain ranges.

Erosion—The Constant Surface Process

The main process going on today on the earth's surface is clearly erosion and is shown by gullies on hillsides, landslides, and similar examples. The agents of erosion, of which rivers are far and away the most important, are constantly wearing down the continents; nevertheless, we still have mountains. The present rate of erosion is such that, theoretically, all topography above sea level will be removed in about 12 million years. However, we know that erosion and deposition have been going on for a few thousand million years, so something must interrupt

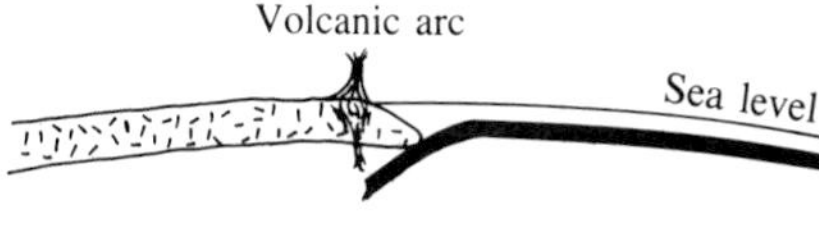

FIGURE 1.24
Continent-ocean plate collision.

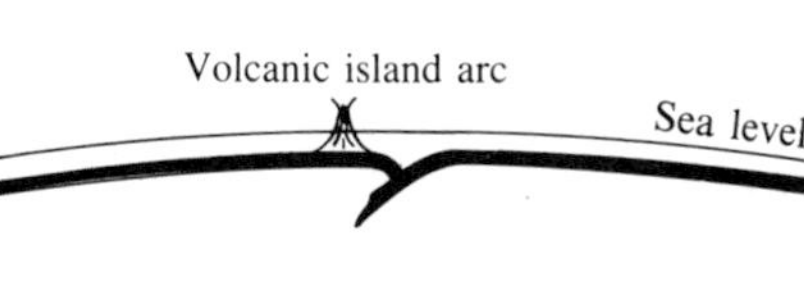

FIGURE 1.25
Ocean-ocean plate collision.

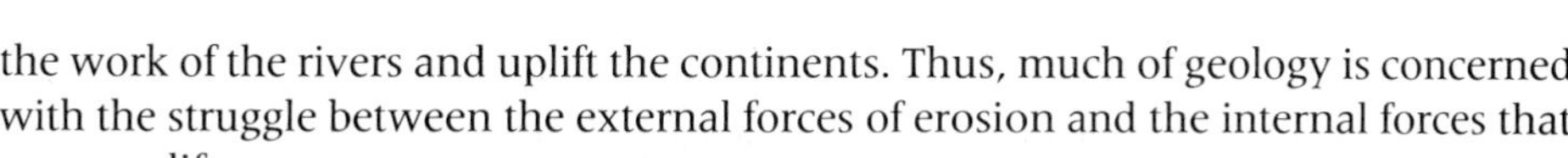

FIGURE 1.26
Continent-continent plate collision.

the work of the rivers and uplift the continents. Thus, much of geology is concerned with the struggle between the external forces of erosion and the internal forces that cause uplift.

Some familiar demonstrations of the internal forces that cause uplift are earthquakes, volcanoes, and folded and tilted sedimentary rocks that are now high on mountains but were originally formed as flat beds at or below sea level. Physical, or dynamic, geology is the study of this battle between the internal force of uplift and the external force of erosion. We will consider these two forces separately to eliminate confusion, but remember that they go hand in hand.

The underlying cause of erosion is downslope movements under the influence of gravity; the agents that cause such movements are water (rivers and so forth), ice (glaciers), and wind. The material that is moved by these agents comes from weathering processes.

The sun's energy is the ultimate source of the energy used in erosion, as shown in Figure 1.27. This diagram shows also that erosion is a necessary byproduct of the distribution of the sun's energy around the earth. Without this distribution of energy by the oceans and the atmosphere, the equatorial regions would become hotter and the poles colder because most of the sun's energy falls near the equator.

Extraterrestrial Effects

The earth is part of the solar system and the universe of stars. Both radiation and solid bodies from space affect the earth and its life. The sun's radiation causes erosion as just described. Other wavelengths of radiation from the sun and more distant stars are lethal; our atmosphere protects us from most of it but may not have in the past. Meteorites (miscalled falling stars) show that solid material from space can impact the earth. In the past such impacts may have caused the extinction of the dinosaurs.

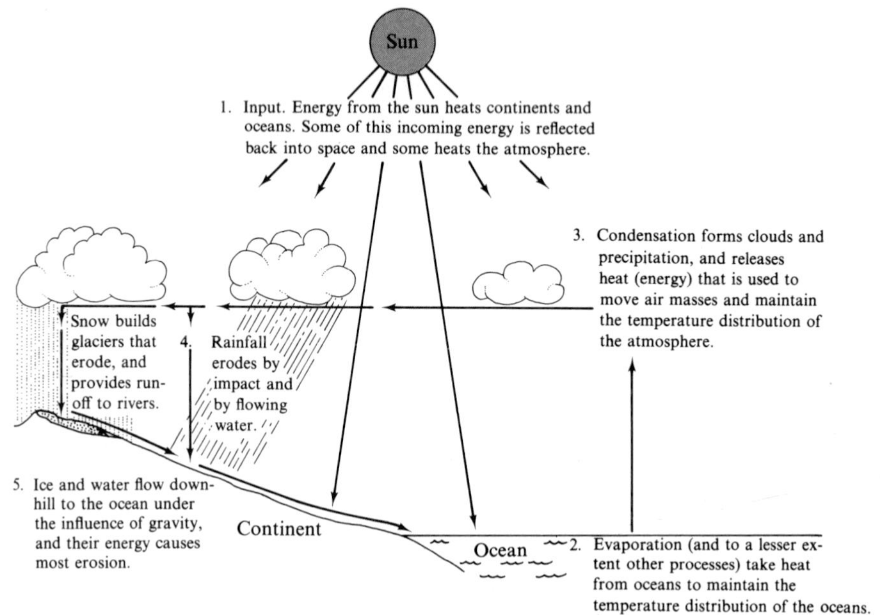

FIGURE 1.27
The water cycle and the earth's energy balance. Incoming energy from the sun is distributed around the earth by the circulation of the oceans and the atmosphere. The water cycle is part of the energy exchange between the oceans and the atmosphere, and it is the immediate source of the energy used in most erosion.

HOW GEOLOGISTS WORK

The crust is the only part of the earth accessible to direct observation. Studying the earth by observations in the crust is like trying to read a book by its covers. Actually, the problem is even greater than this because 71 percent of the surface is covered by oceans.

Geology is based mainly on observations and seeks to determine the history of the earth by explaining these observations logically, using other sciences such as physics, chemistry, and biology. Only a small part of geology can be approached experimentally. For example, although the important use of fossils to date or establish contemporaneity of rock strata is based on the simple, basic principle that life has changed during the history of the earth, this principle could not be established experimentally; it was the result of careful observations and analyses over a long period of time by many people of varied backgrounds. Geologic problems are many, diverse, and complex; almost all must be approached indirectly, and in some cases, different approaches to the same problem lead to conflicting theories. It is generally difficult to test a theory rigorously for several reasons. The scale of most problems prevents laboratory study; that is, one cannot bring a volcano into the laboratory, although some facets of volcanoes can be studied indoors. It is difficult also to simulate geologic time and size in an experiment. All of this means that much of geology lacks exactness and that our ideas change as new data become available. This is not a basic weakness of geology as a science, but means only that much more remains to be discovered; this is a measure of the challenge of geology.

Reasoning ability and a broad background in all branches of science are the main tools of geologists. Geologists use the method of *multiple working hypotheses* to test their theories and to attempt to arrive at the best-reasoned theory. This thought process requires as many hypotheses as possible and the ability to devise ways to test each one. Not always is it possible to arrive at a unique solution—but this is the goal. In the sense that geologists use observation, attention to details, and reasoning, their methods are similar to those of fictional detectives.

The most important method used by the geologist is to plot on maps the locations of the rock types exposed at the earth's surface. The rocks are plotted according to their type and age on most maps. (See Fig. 1.28.) From such maps it is possible to interpret the history of the area. The early geologists had to make their own maps, and work in a remote area was very difficult in many cases. Now, excellent maps produced from aerial photographs are available for most areas. Much geologic mapping is done directly on either black-and-white or color aerial photographs, which have proved to be unexcelled for accurate location of rock units. In addition, the outcrop pattern of the rock units generally shows well on air photos. (See Fig. 1.29.)

RECENT TRENDS

Recent discoveries have expanded the interest of geologists far beyond their classical realm—the rocks of the earth's surface. In addition, geology is becoming more quantitative, although it is still qualitative when compared to a science such as physics. In the early part of the twentieth century, radiometric dating enabled the geologist to work with the very old rocks, and the problem of the origin and early history of the earth began to take form. Since World War II, oceanography has developed rapidly, and the features of the ocean bottoms have been mapped. These

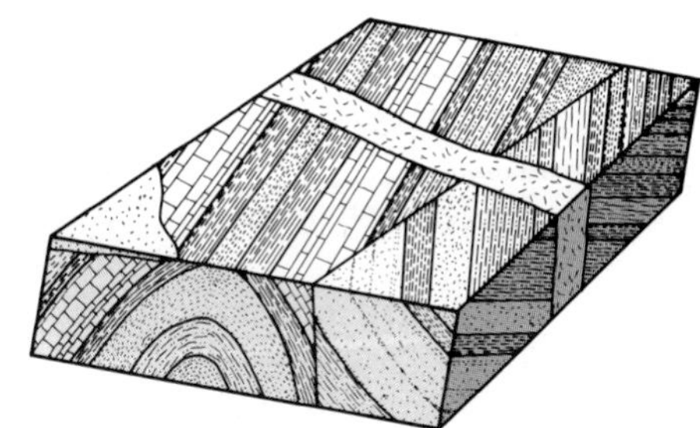

FIGURE 1.28
Block diagram showing a geologic map with cross-sections on the sides. A geologic map shows the distribution of rock types on the surface. From such maps, the history of an area is interpreted.

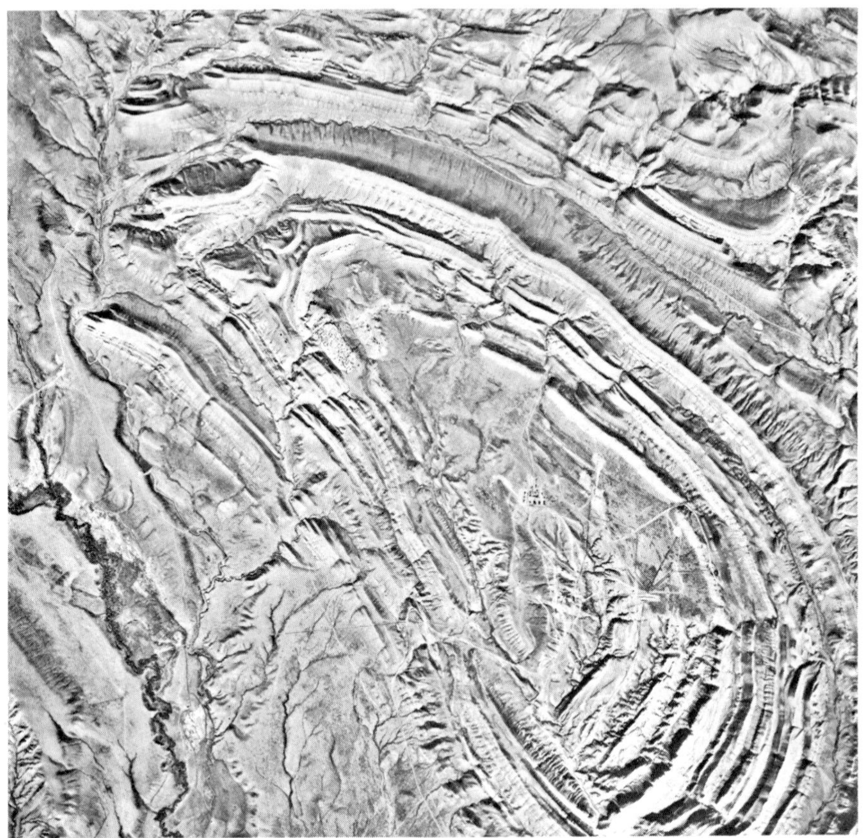

FIGURE 1.29
Vertical aerial photograph of folded rocks in Wyoming. Erosion has etched out the differences in resistance of the various rock layers. Such a photo is essentially a geologic map of the area. Photo from U.S. Geological Survey.

oceanographic discoveries, together with the study of magnetism in ancient rocks, resulted in the theory of plate tectonics.

At present, the space programs are giving us information about other parts of the solar system. Most astronomers believe that all of the solar system formed at the same time, so this new information gives much help to the geologist. The origin of the solar system is, of course, a classic problem of astronomy, but is also part of historical geology.

SUMMARY

Geology is the study of the earth.

Historical geology is concerned with the origin and development of all parts of the earth, including the atmosphere, the oceans, and life, as well as the solid rock body earth.

Geology differs from most other sciences in that it is concerned with absolute time.

The fundamental principle of geology is uniformitarianism, which is simply that the present processes occurring on the earth have always gone on.

Most sedimentary rocks are originally deposited in horizontal beds or layers, and the bottom bed is the oldest.

Earth has a layered structure, with a thin crust overlying the mantle and core.

The continental crust is believed to be granitic, and the oceans are believed to be underlain by basalt.

The rocks of the crust and the upper mantle, above the low- velocity zone, are brittle and are called the lithosphere. The plates that move on the earth's surface are composed of lithosphere.

The layer below the lithosphere is called the asthenosphere.

Geologic studies have revealed that constant change, both physical and biological, has been and is occurring on the earth.

The evidence for continental drift includes paleomagnetic data, apparent fit of the continents, climatic data such as glaciation, apparent rifting, fossils such as *Glossopteris* and *Mesosaurus,* and age of the oceanic sediments.

The earth's magnetic field is nearly a dipole field, but the departures from a simple dipole are enough so that the south magnetic pole is not antipodal to the north magnetic pole.

The dynamo theory explains the earth's magnetic field. In this theory, an electric current moves through the earth's conducting core, creating a magnetic field. The conducting core moves through this magnetic field, and a current is induced. This current starts the whole cycle over again.

Remanent magnetism is the magnetism frozen in a rock at the time the rock forms.

Basalt and other rocks are magnetized by the earth's magnetic field when they cool through a certain temperature (Curie point) that is well below their melting points.

Paleomagnetic (old magnetic) data reveal the north magnetic direction and latitude of ancient rocks, but not where on that latitude the rock formed.

The earth's magnetic field has reversed its polarity numerous times. These changes in magnetic direction are impressed on the new ocean floor created at the mid-ocean ridges and appear as symmetrical bands of alternating magnetic intensity parallel to the mid-ocean ridges.

Sea-floor spreading is the theory that new oceanic crust is being created at the mid-ocean ridges, is then moving away from the ridges, and is being consumed at the submarine trenches.

The bottom sediments on the sea floor are younger and thinner near the mid-ocean ridges and become thicker and older away from the ridges. The oldest sediments so far found are sandstones about 200 million years old.

Plate tectonics is the theory that new lithosphere is created at mid-ocean ridges and consumed at volcanic arcs.

Crustal plates are bounded by mid-ocean ridges, submarine trenches, and transform faults. They are the units of the earth's crust that move in the theory of sea-floor spreading. Plates can be composed of either continent or ocean or both.

Transform faults are the surfaces where one plate moves past another plate. The direction of movement revealed by seismic study is in the opposite sense to the apparent offset of the mid-ocean ridges associated with the two plates involved.

Geology is based mainly on observation and reasoning, together with a knowledge of the other sciences.

The most important method used by geologists is to plot on maps the locations of rock types. Aerial photography and radar images can aid in mapping.

Studies of other planets are contributing to our knowledge of geology.

QUESTIONS

1. What are the basic laws of geology?
2. What is meant by the term uniformitarianism?
3. What is the evidence for continental drift?
4. Outline the theory of sea-floor spreading.
5. What are the three features that identify the boundaries of plates?
6. Describe the three types of convergent plate boundaries.
7. How is the earth's magnetic field believed to be caused?
8. How is remanent magnetism impressed on rocks?
9. What data on the previous position of a continent are obtained from remanent magnetism?
10. What evidence suggests that the southern continents were once joined?
11. What is the evidence for sea-floor spreading?
12. Describe transform faults.
13. What is the geologist's most important method of study of an area?
14. Why is geology not based on experimentation?

SUGGESTED READINGS

Albritton, C. C., ed., *The Fabric of Geology.* Reading, Mass.: Addison-Wesley Publishing Co., Inc., 1963, 372 pp.

Carrigan, C. R., and David Gubbins, "The Source of the Earth's Magnetic Field," *Scientific American* (February 1979), Vol. 240, No. 2, pp. 118–130.

Dewey, J. F., "Plate Tectonics," *Scientific American* (May 1972), Vol. 226, No. 5, pp. 56–68.

Dickinson, W. R., "Plate Tectonics in Geologic History," *Science* (October 8, 1971), Vol. 174, No. 4005, pp. 107–113.

Geikie, Archibald, *The Founders of Geology.* New York: Dover Publications, Inc., 1962, 486 pp. (paperback). Originally published in 1905 by Macmillan and Co..

Hallam, A., "Continental Drift and the Fossil Record," *Scientific American* (November 1972), Vol. 227, No. 5, pp. 56–66.

Heezen, B. C., and I. D. MacGregor, "The Evolution of the Pacific," *Scientific American* (November 1973), Vol. 229, No. 5, pp. 102–112.

Heirtzler, J. R., and W. B. Bryan, "The Floor of the Mid-Atlantic Rift," *Scientific American* (August 1975), Vol. 233, No. 2, pp. 78–90.

James, David E., "The Evolution of the Andes," *Scientific American* (August 1973), Vol. 229, No. 2, pp. 60 –69.

Macdonald, K. C., and B. P. Luyendyk, "The Crest of the East Pacific Rise," *Scientific American* (May 1981), Vol. 244, No. 5, pp. 100–116.

Marsh, B. D., "Island-Arc Volcanism," *American Scientist* (March-April 1979), Vol. 67, No. 2, pp. 161–172.

McKenzie, D. P., and Frank Richter, "Convection Currents in the Earth's Mantle," *Scientific American* (November 1976), Vol. 235, No. 5, pp. 72–89.

Molnar, Peter, and Paul Tapponnier, "The Collision between India and Eurasia," *Scientific American* (April 1977), Vol. 236, No. 4, pp. 30–41.

Short, N. M., and others, *Mission to Earth: Landsat Views the World.* NASA SP-360. Washington, D.C.: U.S. Government Printing Office, 1977, 459 pp.

Valentine, J. W., and E. M. Moores, "Plate Tectonics and the History of Life in the Oceans," *Scientific American* (April 1974), Vol. 230, No. 4, pp. 80–89.

Wilson, J. T., "Some Aspects of the Current Revolution in the Earth Sciences," *Journal of Geological Education* (October 1969), Vol. 17, No. 4, pp. 145–150.

2

Geologic Time—Dating Geologic Events

Historical geology, the history of the earth, is perhaps the most important goal of geology. In this chapter the methods used to determine geologic history will be described. Some of these methods have already been discussed, such as interpretation of the formation of a rock based on study of the rock itself and its relationship to other rocks. Simple examples are the interpretation of a sandstone composed of well-rounded quartz grains as probably the product of deep weathering and long transport, and recognition that an intrusive rock is younger than the rock it intrudes. Geologic history is determined from interpretation of the way rocks were formed and the dating of these events. This chapter is concerned mainly with methods of dating so that rocks and events can be placed on a time scale.

RELATIVE AND ABSOLUTE DATES

Rocks can be dated both relatively and absolutely. **Relative dates** are of two types. The simpler of these is used to determine the sequence of events in a limited area. This method uses *superposition and cross-cutting relationships.* Beds are dated by *superposition from the knowledge that in undisturbed sedimentary rocks, the oldest beds are at the bottom of the sequence and the youngest at the top* (Fig. 2.1A). **Cross-cutting features,** *such as faults and intrusions, like dikes and batholiths, are younger than the rocks they cut* (Fig. 2.1B). This type of relative date is determined from the relationships among the rocks, and these relationships are determined by geologic

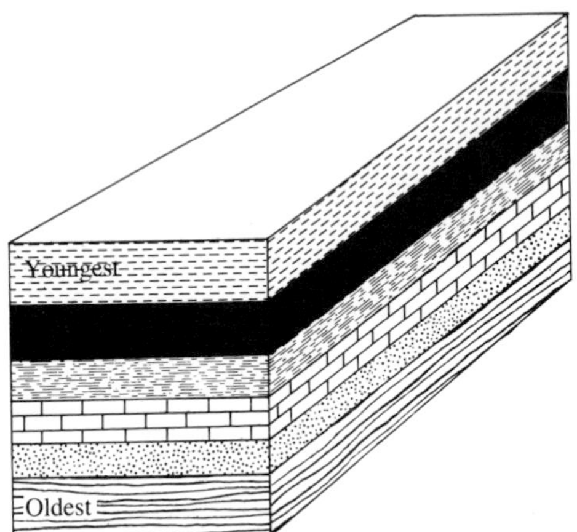

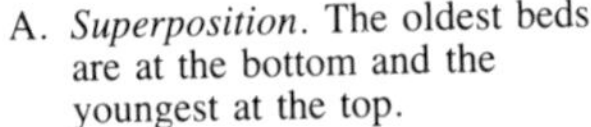

A. *Superposition*. The oldest beds are at the bottom and the youngest at the top.

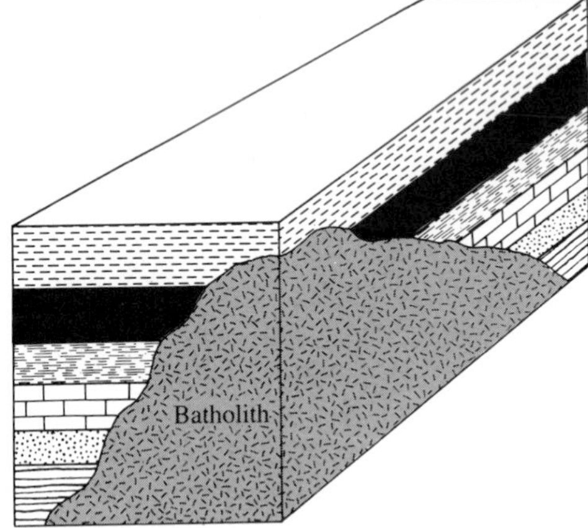

B. *Cross-cutting*. The batholith is younger than the beds that it intrudes.

FIGURE 2.1
Relative dating.

mapping. Geologic mapping is the main method of geologic study and consists of plotting the locations of rock units on an accurate map. *The second way of relative dating is using a geologic time scale based on the study of the fossils in the rocks.* **Absolute dating** is a more recent development and *is based on radioactive decay of elements in rocks;* this method has been used to calibrate the relative geologic time scale developed from the study of fossils.

GEOLOGIC TIME

The recognition of the immensity of geologic time and the development of methods for subdividing geologic time are among the great intellectual accomplishments of humans. It is difficult to discuss the development of ideas in generalities, but a brief outline is presented here. Discoveries made near the end of the eighteenth century led to rapid progress, although there had been earlier studies. The culmination of this early work was the theory now known as *uniformitarianism,* published by James Hutton in 1785 and 1795. Uniformitarianism simply means that the present is the key to the past; that is, the history of old rocks can be interpreted by noting how similar rocks are being formed today. Hutton's book was poorly written, and the theory did not gain wide acceptance until it was popularized by his colleague John Playfair in 1802. Uniformitarianism is a simple idea, but it is the keystone of modern geology. A simple example will show the power of this theory. Many varieties of shell-bearing animals can be seen in the mud along the seashore at low tide. A similar mudstone with similar shells encountered in a canyon wall is interpreted, using uniformitarianism, as a former sea bottom that has been lithified and uplifted. If this seems too elementary, we need only remember that just a few years before Hutton's time, only a few people recognized that fossil shells were evidence of once-living organisms.

Recently uniformitarianism, as just presented, has been criticized as being oversimplified. The objection is that geological processes have modified the earth; therefore, at least in some cases, these processes do not operate now in the same ways that they did in the past. Thus, the present may not be the key to the past. For example, it will be shown that after the origin of life, organisms changed the composition of the atmosphere, making further creation of life impossible. Thus, though uniformitarianism in its simplified version can explain many geologic phenomena, the doctrine has limitations that must be taken into account.

Geologic history is based largely on the study of sedimentary rocks and their contained fossils. Such studies have led to the development of a geologic time scale showing the relative ages of rocks. In recent years it has been possible, using measurements of the radioactive decay of certain elements in rocks, to determine the absolute ages of some rocks. This ability to date igneous and metamorphic rocks has expanded the range of historical geology.

After several false starts, historical geology developed in the last part of the eighteenth century and the first half of the nineteenth. In 1782 the great French chemist Antoine Lavoisier demonstrated that near Paris the quarries dug for pottery and porcelain clay all exposed the same sequence of sedimentary rocks. Georges Cuvier and Alexandre Brongniart published maps in 1810 and 1822 showing the distribution of the various rock types around Paris, and in 1815 William Smith published a geologic map of England. Cuvier and Brongniart studied the fossils in each of the sedimentary layers that they mapped and discovered, as did Smith, that each layer contained a different group of fossils. They discovered that a sedimentary bed could be identified by its fossils. That work set the scene for the next 20 years, for during this period the geologic time scale was developed.

The geologic time scale was the result of individual work by a number of people in western Europe. These men studied the rocks and their fossils at well-exposed places, generally near their homes, and described these rocks and their fossils in books and papers. They called these sequences of rocks systems and named each of the systems. As the studies proceeded in this haphazard manner, the general sequence of older to younger was recognized and the gaps were filled in. It was shown also that rocks in similar stratigraphic position, although far removed from the type area, contained the same or very similar fossils. Thus it took work by many people spread over a wide area to demonstrate that fossils can be used to date rocks, not only in local areas, but all over Europe. In this way a geologic column, or time scale, was developed as a standard of reference; fossils from other areas could be correlated or dated by comparison with fossils on which the time scale was based. The geologic time scale in current use is shown in Figure 2.2. Almost all of the systems or periods were originally defined in the first half of the nineteenth century.

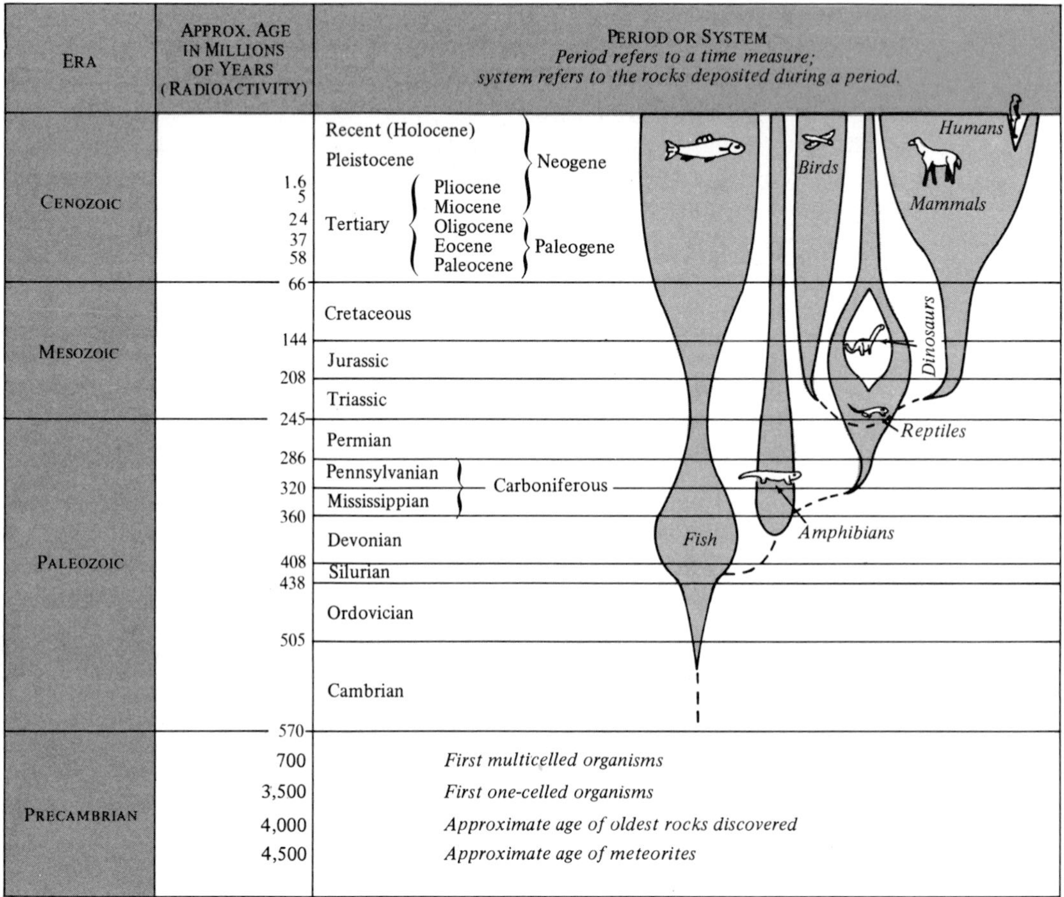

FIGURE 2.2
The geologic time scale. Shown to the right is a very simplified diagram showing the development of life. Not included on the diagram are many types of invertebrate fossils such as clams, brachiopods, corals, sponges, snails, and so forth, which first appeared in the Cambrian or Ordovician and have continued to the present.

The units used in historical geology are of several types. The **systems** *just described are the rocks deposited during a time interval called a* **period.** Thus the Cambrian System was deposited during the Cambrian Period. The systems and periods are arbitrarily defined, and their recognition away from the type area where they were defined is based on interpretation of fossils. A geologist mapping an area locates on the map the occurrence of *easily recognized, objective rock units called* **formations.** Formations are rock units named for a location, a type area, where they are well exposed; examples are the Austin Chalk and the Tensleep Sandstone.

TABLE 2.1
Stratigraphic units.

A. Time and Time-Rock Units

Time Units	*Time-Rock Units*
Eon	—
Era	Erathem
Period	System
Epoch	Series
Age	Stage
(Phase)	Chronozone

B. Rock Units

Group
Formation
Member
Biostratigraphic Units
Biozone (Zone)

All parts of a formation are not necessarily the same age. The sandstone bed in part B of Figure 2.9, for example, might be defined as a formation. A second type of rock unit is the biostratigraphic unit called the **biozone** (zone). A *biostratigraphic unit is defined by the occurrence of one or more types of fossils in the enclosing rocks.* Thus there are three kinds of units: time units such as periods, time-rock units such as systems, and rock units such as formations and biozones. Table 2.1 summarizes the terms used.

The changes in marine invertebrates during geologic history are obvious to even a casual observer; for this reason such fossils can be used to date the enclosing rocks. (See Fig. 2.3.) The changes between periods are not as great but are just as obvious to a trained observer. The most marked changes between adjacent periods are between Cambrian and Ordovician, Permian and Triassic, and Cretaceous and Cenozoic.

The geologic column provided only a relative time scale to which rocks could be referred. It also showed the general development of life and so set the stage for Darwin, who published his theory of evolution in 1859.

UNCONFORMITIES

Once the time scale was set up, geologists traced the systems into previously unmapped areas. One problem that soon appeared was that the systems were

A

B

FIGURE 2.3
Brief comparison of these two fossiliferous slabs of rock shows the great changes in life that have occurred during geologic time. A. Pennsylvanian fossils from Kansas. Bryozoans and brachiopods are most abundant. Photo courtesy Ward's Natural Science Est., Inc. B. Miocene fossils from Virginia. Snails and clams are most abundant. Photo by W. T. Lee, U.S. Geological Survey.

originally defined, in many cases, as the rocks between two important breaks in the geologic record. In other areas there were no breaks, but continuous deposition. Thus, between the original systems were intervals not included in either system; and arguments about the assignment of the neglected intervals are, in some cases, still in progress. The breaks are of several types, such as change from marine to continental deposition, influx of volcanic rocks, and unconformities. (See Fig. 2.4.) The last is the most important and requires discussion.

All breaks in the geologic record that indicate a time interval for which there is no local record are called **unconformities** (see Figs. 2.5 and 2.6); they may take several different forms.

If sedimentary rocks overlie metamorphic or granitic rocks, the sediments were deposited after the time of metamorphism or intrusion. Granitic and metamorphic rocks form deep in the earth, so the erosion that uncovered them required time. In this case, there are no sedimentary rocks representing the time of intrusion or metamorphism and the time required for erosion. This type of unconformity is called a nonconformity. **Nonconformities** *occur where sedimentary rocks are deposited on igneous or metamorphic rocks that formed at some depth in the crust.*

Another type of unconformity, called a **disconformity,** *is marked by a change in fossils, representing a short or a long period of time, that occurs between two parallel beds in the sedimentary section.* Such an unconformity may be due to erosion of previously deposited beds or to nondeposition and may not be at all conspicuous until fossils are studied.

The third type of unconformity is more obvious and consists of folded and eroded sedimentary rocks that are overlain by more sedimentary rocks. Such an unconformity is called an **angular unconformity** *because the bedding of the two sequences of sedimentary rocks is not parallel.* In this case, the time of folding and erosion is not represented by sedimentary rocks.

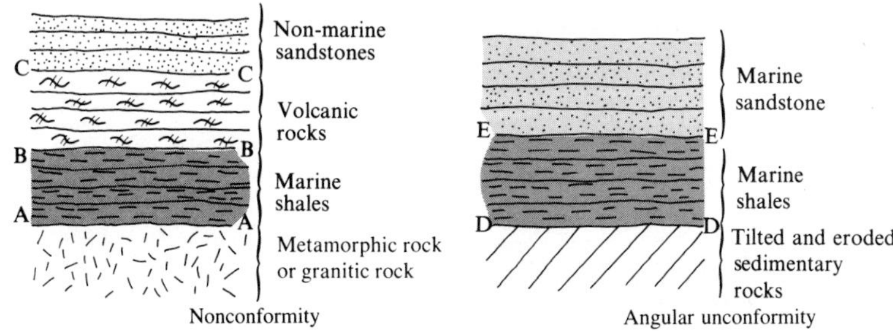

FIGURE 2.4
Natural breaks in the stratigraphic record. A-A and D-D are unconformities because the time necessary for erosion is not represented by sedimentary rocks. The other breaks may or may not be unconformities, depending on whether there is a time interval between the different rock types or continuous deposition. D-D is an angular unconformity.

A

B

FIGURE 2.5
Unconformities. A. An angular unconformity in the Grand Canyon. The rocks beneath the unconformity are Precambrian in age and those just above it are Cambrian. Photo by L. F. Noble, U.S. Geological Survey. B. An erosional unconformity in Silurian rocks in Niagara Gorge, N. Y. Photo by G. K. Gilbert, U.S. Geological Survey.

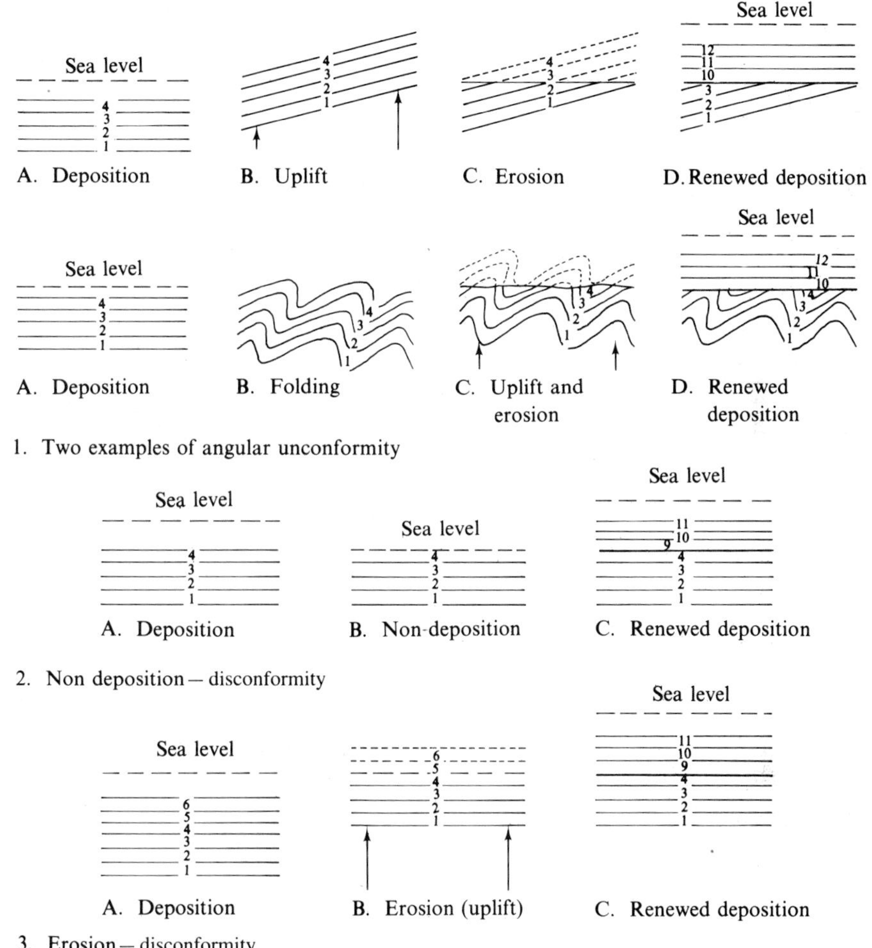

FIGURE 2.6
Development of unconformities. In each case a time interval is not recorded in the bedded rocks. It is commonly difficult to distinguish between cases 2 and 3. In each case, for simplicity, sea level is assumed to be the base level that controls whether erosion or deposition takes place.

CORRELATION AND DATING OF ROCKS

So far we have considered how the geologic systems were defined, and it is obvious that they are recognized in previously unstudied areas by comparing the contained fossils with the fossils from originally defined type areas. In this manner the systems defined in Europe were recognized in North America and the other continents. *This type of correlation by fossils is also most important in establishing the time equivalence of nearby beds.*

Another type of correlation is to establish that two or more outcrops are parts of a once continuous rock body. The simplest method of such rock correlation is actually to trace the beds from one area to the other by walking out the beds. In other cases, distinctive materials in a bed, such as fragments of an unusual rock type or a distinctive or uncommon mineral, or assemblage of minerals, may be used to identify a bed. Another method uses the sequence of beds. (See Fig. 2.7.)

By correlating from one area to another, the total thickness of sedimentary rocks in a region can be determined as shown in Figure 2.8.

Note that the methods of correlation, with the exception of the use of fossils, establish only the continuity of a sedimentary bed. Fossils establish the fact that both beds are the same age. The difference between these two types of correlation can be seen by considering an expanding sea, a not uncommon occurrence in geologic history. As the sea advances, the beach sands, for example, form a continuous bed of progressively younger age. This is shown diagrammatically in Figure 2.9.

This diagram also illustrates another difficulty in the use of fossils. Note that limestone, shale, and sandstone are all being deposited at the same time in different parts of the basin. Each of these different environments will attract and be the home of different types of organisms. Thus, the fossils found in each of these environments will be different, even though they are of the same age. This means that environment must be taken into account in establishing the value of a fossil or group of fossils in age determination. For this reason, free-swimming animals make the best fossils because their remains are found in all environments. *Fossils that establish clearly the age of the enclosing rocks are very useful and have been called* **index fossils.** Ideally, an index fossil should be widespread in all environments, be abundant, and have a short time span. Very few fossils meet all of these

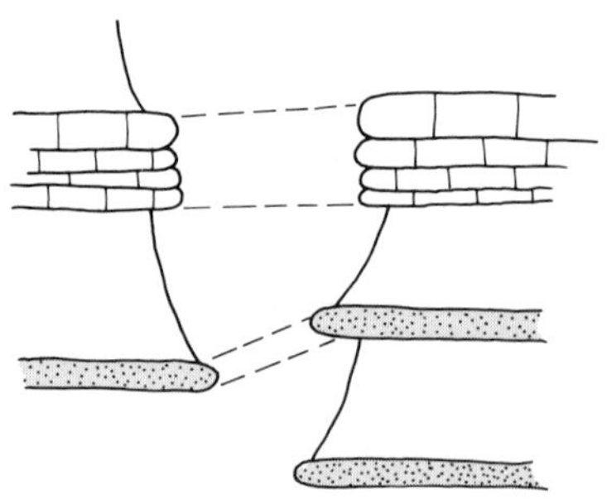

Correlation by sequence of beds. Note the changes in thickness in the two areas.

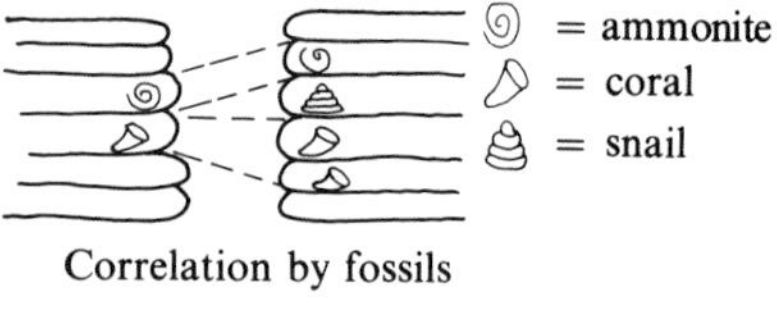

Correlation by fossils

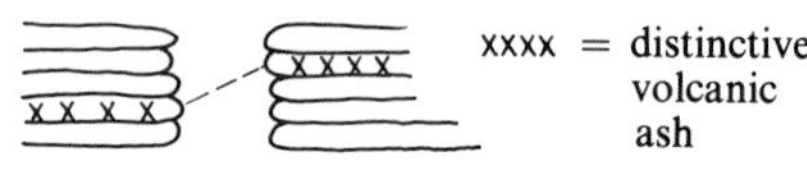

Correlation by lithologic similarity. Also shows use of a keybed (an easily recognized bed)

FIGURE 2.7
Types of correlation.

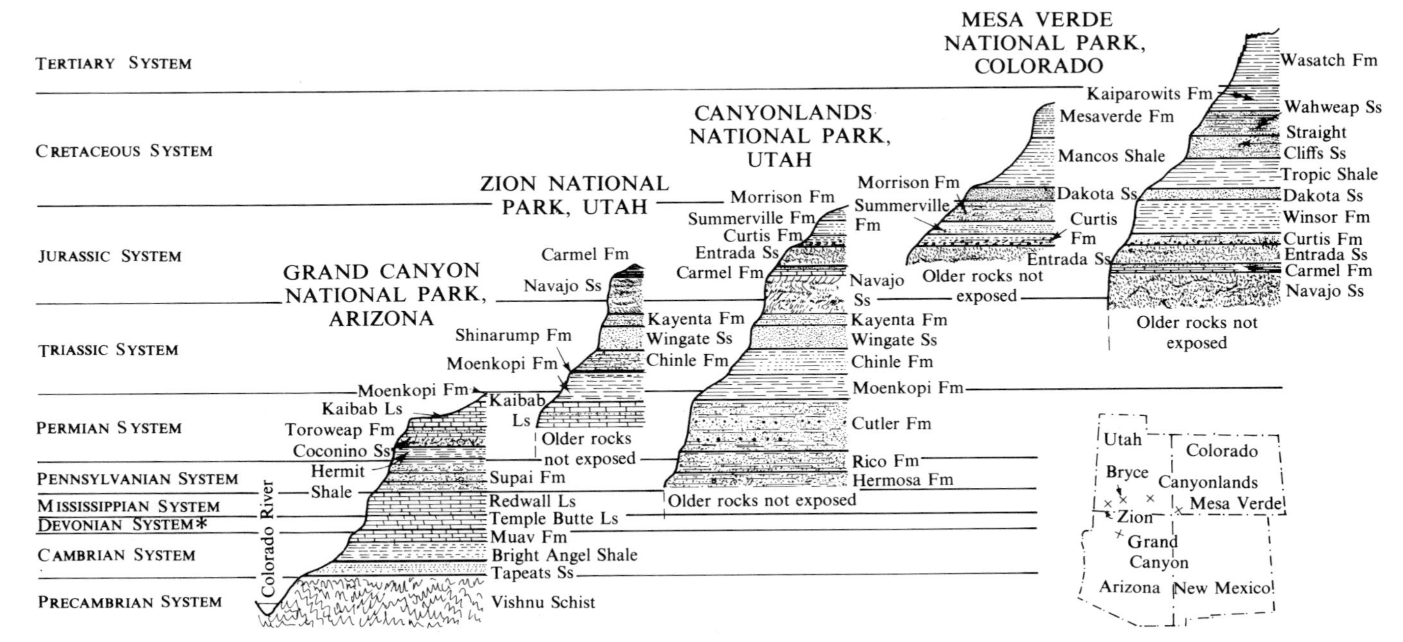

FIGURE 2.8
Correlation of the strata at several places on the Colorado plateau reveals the total extent of the sedimentary rocks. After U.S. Geological Survey.

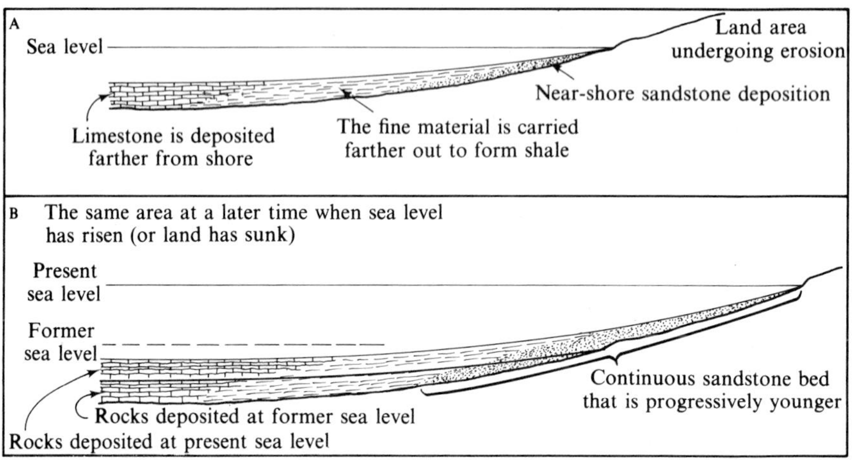

FIGURE 2.9
Deposition in a rising, or expanding, sea. A. The deposition at the start. B. Deposition at a later time after sea level has risen.

requirements, and generally, groups of fossils are used to establish the age of a bed. (See Figs. 2.10 and 2.11.)

Time-lines, or **surfaces,** can be established in sedimentary rocks by physical means alone in some rare instances. For example, a thin ash bed from a volcano may be spread over a very large area by wind. The deposition of such an ash bed may occur over a number of days or even weeks, but in terms of geologic time it is instantaneous. These rare occurrences help in local correlation problems and show the validity of the methods of dating and correlation described here.

The only rocks that have reasonably abundant fossils are marine sedimentary rocks, but not even all of these have enough fossils to establish their age clearly. On the previous pages, we have discussed dating of fossiliferous marine rocks. Similar techniques can be used to date continental sediments, but here the number of fossils is generally much smaller. The fact is easy to understand if one compares the number of easily fossilized animals exposed at low tide at the seashore with the

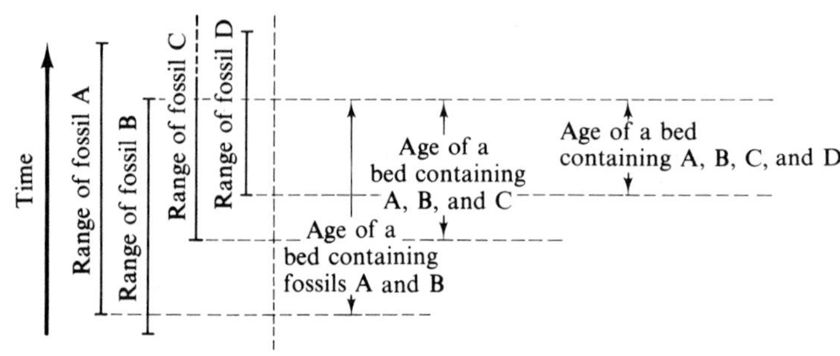

FIGURE 2.10
The use of overlapping ranges of fossils to date rocks more precisely than can be accomplished by the use of a single fossil.

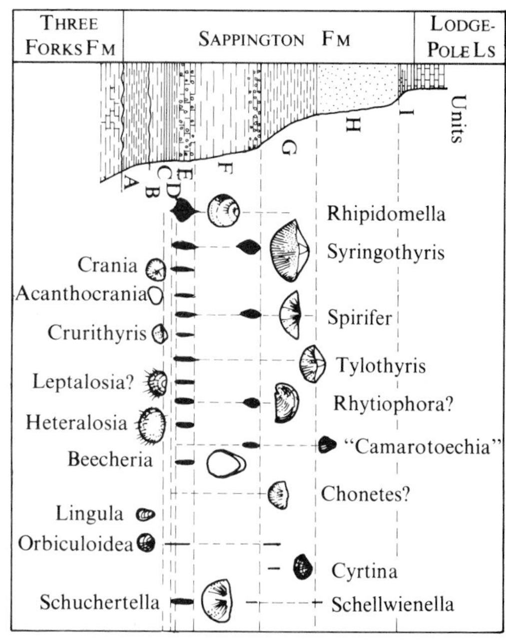

FIGURE 2.11
An example of the distribution of brachiopods in the Sappington Formation in western Montana. The relative number of each fossil found in the beds is indicated by the thickness of the black lines. The occurrence of these animals was at least in part controlled by the sedimentary environment. From Gutschick and Rodriquez, 1967.

much smaller number of easily fossilized animals living in a similar-sized area of forest or grassland, and if one considers how easy it is for a clam shell to be buried and so preserved, especially when compared to the chance that a plant leaf or a rabbit skeleton will be buried and so preserved as a fossil. A land organism, even if buried, may not be preserved because oxidizing conditions that destroy organic material exist even at some depth. Also, a dead rabbit is apt to be eaten and dismembered by scavengers. Another problem in the use of continental fossils is that they are controlled even more by climate than are marine organisms. The progression of types of plant and animal life seen during a mountain climb will illustrate this point.

Dating igneous and metamorphic rocks presents even more problems because such rocks almost never contain fossils. They must be dated by their relationships to fossil-bearing rocks. Volcanic rocks are dated by the fossiliferous rocks above and below and, in some cases, by interbedded sedimentary rocks. Intrusives must be younger than the rocks that they intrude; they cannot be dated more closely unless the intrusive has been uncovered by erosion and, then, unconformably covered by younger rocks. (See Fig. 2.12.)

Metamorphic rocks are even more difficult to date because both the age of the original rock and the date of the metamorphism should be obtained. The age of the parent rock may be found by tracing the metamorphic rock to a place where fossils are preserved. Finding the date of metamorphism may be more difficult because, in general, an unconformable cover, if present, gives only the upper limit.

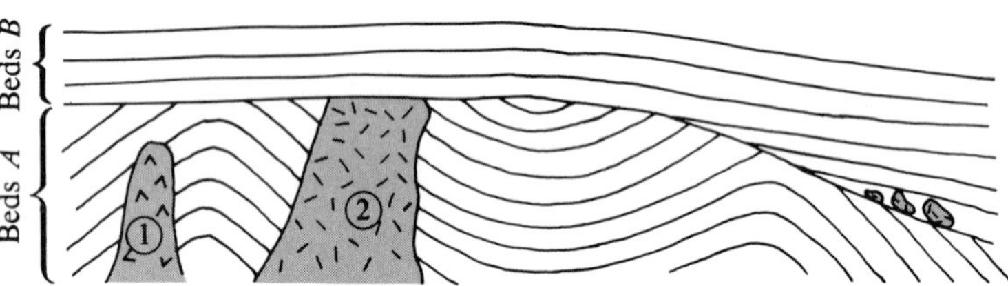

FIGURE 2.12
Dating of intrusive rocks. Intrusive body ① is younger than beds A; no upper limit can be determined. Intrusive ② is younger than beds A and older than beds B. Closer dating may be possible if the conglomerate composed of pebbles of ② in the basin at the right contains fossils. The contact between beds A and B is an angular unconformity.

RADIOMETRIC DATING

Some of these problems in dating igneous and metamorphic rocks can be overcome by modern radiometric methods. Such methods produce an *absolute age* in years and so have been used, in addition, to date the geologic time scale in years. (See Fig. 2.13.) In spite of the seeming precision of radiometric dating, it is not yet possible, with a few exceptions, to date a rock as accurately this way as with fossils. This is because of inherent inaccuracies in the measurements of the amounts of the elements produced by radioactive decay. These inaccuracies produce an uncertainty in the radiometric date that is more than the span of zones based on fossils, especially in those parts of geologic time that contain abundant fossils of short-lived organisms. As methods of analysis are improved, this limitation may be overcome.

Not all atomic nuclei are stable. Radioactive nuclei disintegrate spontaneously, releasing energy in the process. (See Fig. 2.14.) Understanding why and how this takes place was one of the great advances in physics during the first half of the

FIGURE 2.13
Determination of absolute age of one of the geologic periods. The sandstone contains Cambrian fossils. Dikes A and B have been dated by radiometric methods. The Cambrian Period is younger than dike A and older than dike B. Can the absolute age of the Cambrian be determined from a single occurrence such as this?

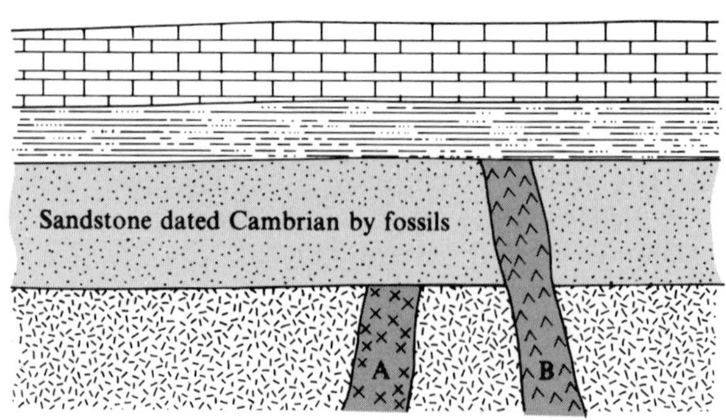

FIGURE 2.14
The mineral uraninite. A. An ordinary photograph. B. An autoradiograph made by placing photographic film next to the mineral in a darkroom. The radiation from the uraninite exposes the film even though no visible light reaches the film. This can be done without a darkroom by wrapping the film in opaque paper and laying the flat mineral on the film. Exposure times are generally a few weeks. Photo courtesy Ward's Natural Science Est., Inc.

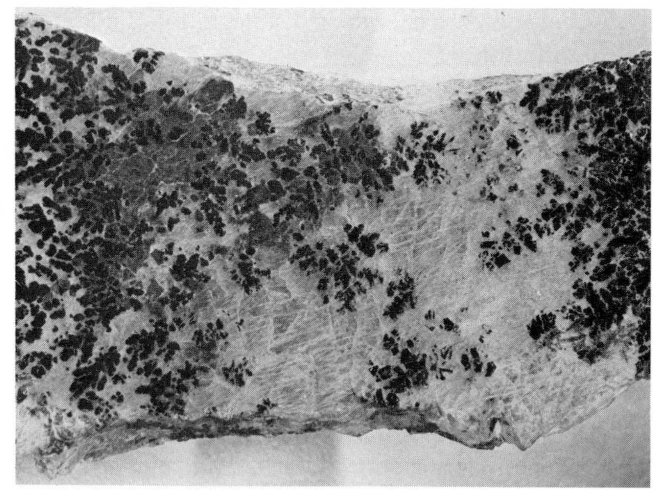

A

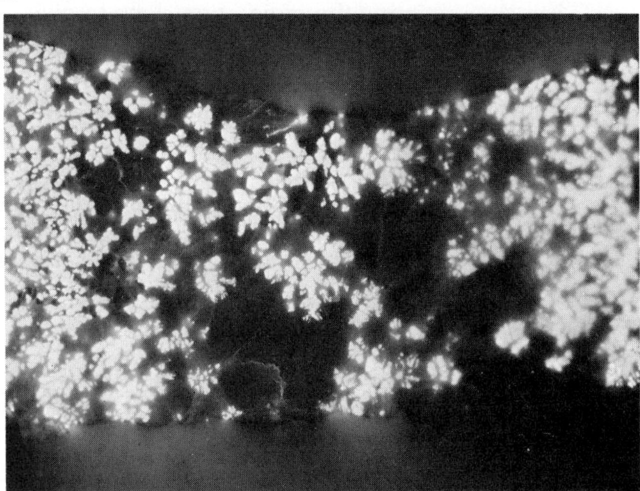

B

twentieth century. In natural radioactivity, the unstable nucleus emits several types of high-energy particles and also releases energy in the form of electromagnetic waves similar to light energy.

In radioactive decay, an atom is changed to an atom of another element. This means that the number of atoms of a radioactive element decreases with the passage of time. *If the rate of disintegration is known, then by measuring the amounts of the parent and daughter elements, the age of crystallization of the mineral containing the parent element can be found.* This is the principle of radiometric dating, which works only because the rate of disintegration is not affected by temperature, pressure, and so forth.

The rate of radioactive decay is expressed in terms of the **half-life.** *In one half-life, one-half of the radioactive atoms present will decay to the daughter element.* In the next half-life, one-half of the remaining radioactive atoms will decay, and so on. (See Fig. 2.15.)

Radiometric dates are obtained by studying the daughter elements produced by a radioactive element in a mineral. Sometimes more than one element is studied in an effort to provide a check on accuracy. A radiometric date tells the time of formation of that mineral and so tells the time of crystallization of an igneous rock or the time of metamorphic recrystallization in the case of a metamorphic rock. In some cases, in which microscopic studies show that different minerals formed at different times in a rock that has undergone more than one period of metamorphism, it is possible to date the periods of metamorphism by dating elements in minerals formed during each period of metamorphism. In many radioactive decays, a gas is one of the products, and heating during metamorphism may drive off the gas, thus resetting the radioactive clock to give the time of metamorphism.

A radioactive element changes to another element by spontaneously emitting energy. The rate of this decay is unaffected by temperature or pressure; hence, if we know the rate of formation of the daughter products, all we need to find is the ratio of original element to daughter product to be able to calculate the time of crystallization of the mineral containing the original radioactive element. The main assumptions, then, in using radiodates are that the decay rate is known and constant and that no daughter or parent elements are lost. The latter assumption is the least sound because of weathering and metamorphism.

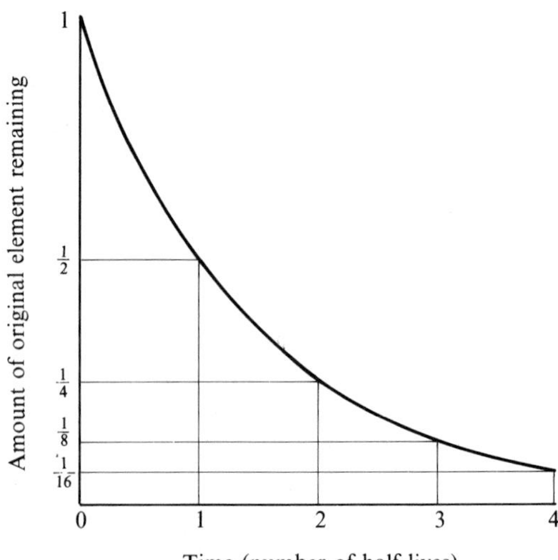

FIGURE 2.15
Rate of radioactive decay. During each half-life, one-half of the remaining amount of the radioactive element decays to its daughter element.

Several radioactive elements are used to date rocks, but only the most important are reviewed here. A few minerals contain uranium, and these generally also contain thorium, which is about the same atomic size. These minerals are rare and with a few exceptions are almost completely confined to pegmatite veins in batholiths. Because such veins are probably the last to crystallize, these occurrences give only the upper limit of the age of the batholith. Each of the three radioactive isotopes that these minerals contain—the two of uranium as well as that of thorium—produces a series of radioactive daughter products; each member of each series decays to the next daughter until, finally, a stable isotope of lead is produced. These many steps constitute a possible source of inaccuracy, as it is possible for any one of the daughters to be removed by leaching or some other process. The elements and the final products are

Uranium 238 → Lead 206
Uranium 235 → Lead 207
Thorium 232 → Lead 208

Thus, in this case, it is possible to measure three ratios, which should all agree in age.

A problem with age determinations involving lead is that many rocks contain some lead of nonradioactive origin. The amount of nonradiogenic lead is quite small but may require an appreciable correction in calculating the age of very old rocks. This ordinary lead contains the isotope lead 204, so its presence can be easily recognized. The problem is that along with the minor amounts of lead 204 are the three isotopes lead 206, lead 207, and lead 208, which also occur radiogenically, and the ratio of all four isotopes varies. Thus to correct for the preexisting ordinary lead, one must analyze nonradioactive rocks in the area to find the ratio between lead 204 and each of the other isotopes. In this way the nonradiogenic lead 206, lead 207, and lead 208 can be excluded from that due to radioactive decay. Meteorites contain lead 204, and the correction for the original lead is a problem. Meteorites are used to date the formation of the earth, so this problem results in some uncertainty in this important date.

Rubidium is a radioactive element that is generally present in small amounts in any mineral that contains potassium because both elements are about the same size. Rubidium decays in a single step to strontium (rubidium 87 → strontium 87). Because potassium minerals (feldspar and mica) that contain rubidium are so common and only a single radioactive decay is involved, this process may become the most useful method of radiometric dating. However, at present the half-life of this decay cannot be measured accurately because the low energy of the beta particles emitted makes them difficult to detect. Some studies have suggested that the rubidium and strontium isotopes in some rocks may be inherited from their parent materials; thus the rubidium-strontium ages determined for such rocks would tend to be too old.

Potassium also has a radioactive isotope that is useful in dating, especially because it is such a common element. The small amount of radioactive potassium decays to form two daughters:

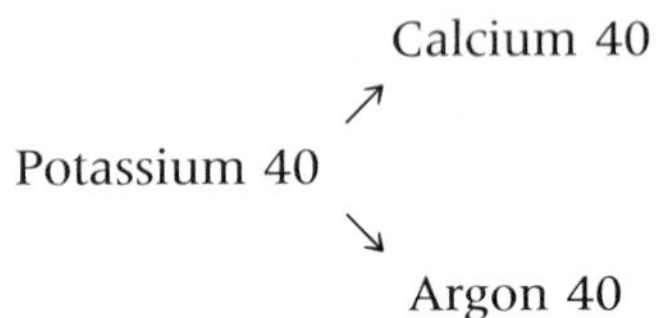

Calcium 40 is ordinary calcium and so is present in most minerals. Therefore, the method used is to measure the amounts of potassium 40 and argon 40. The main problem is finding minerals that retain the argon gas. It appears that micas, in spite of their cleavage, retain most of the argon but feldspar loses about one-fourth of the argon. Potassium 40 is the most commonly used method of radiometric dating.

Another type of radiometric dating is to measure the damage done to a crystal lattice. The crystal structure of minerals containing radioactive elements can be damaged or destroyed by the radiation, especially if the mineral has weak bonding. In other cases radiation produces discoloration, and this can cause gray (smoky) quartz, violet fluorite, and blue or yellow halite. Another common example of radiation damage is the halos, or rings, that commonly surround tiny crystals of radioactive minerals. These small crystals generally occur as inclusions in later-formed minerals; dark halos are common, for example, in biotite. Uranium and thorium minerals emit particles with various amounts of energy, and because the particles do most of their damage near the end of their movement, traveling a distance determined by the energy of each particle, they form a series of concentric halos. (See Fig. 2.16.) The amount of damage done to a mineral (the thickness and darkness of the halo) can be used to calculate the age of the mineral, but so far such dating methods have met with only limited success. Uranium 238 also decays very slowly by spontaneous fission. The high-energy fission particles released in this way do much damage to the enclosing crystal. Under high magnification with an electron microscope, the paths of such particles can be seen in some minerals after etching with acid. The path is a small tube, not unlike a bullet hole. (See Fig. 2.17.) By counting the tubes, the number of fission disintegrations that have occurred can

FIGURE 2.16
Radiation halo. Such halos are commonly seen in biotite crystals, where they surround tiny crystals of radioactive minerals such as uraninite. Typical size is less than one-tenth of a millimeter.

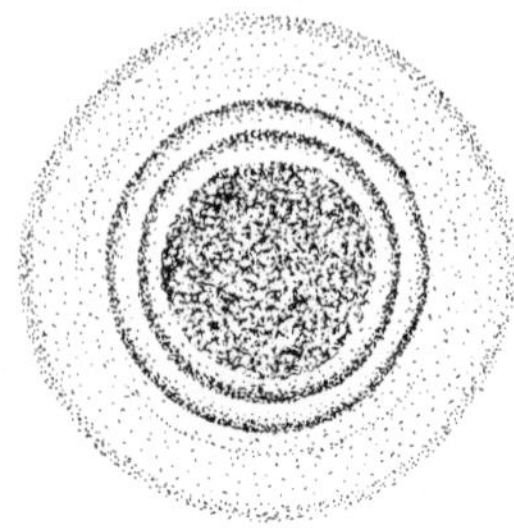

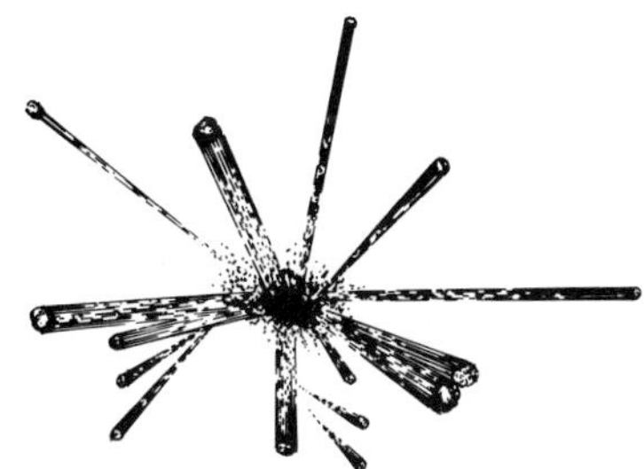

FIGURE 2.17
Fossil fission tracks. The tubes are caused by high-energy radioactive particles escaping from the mineral at the center. The tubes are much like bullet holes. The longer tubes are about one one-hundredth of a millimeter in length.

be determined. The number of remaining uranium atoms, the source of the particles, can be determined by instrumental analysis, and thus the age of the mineral can be calculated. This method is useful for dating material that is a few tens of years old to the oldest objects in the universe.

Radiation can increase the energy of an atom by moving its electrons into different orbital shells and by trapping within the atom other electrons from beta particles or from other atoms. Remember that each electron belongs in a certain shell. This energy can be released by freeing the electrons so that they can escape or move back to the original shell. One way to do this is to excite the atom by heating it. The energy is released in the form of light. This phenomenon can be seen easily by heating a mineral such as fluorite in a frying pan in a dark room. The complete explanation of this phenomenon would require much deeper knowledge of atomic structure. Because the amount of energy stored depends on the length of time a mineral has been exposed to radiation, the age of some minerals can be determined by measuring the amount of stored energy. So far this has not proved to be an accurate method of dating minerals.

Carbon-14 dating is somewhat different. Cosmic rays come to the earth from space and produce neutrons in the atmosphere. A neutron colliding with a nitrogen atom knocks a proton out, converting the nitrogen to carbon 14:

$$^{14}_{7}\text{N} + {}^{1}n \rightarrow {}^{14}_{6}\text{C} + {}^{1}_{1}\text{H}$$

Carbon 14 is radioactive and decays by beta emission to nitrogen 14:

$$^{14}_{6}\text{C} \rightarrow {}^{14}_{7}\text{N} + \text{B}^{-}$$

These reactions are used in the carbon-14 method of dating, which is effective in dating carbonaceous material, such as wood or coal, that is less than 40,000 to 50,000 years old. This method depends upon the fact that all living matter contains a fixed ratio of ordinary carbon 12 and radioactive carbon 14. Cosmic rays react with nitrogen in the atmosphere to form the carbon 14, which ultimately gets into all living matter via the food cycle. The carbon 14 decays back to nitrogen. Therefore, the ratio of carbon 12 to carbon 14 can be used to calculate how long an organism has been dead because a dead organism takes in no more carbon 14. Extensive use of this method is made in archeology and in glacial studies.

SUMMARY

Rocks can be dated both relatively and absolutely.

Relative dates can be determined by the relationships of the rock units to each other by using superposition and cross-cutting relationships and by referring the rocks to a geologic time scale based on the fossils in the rocks.

Absolute dating is based on radioactive decay of elements in the rocks and has been used to calibrate the relative geologic time scale on the basis of fossils.

Uniformitarianism is the basis of geological interpretation and means that ancient rocks are interpreted as having formed in the same ways that similar rocks are presently forming, or "the present is the key to the past."

Systems are the rocks deposited during a time interval called a period.

Formations are rock units that can be mapped easily. The other stratigraphic units are summarized in Table 2.1.

The most marked changes in life occurred at the ends of the Cambrian, Permian, and Cretaceous systems.

The development of life through geologic time was important evidence for Darwin's evolution theory and the recognition of the immensity of geologic time.

Unconformities are breaks in the geologic record that indicate a time interval for which there is no local depositional record. If the older rocks were deformed so that their bedding planes are not parallel to the younger rocks, an angular unconformity results. If the bedding planes are parallel above and below the unconformity, a disconformity exists. Where sedimentary beds overlie granitic or metamorphic rocks, a nonconformity exists.

Correlation and dating of rocks can be accomplished in many ways. Comparison of fossils can establish the time equivalence of beds. Rock correlation can establish that two outcrops are parts of a single rock body. This can be accomplished if the bed has distinctive properties that can be recognized, from a similar sequence of beds, or by actually tracing the beds on the ground.

A lithic body, such as a sandstone unit, is not necessarily the same age throughout.

Index fossils should be found in all environments, be abundant, and have a short time span. Very few fossils meet these requirements, so most dating uses groups of fossils. Fossils are most abundant in marine sedimentary rocks.

Igneous rocks are dated by their relationships to sedimentary rocks or by radiometric methods. An intrusive rock is younger than the rocks that it intrudes.

In dating metamorphic rocks, the age of the parent rock and the age of metamorphism should be determined. They are dated by tracing them to places where they are not metamorphosed, by their relationships to sedimentary rocks, and by radiometric methods.

A radioactive element spontaneously changes into another element at a rate that is unaffected by temperature and pressure.

The more important methods of radiometric dating are

Uranium 238 → Lead 206

Uranium 235 → Lead 207

Thorium 232 → Lead 208

Rubidium 87 → Strontium 87

Calcium 40
↗
Potassium 40
↘
Argon 40

Carbon 14

Neutrons from cosmic rays colliding with nitrogen in the atmosphere produce carbon 14.

QUESTIONS

1. Might you find fragments of either the underlying bed or the overlying bed in a layer of coarse sandstone? Why?
2. Describe in general terms how the geologic column was established and discuss whether it consists of natural subdivisions in North America.
3. What are unconformities? Describe the various types.
4. How was the value of fossils in determining geologic age established?
5. What is meant by correlation?
6. List and briefly describe the methods of correlating rocks.
7. Why is it generally easier to date marine rocks than continental sediments?
8. How are metamorphic rocks dated? What dates are required?
9. Which method of radiometric dating is best and why?
10. What are index fossils?
11. What effect would a small amount of modern root material mixed with a sample to be dated by the carbon-14 method have?
12. How can an angular unconformity be distinguished from a low-angle thrust fault?
13. Fossils aid in determining the age of a rock unit. How else do they aid in geologic interpretation?
14. Prove to your friend, who is an art major and has not taken any science, that the earth is very old.
15. Distinguish between relative and absolute dating in geology.

SUGGESTED READINGS

Cohee, G. F., M. F. Glaessner, and H. D. Hedberg, eds., Contributions to the Geologic Time Scale. *Studies in Geology,* No. 6. Tulsa, Okla.: American Association of Petroleum Geologists, 1978, 388 pp. (paperback).

Eiseley, L. C., ''Charles Lyell,'' *Scientific American* (August 1959), Vol. 201, No. 2, pp. 98–106. Reprint 846, W. H. Freeman and Co., San Francisco.

Emiliani, Cesare, ''Ancient Temperatures,'' *Scientific American* (February 1958), Vol. 198, No. 2, pp. 54–63. Reprint 815, W. H. Freeman and Co., San Francisco.

Engel, A. E. J., ''Geologic Evolution of North America,'' *Science* (April 12, 1963), Vol. 140, No. 3563, pp. 143–152.

Faul, Henry, *Ages of Rocks, Planets, and Stars.* New York: McGraw-Hill Book Co., 1966, 109 pp. (paperback).

Hay, E. A., ''Uniformitarianism Reconsidered,'' *Journal of Geological Education* (February 1967), Vol. 15, No. 1, pp. 11–12.

Macdougall, J. D., ''Fission-track Dating,'' *Scientific American* (December 1976), Vol. 235, No. 6, pp. 114–122.

Nichols, R. L., ''The Comprehension of Geologic Time,'' *Journal of Geological Education* (March 1974), Vol. 22, No. 2, pp. 65–68.

Ralph, E. K., and H. N. Michael, ''Twenty-five Years of Radiocarbon Dating,'' *American Scientist* (September 1974), Vol. 62, No. 5, pp. 553–560.

3
Reading the Rocks—Environment

INTERPRETING GEOLOGIC HISTORY

Deciphering the history of the earth or any part of it is a difficult reiterative process. It has many steps, and each affects every other step. Multiple working hypotheses must be used to test each step. The process begins with study of the rocks themselves and their relationships to other rocks. This leads to some ideas about their mode of origin. The rocks must be mapped, dated, and correlated, using the methods described in this chapter. If the rocks have been deformed by folding and faulting, their original relationships and positions must be determined. Each layer or rock unit can now be reconstructed to show the geography at its time of formation. Remember that the rock record is incomplete because of erosion and because many rock layers are exposed at only a few places. The final step is to reconstruct the events that caused the changes between each of the steps.

This brief outline shows how the incomplete rock record is interpreted. Note the various levels of abstraction that are involved. It is no wonder that the history of the earth is not completely known. Interpreting the history of the earth or a small area is difficult, and revision is constantly necessary as more data are obtained. Thus geology is an ongoing process that offers great challenge; many fundamental problems remain to be solved.

INTERPRETING SEDIMENTARY ROCKS

A sedimentary rock has the features of the source area, the mode of transportation, the mode of deposition, and the features of the depositional site all impressed on it. Even a beginning student can recognize the excellent sorting, rounding, and frosting of wind-deposited (dune) sand or the mixture of all sizes from boulders to silt of angular, scratched fragments in a glacial deposit (till). Many rocks have features that reveal much about their origins, but the features are not as obvious as in these examples.

A sandstone composed almost entirely of quartz grains can imply that weathering was complete and no grains of other minerals survived or that the source that underwent erosion was itself a quartz sandstone. If the grains in the sandstone are well rounded, it may imply that the other less-resistant mineral grains did not survive long transportation.

A sandstone composed of quartz and feldspar grains (arkose) may be the result of little chemical weathering at the source area, perhaps because of cold temperatures or because of rapid deposition and burial.

Less obvious examples require consideration of other facets of sedimentary rocks.

Features of Sedimentary Rocks

The most noticeable feature of an outcrop of sedimentary rocks is the **bedding,** *which records the layers in the order of deposition, with the oldest at the bottom.* In some instances the beds are too thick to show in a small outcrop, and, for the same reason, many hand specimens do not show obvious bedding.

Study of the types of fossils and of sedimentary features (see Table 3.1) can tell much about the environment of deposition and, in some cases, the postdepositional history. **Mud cracks** on bedding planes record periodic drying and suggest shallow water and, perhaps, seasonal drying (Fig. 3.1). **Ripple-marks,** too, suggest shallow water with some current action, but are also known to form in deep water. Detailed study of ripple-marks can show type and direction of current. (See Figs. 3.2 and 3.3.) Current action can also cause localized scouring or cutting to produce **cut-and-fill features.** (See Fig. 3.4.) Another type of current feature is **cross-bedding.** (See Figs. 3.5 and 3.6.) Sand dunes likewise exhibit cross-bedding. Currents can also align mineral grains and fossils and so impart linear internal structures to rocks that can be used to determine the current direction.

If a mixture of sediments is suddenly deposited into a sedimentary basin, then the large fragments sink faster than the small ones. This does not violate Galileo's famous experiment because here we are concerned with falling bodies in a viscous medium in which velocity of descent is controlled mainly by the size. The type of bedding produced by this kind of sedimentation is called **graded bedding** (Fig. 3.7); we can demonstrate this experimentally by putting sediments of various sizes in a jar of water, shaking, and then allowing the contents to settle. Graded beds can

TABLE 3.1

Physical features of some common environments. All the features indicated should not be expected in every case.

	Rock Type					Bedding			Features				
	Silt, mud	Sand	Gravel	Carbonate	Evaporite	Laminated	Thin	Thick	Sorting	Mud cracks	Cross-beds	Ripple-marks	
MARINE													
Littoral (tidal zone)	X	X	X			X	X		Good	X		X	
Continental shelf	X	X					X	X	Good		X	X	
Carbonate platform (shallow, off-shore)				X			X	X					Oolites
Continental slope	X	*						X	Fair				*Submarine fans with sole marks and graded bedding
Abyssal (deep water)	X*						X		Good				*Ooze
TRANSITIONAL													
Bay, estuary	X						X			X		X	
Lagoon	X				X	X	X			X		X	
TERRESTRIAL													
River	X	X	X						Poor		X	X	Rounded pebbles
Alluvial fan		X	X						Poor				Angular fragments
Lake	X					X			Fair	X			
Swamp	X												Organic material; peat, coal
Glacial	X	X	X						Very poor				Striated pebbles
Glacial lake	X					X*							*Varved (yearly) layers
Dunes		X							Very good		X		Frosted, rounded grains
Playa (desert lake)	X				X	X	X		Fair	X	X		

A

B

FIGURE 3.1
A. Mud cracks at Death Valley, California. B. Ancient mud cracks preserved in a sandstone in New Mexico. Photos from U.S. Geological Survey; A by J. R. Stacy, B by E. F. Patterson.

form in many ways, such as a sudden influx of sediment caused by a storm or seasonal flow of rivers. Their formation at the base of the continental slope is described in the next section.

The **sorting** *of a sediment is a measure of the range in size of its fragments.* If the fragments are all near the same size, the sediment is well-sorted.

Slump features, as shown in Figure 3.8, form if sediments are deposited on a slope or if they are tilted while still soft. Another type of deformation of soft rock is the vertical movement of sediment shown in Figure 3.9.

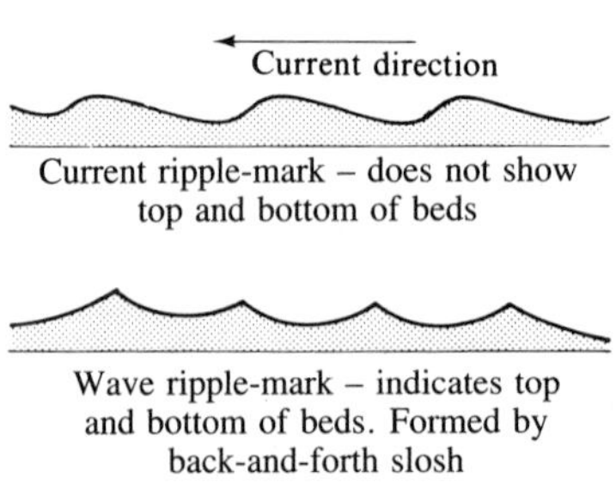

FIGURE 3.2
Ripple-marks form where currents can act on sediments. Ripple-marks form at right angles to the current direction, and asymmetrical ripple-marks indicate which way the current flowed. To see why only wave ripple-marks can be used to determine top and bottom of beds, turn the page upside down.

FIGURE 3.3
Oscillation ripple-marks, Erie County, Ohio. Note the presence of ripple-marks in successive layers. Photo by D. F. Demarest, U.S. Geological Survey.

Deposition

Most sedimentary rocks are deposited in oceans, and these marine rocks, easily dated by their contained fossils, record much of the earth's history. In the present section a few of the rock types found in the more important marine environments will be considered.

Clastic Rocks One of the interesting observations of marine sedimentary rocks is that at many places individual sedimentary units, such as sandstones, can be traced for hundreds of miles. At other places the beds are irregular and have limited horizontal extent. To see how this comes about we will consider how marine sedimentary rocks are formed. We will begin with the clastic rocks (conglomerates, sandstones, and shales).

Because most marine sediments consist of material delivered by rivers, most sedimentary rocks form near the continents. Sampling and drilling in the oceans show this to be true. If we wish to interpret the thick accumulations of old sedimentary rocks exposed in today's mountain ranges, we should study the continental shelves and the continental slopes. (See Fig. 3.10.)

The continental shelves are covered with clastic sediments for the most part. One would expect that, in most cases, the coarsest material should be found near shore or near river mouths and the finer silt and clay should be farther from the

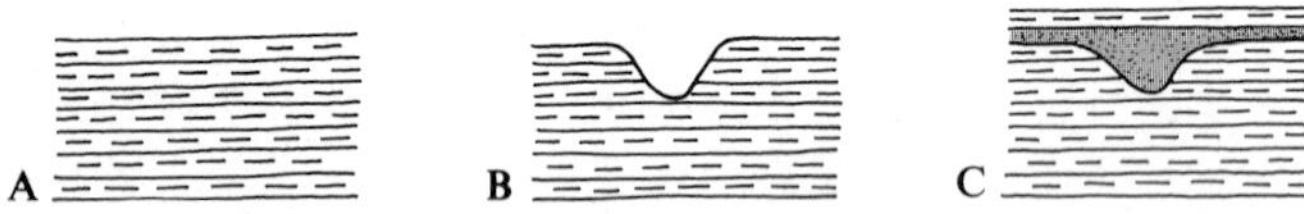

D

FIGURE 3.4
Scouring produced by localized currents. A. Initial deposition. B. Channel cut by current. C. Deposition is resumed, filling the channel. Such a feature records an interruption in deposition and can be used to distinguish top and bottom. D. Channel filled with conglomerate. Note that the conglomerate is more resistant to weathering than the sandstone. Channel is about ten feet deep. Daggett County, Utah. Photo by W. R. Hansen, U.S. Geological Survey.

FIGURE 3.5
Development of one type of cross-beds. A. Initial deposition. B. Channel cut by current. C. Channel filled by deposition from one side. D. Normal deposition after channel is filled. Cross-beds can also be used to distinguish top and bottom.

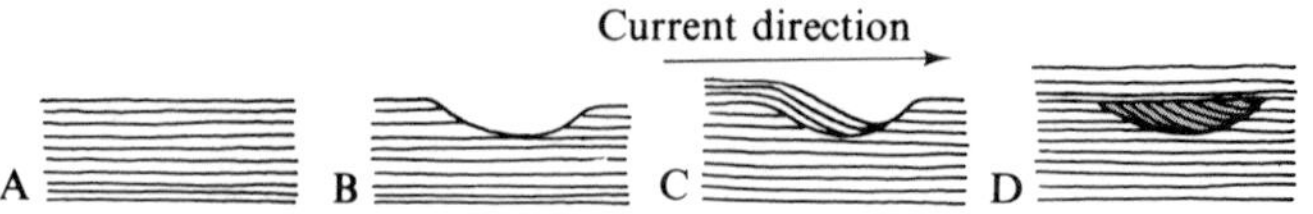

FIGURE 3.6
Cross-bedding. A. In river-deposited coarse sandstone. Photo by H. E. Malde, U.S. Geological Survey. B. In wind-deposited sandstone in Zion National Park, Utah. Photo by H. E. Gregory, U.S. Geological Survey.

A

B

FIGURE 3.7
Graded bedding. Produced by the more rapid settling of coarse material than fine material.

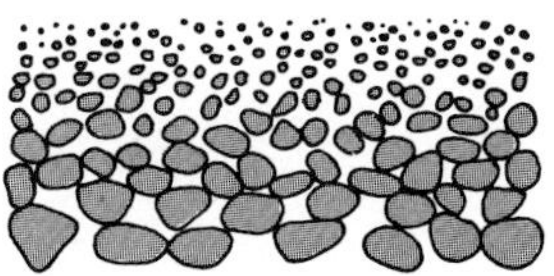

FIGURE 3.8
Soft sediments can slide down even very gentle slopes, developing slump features.

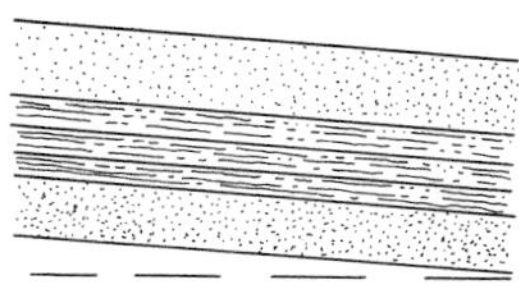

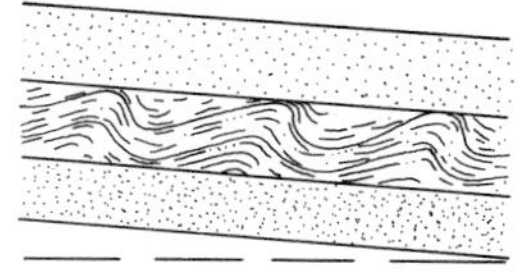

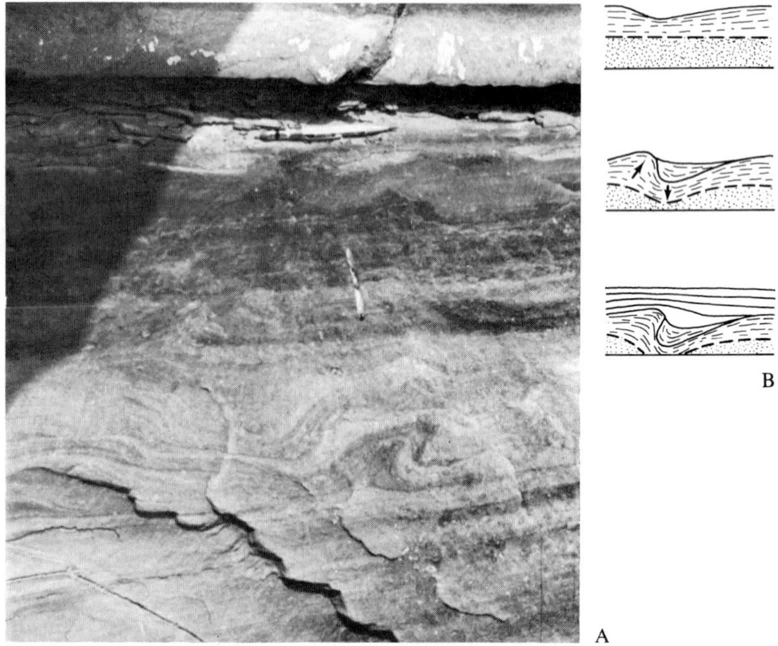

FIGURE 3.9
A. Photo of deformed bedding. B. Diagram showing how deformation of soft rock forms the structure shown in photo.

source. (See Fig. 3.11.) This is probably true in a very general way, but the many departures suggest that the situation is not simple. (See Fig. 3.12.) One very obvious reason is that during the recent glacial time the oceans were lower because much water was held in the glacial ice. Apparently, too, tidal and other currents are fairly effective in moving and sorting clastic material on the relatively shallow shelves. This type of reworking of the sediments may possibly account for the

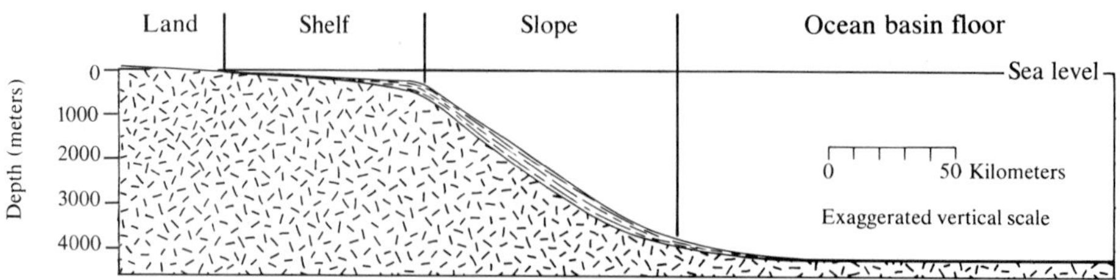

FIGURE 3.10
Continental margin showing typical continental shelf, continental slope, and ocean basin floor. The vertical scale is exaggerated.

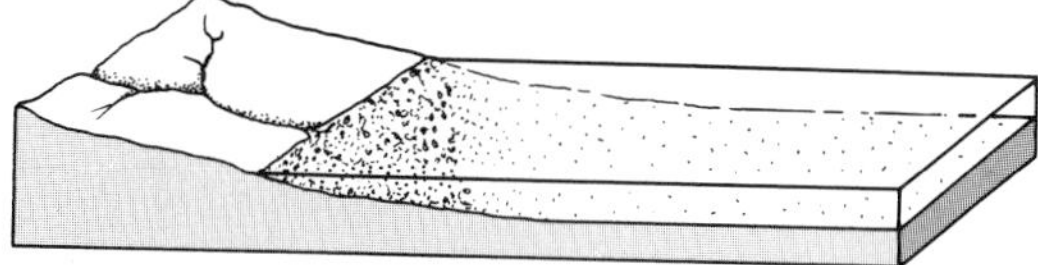

FIGURE 3.11
Idealized diagram showing coarse material being deposited near the shore and finer material farther out.

uniform beds extending over large areas that are found on the present continents. In any event, pebbles and conglomerate are not uncommon far out on the shelf.

The present coastal plains on the Atlantic and Gulf coasts are underlain by what are apparently continental shelf deposits. Uplift has exposed the landward parts of these sedimentary accumulations, and seismic studies and deep drilling indicate that they extend onto and thicken on the present continental shelf. On the Gulf Coast these deposits are several miles thick, and their features and fossils indicate shallow-water deposition. The latter suggests slow subsidence.

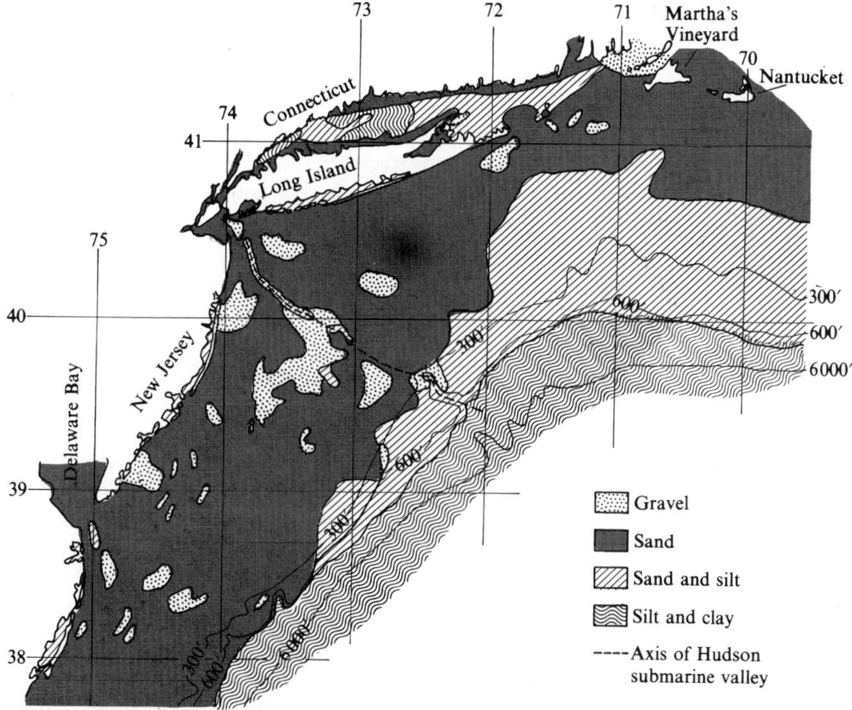

FIGURE 3.12
Bottom sediments on part of the continental shelf in northeastern United States. The many departures from the idealized case shown in Figure 3.11 are probably caused by bottom currents and by deposition when sea level was lowered during glacial times. Adapted from Shepard and Cohee, 1936.

Much less is known about the structure and formation of the continental slopes. Limited sampling shows the surface, at least, to be covered mainly with fine sediments, presumably swept off the shelves. A more important process in the building of the slopes may be submarine landslides or mudslides. Following earthquakes, submarine telephone cables may be broken. The cables do not all break at the time of the earthquake but at later times, farther down the slope. In some cases many cables have been broken and the time of breaking is accurately known. This, plus the burial of some of the cables reported by the repair ships, strongly suggests that submarine landslides move down the continental slopes.

Another feature of continental shelves and slopes is the *submarine canyons.* These submarine canyons have many of the features of river canyons, and some are larger than such river canyons as the Grand Canyon. The origin of such features has been the subject of lively speculation, and not everyone is yet satisfied with the explanations offered. At first they were thought to have formed by ordinary river erosion at a time when sea level was lower than it is at present. Many of the submarine canyons have heads near the mouths of large rivers, making this a reasonable hypothesis. However, many of them extend to depths of over 3.2 km (2 mi), and sea-level changes of this magnitude in the recent past seem impossible. The submarine landslides mentioned above offer the most likely explanation. Such landslides consist of sediment-laden water. Because they are heavier than the surrounding water, these landslides or mudflows move along the bottom and are capable of eroding the soft bottom material. *These muddy waters are called* **turbidity currents.** The rocks formed by them are called **turbidites.** They can move down very gentle slopes and can transport fairly large pebbles. Landsliding of unconsolidated material on the continental slopes can account for the canyons there; and turbidity currents initiated by stream discharge, storms, or other currents can account for the submarine canyons on the continental shelves. As well as explaining the canyons, these processes can account also for the building of the continental slopes—seismic studies show the slopes to be composed of sedimentary rocks.

In addition, turbidity currents and submarine landslides can account for the smooth ocean bottoms at the foot of the continental slope at many places. These smooth surfaces are apparently similar to the fans that develop where rivers enter broad valleys. These submarine fans contain pebbles and shallow-water shells at some places. The fans do not occur everywhere at the foot of the continental slope.

To further evaluate the possible existence and importance of turbidity currents in the geologic past, we will look at their depositional features. Turbidity currents, because of their density and speed, can transport large fragments. These larger fragments and the main mass of fine material are deposited fairly rapidly. The larger, heavier fragments settle first, forming graded beds. The eroding ability of turbidity currents suggests that they may erode or otherwise deform the top of the underlying bed. This is shown in Figure 3.13.

Carbonate Rocks *The carbonate rocks are limestone and dolomite.* Limestones are being deposited at present in warm shallow seas, and most of the features of

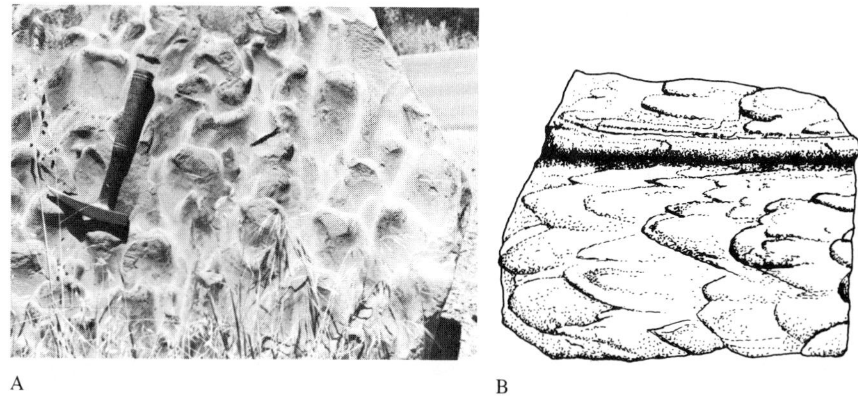

FIGURE 3.13
Depositional features of turbidity currents. A. Photo. The bottom of a bed is shown. B. Diagram. The lobelike casts point in the opposite direction to current movement. The feature that crosses the specimen is the cast of a groove, probably caused by a boulder carried by the turbidity current. The specimen shown is about two feet in diameter.

ancient limestones suggest that they, too, were formed in such an environment. Calcite, the mineral of which limestone is composed, is soluble in water containing dissolved carbon dioxide. To precipitate calcite, the amount of dissolved carbon dioxide must be reduced because calcite is only slightly soluble in pure water. Warm water can dissolve much less of any gas, including carbon dioxide; this is one reason why limestones form in warm water. Marine plants are also common in warm water, and they extract carbon dioxide from the water as part of their life process. This, too, aids in the precipitation of calcite. Many lime-secreting organisms live in this environment, and their shells and other structures also accumulate on the sea floor. Fossils of organisms that lived in warm shallow waters are present in most limestones, showing that this is the environment in which most limestones formed. A very few limestones form by chemical precipitation in restricted parts of the oceans and in lakes.

Most dolomite appears to form by replacement of calcite in limestone by the mineral dolomite. (The term dolomite is used to refer to both the mineral and the rock.) The replacement may be partial or complete. The process is not well understood, but old rocks are more apt to be dolomite than younger ones.

Oozes In the deep ocean, far from land, the sediments consist mainly of wind-carried dust, largely of volcanic origin, and the fallen shells of tiny animals that live near the surface. *Sediments of this type that are composed of more than 30 percent biologic material are called* **oozes.** (See Fig. 3.14.) The rate of accumulation of oozes is very slow compared with that of other sedimentary rocks. It is estimated that it requires between 1000 and 10,000 years to deposit one millimeter of ooze. The volcanic dust settles so slowly that most of it is oxidized while sinking to the bottom, and so it is termed *red clay.*

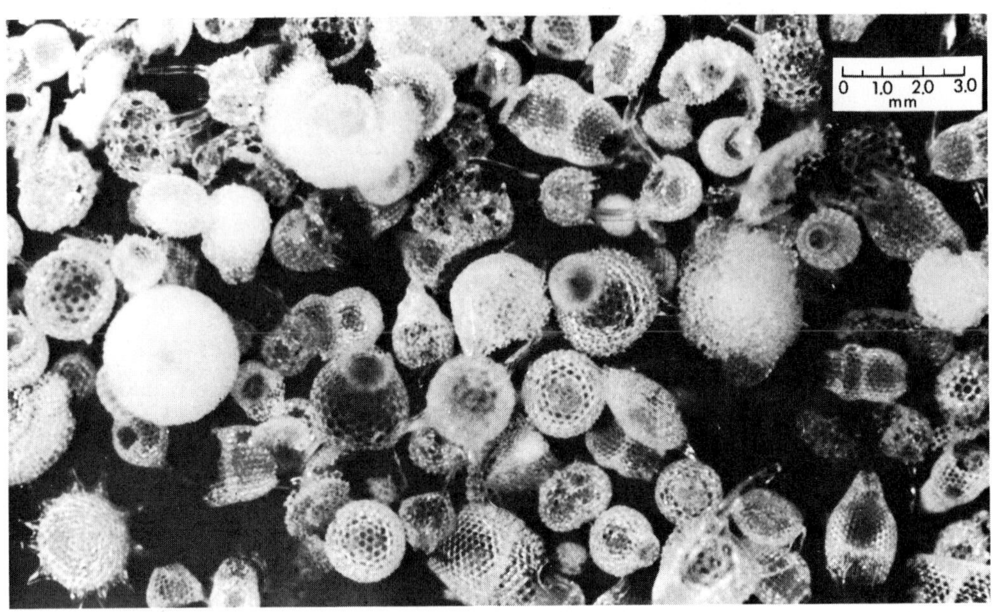

FIGURE 3.14
Radiolarian ooze from a depth of 5205 m (17,066 ft), about 970 km (600 mi) south of Hawaii. Photo from D. R. Horn, Lamont-Doherty Geological Observatory.

Geosynclines and Plate Tectonics

It was noted long ago by geologists that many mountain ranges are formed of bent and broken layers of sedimentary rocks. The sedimentary rocks in mountain ranges generally are much thicker than nearby rocks of similar age. These sedimentary rocks had been deposited as flat layers in shallow seas. We know that the seas were shallow because the rocks have features that commonly form in shallow water, such as mud cracks, and the fossils that they contain are of organisms that live in shallow waters. *The mountain ranges were thought to have begun as shallow, elongate areas of deposition, and these areas of sedimentation were named* **geosynclines.** The term geosyncline is used in this book to mean an area of deposition that is later deformed, generally forming a mountain range. We now relate the classic concept of the geosyncline to plate tectonics.

Many mountain ranges are near the margins of continents, and so the geosyncline, or at least part of it, may have been the continental shelf–continental slope area just described. The sediments of continental shelves tend to be sandstones near the coast, with shale or limestone accumulating farther from shore, although many exceptions are found (Fig. 3.15). The rocks that underlie the continental slope are generally clastics—shale and sandstone—commonly intermixed. Few or no volcanic rocks are found in the continental shelf–slope area.

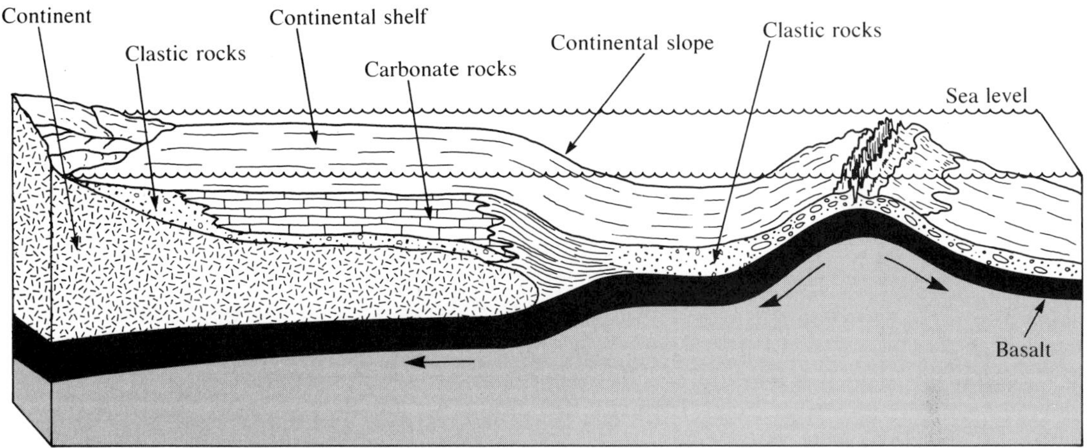

FIGURE 3.15
Development of thick sections of sedimentary rocks at a continental shelf and slope and much thinner layers of sediments and volcanic rocks on the ocean floors and at mid-ocean ridges.

In parts of most geosynclinal mountain ranges, volcanic rocks are found interbedded with sedimentary rocks. This suggests that in these cases convergent plate boundaries may be involved. Volcanic island arcs form where oceanic plates collide with continental plates and so are also at the margins of continents (Fig. 3.16). The volcanic borderlands can be the source of both volcanic rocks and clastic sediments. Clastic sediments containing volcanic rocks generally are very poorly sorted and so have a wide variety of sizes and types of fragments. Such rocks contrast with the sediments on continental shelves, which tend to be well sorted. Later plate movements are believed to crumple the geosynclinal sediments.

Geosynclines at plate boundaries are more complex because of the plate motion and volcanic activity that can result in a number of forms (Fig. 3.17). The configuration closest to the classical geosyncline is the offshore volcanic island arc. This configuration can develop in at least two ways. *The sea between the island arc and the continent is called a* **backarc basin.** On the continent side of this basin, typical continental shelf and slope sediments, such as sandstone, shale, and some limestone, will accumulate. Near the active volcanic island arc, volcanic rocks and sediments composed of volcanic fragments will accumulate rapidly. On the ocean side of the volcanic islands, similar sediments will be trapped in the trench. The ocean-floor sediments on the descending plate may be scraped off the plate and so also accumulate near the trench.

Deformation occurs at converging plate boundaries. If one of the plates is composed of oceanic lithosphere, that plate will descend and a subduction zone, with its associated magmatic activity, will be initiated.

The volcanic island arcs just considered can form in at least two ways. The

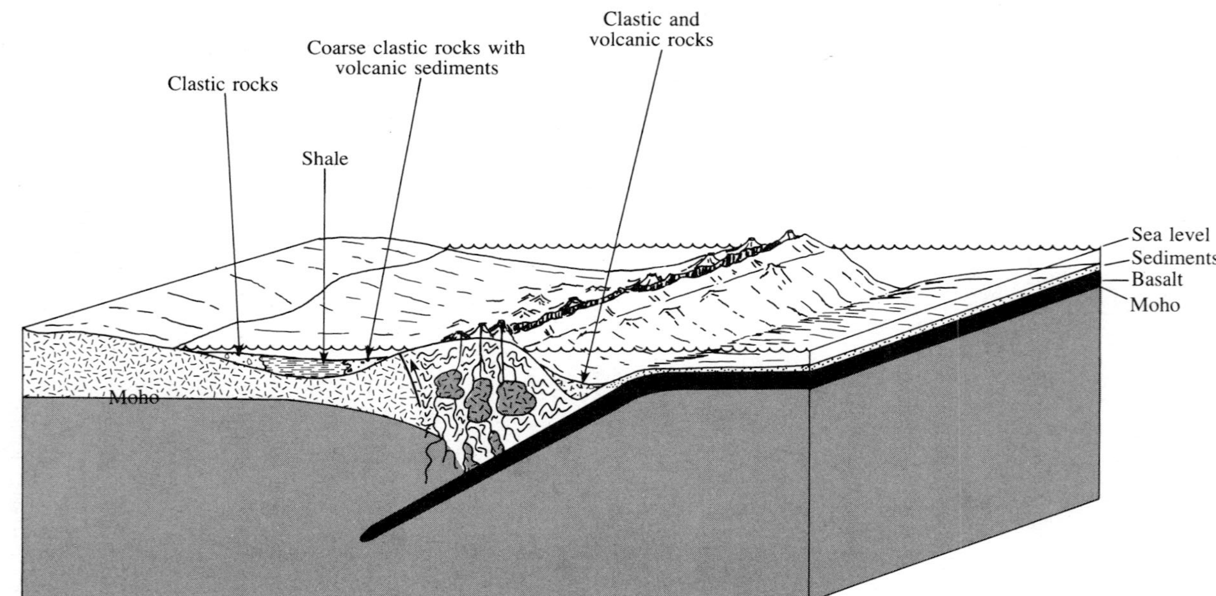

FIGURE 3.16
Development of thick sections of sedimentary and volcanic sedimentary rocks at a convergent plate boundary.

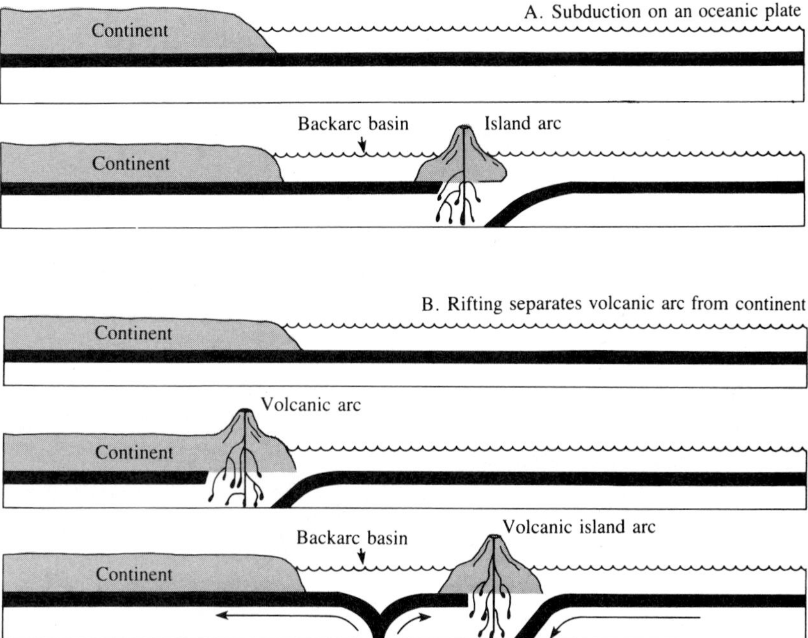

FIGURE 3.17
Origin of backarc basins. A. Subduction zone forms on oceanic plate. B. Subduction zone forms at continent-ocean boundary, and later rifting forms backarc basin.

simplest is if the subduction zone initially forms well offshore (Fig. 3.17A). Why subduction would occur at such a location is not understood. The Aleutian Islands appear to have formed in this way because the islands are younger than the floor of their backarc.

The other way that some volcanic island arcs formed requires two steps. The subduction zone initially forms at the continent-ocean border. Then a rift develops that separates the volcanic arc from the continent (Fig. 3.17B). Honshu Island and the Sea of Japan probably formed in this way because parts of Honshu are much older than the volcanic arc. Unlike Honshu, however, the crust under the Sea of Japan is younger than the volcanic arc.

Along the west coast of South America, a subduction zone has formed at a plate boundary separating oceanic and continental lithosphere. The subduction zone slopes down under the continent, and the volcanic arc has formed on the continent (Fig. 3.18). The volcanoes are mainly andesite; this rock type was named for the Andes Mountains. The magmatic rocks, both volcanic and intrusive, that are associated with most subduction zones are generally of this composition. Here the sediments initially resemble those of a classical geosyncline, with continental-shelf sediments at the continent and deeper-water sediments in the trench. Material from the volcanic arc and sediment scraped from the descending slab of lithospheric

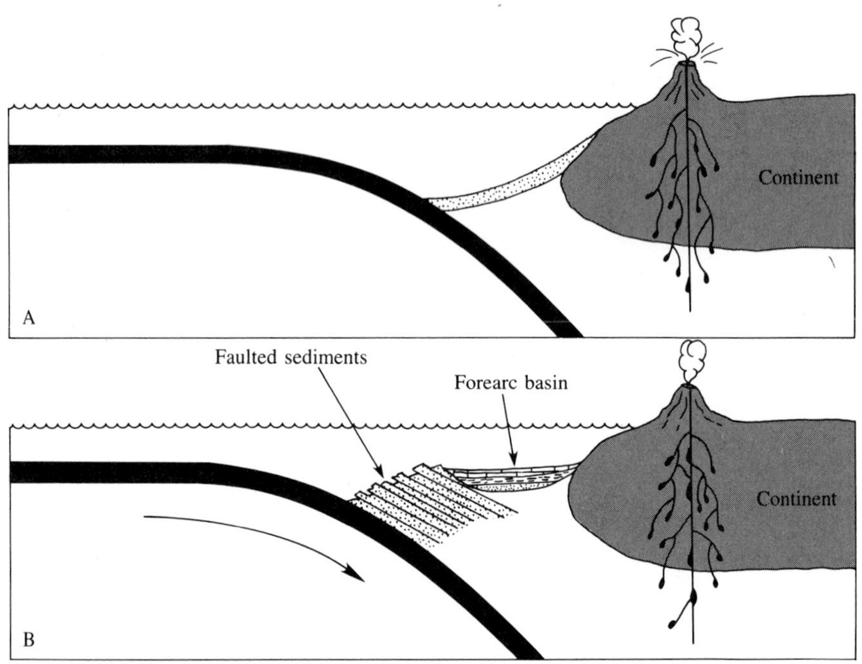

FIGURE 3.18
Volcanic arcs at continent-ocean boundary. A. Volcanic arc forms on continent edge. B. As subduction proceeds, sediments are scraped off the oceanic plate, forming a ridge and forearc basin.

plate soon overwhelm the initial sediments (Fig. 3.18B). Many earthquakes, including some very deep and some of high magnitude, are associated with this South American subduction zone.

At some subduction zones, the sediments scraped off the oceanic slab as it descends may accumulate and form a submarine ridge or an island (Fig. 3.18B). *The sedimentary rocks are broken into thin slices in thrust contact with each other, or they may be a heterogeneous mixture of blocks and slabs enclosed in a matrix of rock that was soft sediment at the time of formation.* The term **mélange** has been applied to such mixed rocks. At some subduction zones (Indonesia is an example) islands formed in this way create *a basin between the trench and the volcanic arc. Such basins are termed* **forearc basins.** Where exposed on continents, the rocks in the ridges formed by this tectonic accumulation are difficult to study because of their broken nature. The Franciscan rocks of the California coast, which have puzzled geologists for decades, probably formed in such an environment.

Igneous and metamorphic activity also occur near subduction zones. The descending slab is partially melted and is the source of the magma. The rising magma and probably the active hot fluids derived from moist sediments on the melting plate cause metamorphism of the rocks above the subduction zone. In this way rocks of continental composition are formed.

Continent-continent collisions cause more intense compressional deformation. The collision of Africa and Europe produced the thrust structures in the Alps and in North Africa. The collision of India with Asia produced the Himalaya Mountains.

Where two continents collide they are welded together, forming a new, larger continent. In such cases, the youngest part of the new, larger continent will be the area of collision. Generally the oldest parts of continents are in their interiors. Collisions, then, can account for many places where younger rocks in continental interiors are surrounded by older rocks. **Rifting,** *or the breaking-up of a continent by the formation of a new spreading center,* can also change the arrangement of deformed belts. So far we have seen that the magmatic and metamorphic activities that accompany subduction and continental collision are the processes by which continents are formed. Continent-continent collisions and rifting of continents have changed the continents throughout geologic time. Collisions between North America and both Africa and Europe have caused orogeny that produced parts of the present Appalachian Mountains. After these collisions, rifting reopened the Atlantic Ocean; at some places, however, the separation did not occur at the old shoreline. As a result, some areas in New England and Newfoundland appear to have once been parts of other continents. This conclusion is based on rock types and fossils. At other places on both coasts of North America and elsewhere, parts of the present continents also appear to be foreign, but their sources are problems.

Interpreting Sedimentary Rocks—Some Simple Examples

Interpreting geologic history from the study of sedimentary rocks is an important activity carried on by many geologists. These studies are conducted at many levels, both in the field and in the laboratory. The few examples given here may help you to interpret the rocks in your area. These examples are simple, but a little imagination will reveal the many other possibilities and the many complexities that are possible. The following examples will be limited to the clastic sedimentary rocks, especially sandstones.

In studying a hand specimen of sandstone, the composition of the grains may reveal the rock type that was eroded to form the sandstone. The amount of weathering influences the mineralogy of the grains so that the grains of rock that could reveal the source are, in general, present only if very little weathering has occurred. Such a lack of weathering could be caused by a dry or cold climate, rapid or short transportation, or rapid burial, perhaps caused by rapid deposition. A quartz sandstone could reveal the opposite conditions. In general, a quartz sandstone implies complete weathering, and a feldspathic sandstone (arkose) implies less weathering and/or rapid burial. The silt in shale may also reveal the degree of weathering or reworking. The shale associated with quartz sandstone generally has mainly quartz in its silt fraction. Shales that are rapidly deposited with little weathering contain much feldspar in their silt fraction. The color of a sedimentary rock may also reveal information about its formation. Red color, as in red beds, reveals oxidation because the red color comes mainly from hematite (red iron oxide). Seasonal drying, as commonly occurs in arid climates, is the way that most red beds form. They can, however, form in other ways. Black color in sedimentary rocks is generally caused by carbon of organic origin. This implies that organisms were deposited faster than they could be destroyed by bacteria and oxidation. Such situations are common in stagnant water.

Study of the size, shape, and sorting of the grains may help in deciding which conditions were most important in the formation of the sandstone. Large grains suggest nearness to the source or short distance of transportation. Angular fragments, unless very resistant to abrasion, have not traveled far. The sorting can reveal, among other things, the mode of transportation. Wind-transported sediments are very well sorted, and glacially deposited rocks have an extreme size range. All of these features can be seen with a hand lens, but the complete study of a hand specimen in the laboratory requires the use of the microscope and other instruments.

Much more can be learned at the outcrop of the rock layer. Here the bedding can be seen. The thickness and uniformity of the beds can reveal much about their origin. The irregular bedding of river deposits contrasts with the regularity of many shallow-water marine sedimentary rocks. The many features of sedimentary rocks, such as ripple-marks, cross-beds, and graded bedding, can also be studied at the outcrop. These features and their interpretation were described earlier. Fossils can be collected at the outcrop. The environment of deposition can, in many cases, be determined from study of fossils because the types of animals that live near shore are different from those in deep water; they also differ between warm and cold water. Perhaps the most important use made of fossils is to determine the age of the sedimentary layers. The deformation, if any, of the beds can also be seen in many outcrops.

The next level of study of a sedimentary layer is to trace its areal extent by mapping its outcrops. By combining the interpretations made at each outcrop, it is possible to reconstruct the basin in which the beds were deposited. An example of such an interpretation is shown in Figure 3.19. In this example, the source area for the Morrison Formation is revealed from the current directions shown by the cross-bedding, the increase in thickness near the source, and the coarseness of the grains near the source. Note also that the Morrison Formation is exposed in only a small part of the basin; everywhere else it has either been removed by erosion or is covered by younger rocks. Most geologic interpretations must be made on the basis of such fragmentary evidence.

INTERPRETING FOSSILS

Up to this point, fossils have been considered only as a means of dating the rocks that enclose them. If some of the other information that fossils can reveal is considered, we will not only learn more about interpreting rocks but also see some of the problems of dating. If you were to walk a few miles along most coasts, you would probably see several different habitats, such as beach, rocky coast, or marsh. Different plants and animals live in each environment but, of course, all these organisms are contemporary. So it is with ancient rocks—rocks of different environments may have different fossils even though the rocks are the same age. Another problem arises if you compare, say, the invertebrates found on a sandy beach on the Gulf Coast to those on a sandy beach in Maine; they are different. So with ancient rocks one should not expect sandstones of the same age to have the same fossils unless the sandstones came from the same environment, or, said another way, all sandstones are not formed in the same environment. Again, if you compare the Gulf Coast organisms with those on a similar beach in California they are different, even though the physical environment may be the same, because the Pacific faunal province is different from the Gulf faunal province. These points are obvious when we think about the present coast, but imagine the confusion they can cause in dating and interpreting ancient rocks.

To summarize, life, and therefore the fossil evidence of it, changes not only with time, but also with environment and province. Some of the environmental factors that affect life are type of bottom (sand, mud, carbonate, etc.), temperature, depth, salinity, wave and current action, and clear or cloudy water. This list is by no means complete. It should come as no surprise that uniformitarianism is used to interpret the environment in which the fossil organism lived. We compare the fossils to organisms living in present-day communities. This approach works better with young fossils whose relatives are still living than with organisms with no close living relatives. An obvious possibility with fossil assemblages is that some of the shells may have been transported to their final resting place by waves or currents; that is, they were not part of the assemblage during life.

ISOTOPES AND ENVIRONMENT

Isotopes can be used to determine ancient temperatures. Oxygen isotopes are used in temperature determination. Oxygen has three isotopes: oxygen 16, oxygen 17, and oxygen 18. When water evaporates, the lightest isotope—common oxygen 16—evaporates at a slightly higher rate than do the heavier isotopes. This difference in evaporation rates is proportional to temperature. The ratio of oxygen 16 to oxygen 18 can be measured and used to determine the temperature. In practice, this ratio is measured in fossil shells made of calcium carbonate ($CaCO_3$). There are many pitfalls in this method, but so far the results have agreed fairly well with other geological evidence.

Other features used for estimating temperatures are coral reefs, evaporites, desert deposits, and glacial deposits. Fossils, both plant and animal, have also been used very successfully to reconstruct climates and ecology.

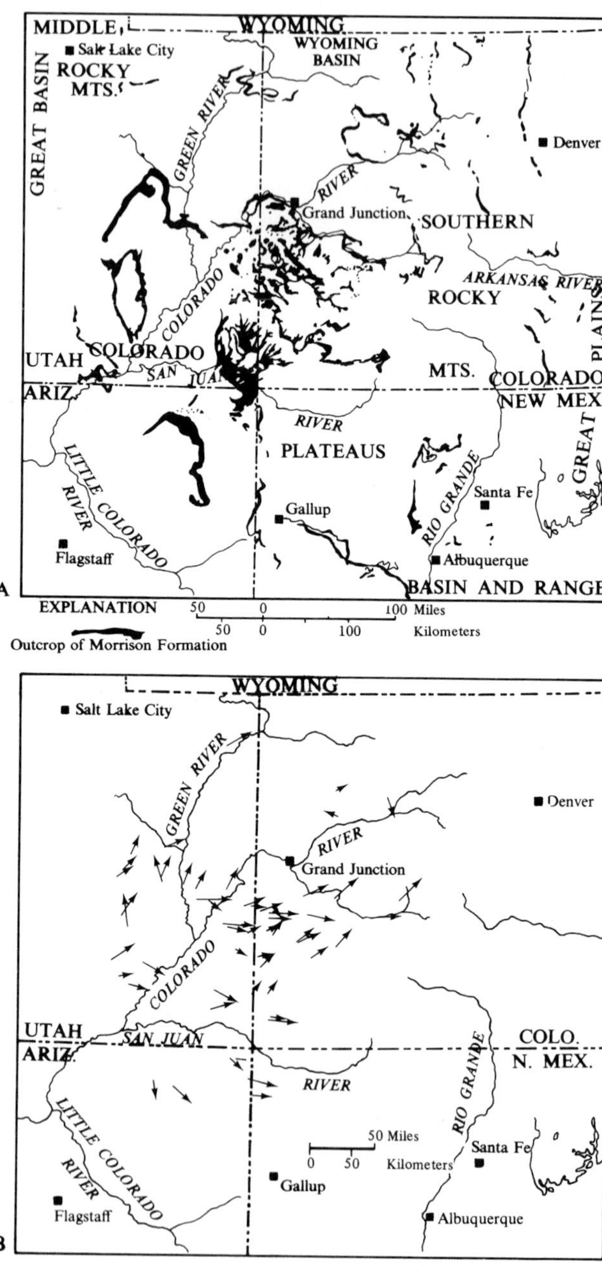
A
MIDDLE ROCKY MTS.
Salt Lake City
WYOMING
WYOMING BASIN
GREAT BASIN
GREEN RIVER
Denver
RIVER
Grand Junction
SOUTHERN
COLORADO
ARKANSAS RIVER
ROCKY
GREAT PLAINS
UTAH
COLORADO
SAN JUAN
MTS.
COLORADO
ARIZ.
NEW MEX.
RIVER
PLATEAUS
LITTLE COLORADO RIVER
RIO GRANDE
Santa Fe
GREAT
Gallup
Flagstaff
Albuquerque
BASIN AND RANGE
EXPLANATION
Outcrop of Morrison Formation
50 0 100 Miles
50 0 100 Kilometers
B
WYOMING
Salt Lake City
GREEN RIVER
Denver
RIVER
Grand Junction
COLORADO
UTAH
SAN JUAN
COLO.
ARIZ.
N. MEX.
RIVER
RIO GRANDE
LITTLE COLORADO RIVER
50 Miles
0 50 Kilometers
Santa Fe
Gallup
Flagstaff
Albuquerque

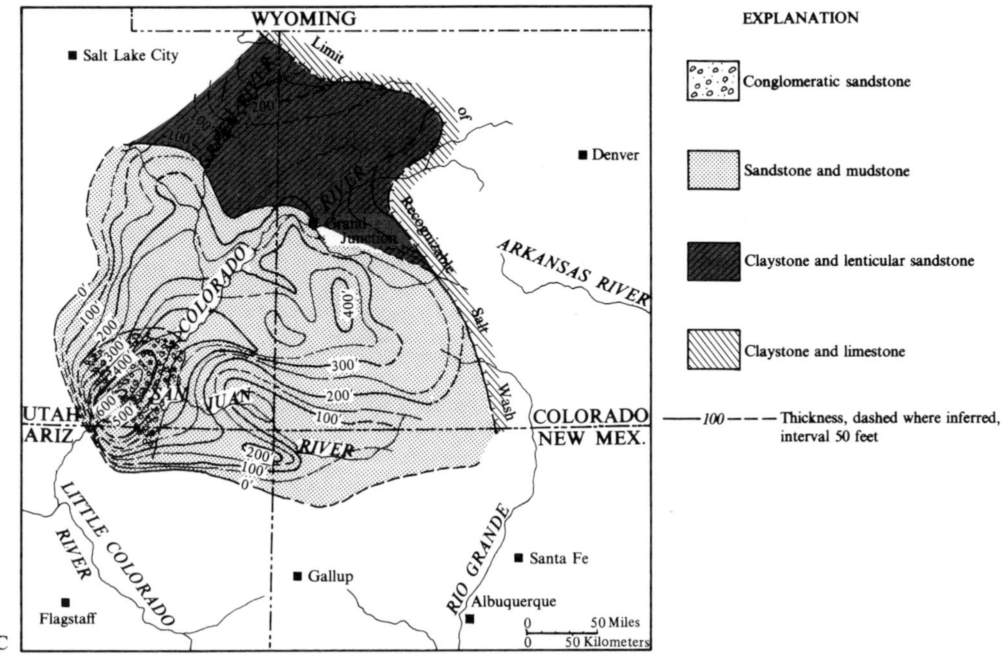

FIGURE 3.19
Reconstruction of an ancient sedimentary basin from interpretation of the features in scattered outcrops. A. Map of outcrops of the Morrison Formation. B. Current direction inferred from cross-bedding in the Salt Wash Member of the Morrison Formation. The arrows show the downslope direction of cross-beds. Length of the arrows is proportional to the consistency. C. Thickness and sediment type in the Salt Wash Member. Note that the coarser sediments are closer to the source area inferred from both thickness and cross-bedding. From Craig and others, 1955.

SUMMARY

Sedimentary rocks are important in the interpretation of geologic history through study of grain size and shape, depositional features, size and shape of depositional area, fossil content (including dating), and deformation of the area.

A sedimentary rock has the features of the source area, the mode of transportation, the mode of deposition, and the characteristics of the depositional site all impressed on it.

Bedding of sedimentary rocks records the layers in order of deposition, with the oldest at the bottom.

Mud cracks record periodic drying and can show top and bottom of deformed beds.

Ripple-marks can show type and direction of currents, and wave ripple-marks can show top and bottom of beds.

Cut-and-fill features are caused by localized scouring and can show top and bottom of beds.

Cross-bedding, another current feature, also shows top and bottom of beds.

Graded bedding is formed by larger fragments settling faster than smaller fragments; bottom is indicated by the larger fragments.

The sorting of a sediment is a measure of the range in size of its fragments.

Continental shelves, the almost flat portions of the continent below sea level, are generally covered with clastic sediments.

Continental slopes, the steep slopes between the continental shelves and the deep ocean platforms, are mainly covered with fine sediments.

Slopes and shelves have some large submarine canyons. These may be caused by turbidity currents—underwater landslides consisting of heavy, sediment-laden water that can erode the soft bottom material.

The carbonate rocks are limestone and dolomite. Limestones are being deposited at present in warm shallow seas, and most of the features of ancient limestones suggest that they, too, were formed in such an environment.

In the deep ocean, the sediments are composed of slowly settled, wind-carried dust, largely of volcanic origin, and fallen shells of tiny animals. Sediments that have more than 30 percent biologic material are called oozes.

Many mountain ranges are formed of bent and broken layers of sedimentary rocks. These sedimentary rocks are much thicker than nearby rocks of the same age. Geosyncline is the name given to the elongate depositional areas in which the rocks accumulated.

Geosynclines generally form at continent-ocean borders and may form at continental shelf–slopes in the interior of a plate.

At convergent plate boundaries a geosyncline may form between a volcanic island arc and the continent. This area is called a backarc basin.

The deformation that occurs at convergent plate boundaries may deform the sedimentary and volcanic rocks, producing the structures seen in mountain ranges.

At some subduction zones, the sediments scraped off the oceanic slab as it descends may accumulate and form a submarine ridge or an island. Islands formed in this way may create a basin, termed a forearc basin, between the trench and the volcanic arc. The sedimentary rocks in such zones may be broken into thin slices, or they may be a heterogeneous mixture of blocks and slabs enclosed in a matrix of rock that was soft sediment at the time of formation. The term mélange has been applied to such rocks.

Fossils can be used not only to date rocks but also to reveal much information about the environment and province in which the rocks formed. Fossils can also be transported after the death of the organism, so care must be exercised in their interpretation.

Rifting is the breaking-up of a continent by the formation of a new spreading center.

Geologic history is interpreted by determining how the rocks were formed and when, how they have been deformed and when, and how and when they have been eroded. Multiple hypotheses must be used to test each step in the reconstruction of events.

Ancient environments are interpreted from the rocks themselves, the fossils, and the ratio of oxygen 16 to oxygen 18.

QUESTIONS

1. How can top and bottom of the beds be determined in a group of vertical sedimentary beds?
2. How might sediments deposited in shallow water differ from those deposited in deep water?
3. What features could be used to distinguish turbidity-current–deposited sediments?
4. What geologic history might be inferred from each of the following rocks?
   a. A feldspar-rich sandstone.
   b. A well-rounded conglomerate.
   c. A sandstone composed almost entirely of quartz.
   d. A limestone with abundant fossils.
5. What kinds of fossils would you expect to find in rocks deposited in the following environments?
   a. Tropical shallow water.
   b. Deep water.
   c. Temperate near-shore.
   d. Marsh.
6. What are the characteristics of geosynclinal sediments?
7. Where might geosynclines form?
8. Why might two occurrences of the same fossil not be of the same age?
9. Why might two rocks of the same age have different fossils?
10. What isotope can be used to determine the temperature of deposition?

SUGGESTED READINGS

Cook, F. A., L. D. Brown, and J. E. Oliver, "The Southern Appalachians and the Growth of Continents," *Scientific American* (October 1980), Vol. 243, No. 4, pp. 156–188.

Courtillot, Vincent, and G. E. Vink, "How Continents Break Up," *Scientific American* (July 1983), Vol. 249, No. 1, pp. 43–52.

Dickinson, W. R., "Plate Tectonics in Geologic History," *Science* (October 8, 1971), Vol. 174, No. 4005, pp. 107–113.

Dietz, R. S., "Geosynclines, Mountains and Continent-Building," *Scientific American* (March 1972), Vol. 226, No. 3, pp. 30–38.

Hallam, A., "Continental Drift and the Fossil Record," *Scientific American* (November 1972), Vol. 227, No. 5, pp. 56–66.

Heezen, B. C., "The Origin of Submarine Canyons," *Scientific American* (August 1956), Vol. 195, No. 2, pp. 36–41. Reprint 807, W. H. Freeman and Co., San Francisco.

Ingersoll, R. V., "Tectonics of Sedimentary Basins," *Bulletin Geological Society of America* (November 1988), Vol. 100, No. 11, pp. 1704–1719.

James, David E., "The Evolution of the Andes," *Scientific American* (August 1973), Vol. 229, No. 2, pp. 60–69.

Land, L. S., "The Origin of Massive Dolomite," *Journal of Geological Education* (March 1985), Vol. 33, No. 2, pp. 112–125.

Molnar, Peter, and Paul Tapponnier, "The Collision between India and Eurasia," *Scientific American* (April 1977), Vol. 236, No. 4, pp. 30–41.

Pettijohn, F. J., and P. E. Potter, *Atlas and Glossary of Primary Sedimentary Structures.* New York: Springer-Verlag, 1964, 370 pp.

Potter, P. E., and F. J. Pettijohn, *Paleocurrents and Basin Analysis.* Berlin: Springer-Verlag, published in United States by Academic Press, Inc., New York, 1963, 296 pp.

Reading, H. G., ed., *Sedimentary Environments and Facies.* New York: Elsevier North Holland, Inc., 1978, 557 pp.

Valentine, J. W., and E. M. Moores, "Plate Tectonics and the History of Life in the Oceans," *Scientific American* (April 1974), Vol. 230, No. 4, pp. 80–89.

Zenger, D. H., "Dolomitization and Uniformitarianism," *Journal of Geological Education* (May 1972), Vol. 20, No. 3, pp. 107–124.

4

Evolution and Fossils

In the following chapters the development of life is traced from its origin to the present. In this chapter the way that people recognized the meaning of the diversity of both fossils and present life is described. Indeed, it was not always recognized that fossils are the remains of once-living organisms. Even after this was generally conceded, separate creations were invoked to account for the diversity and for the numerous gaps in the record.

EVOLUTION

Evolution is a biologic process, but it is also important to geology. The evidence for evolution comes both from fossils—the province of the geologist—and from living organisms—the province of the biologist. *Biological evolution is the continuing change in populations of organisms that is now occurring and has occurred in the geologic past.* In a sense, evolution is the biologic counterpart of the continuing changes that have occurred and continue to occur on the earth. Because of its importance in human intellectual development, evolution will be discussed historically.

Linnaeus—Species and Genus

An important step occurred in 1737 when Carolus Linnaeus published his book *Systema Naturae.* In this book he attempted to classify all living organisms in a logical system. He used two names, genus and species, for each kind of organism, and this system is still in use. The tenth edition of this book, published in 1758, is the generally accepted start of this system of classification. The value of this system

TABLE 4.1
Classification of the more important fossil-forming organisms.

A. Classification scheme and an example.

Kingdom	Animalia
Phylum	Chordata
Class	Mammalia
Order	Primate
Family	Hominidae
Genus	*Homo*
Species	*sapiens*
Individual	John Doe

is that it indicates that there are degrees of similarities and differences among organisms rather than random variations.

Although there is no completely satisfactory definition of **species,** *it can be considered as a group of similar individuals that can interbreed to produce fertile offspring.* A **genus** *is a group of closely related species.* Thus the cats belong to the genus *Felis;* the common house cat is *Felis catus,* and the ocelot is *Felis pardalis* (which is written *F. pardalis* if the genus has been mentioned nearby). Note that the first letter of the generic name (genus) is capitalized, but the specific name (species) is in lower case, and that both genus and species are in italics. The other terms in the system, *family, order, phylum,* and so on, are shown in Table 4.1. To biologists working with living organisms, a species is a group that can interbreed, and the whole system attempts to show the genetic relationships among organisms. Geologists working with fossils have the same goal, but they must work with dead remains of organisms. A typical fossil is a clam or a snail shell, so geologists must define a species and deduce the relationships to other species on the basis of shell morphology alone. As will be described in later chapters, in rare cases they may have more evidence on which to base their conclusions. Because they work with form and structure alone, geologists define a species on the basis of a single fossil or a group of fossils called the **type specimen(s).** All other fossils referred to this species must be compared to the type(s). This is done by using descriptions, pictures, plaster casts, other specimens from the same location as the type, and direct comparison with type(s). Linnaeus laid this foundation for Darwin about one hundred years before Darwin left on the voyage of the *Beagle,* and in this intervening time the classification was refined and expanded.

Lamarck and Cuvier

Geology developed rapidly near the beginning of the nineteenth century, when the geologic time scale was set up. Evolution was an idea mentioned by a few workers. In 1809 Jean-Baptiste Pierre Antoine de Monet, Chevalier de Lamarck, described his theory of evolution. He believed that all organisms evolved from a single ancestor and that the environment created the need for change or evolution. His followers interpreted this to mean that the offspring inherited the acquired characteristics of the parents and, in turn, passed them on to their children. For example, Lamarckists believed that a blacksmith who developed large biceps would pass these muscles on to his children. We now know that traits are not inherited in this way.

One of Lamarck's critics was Georges Léopold Chrétien Frédéric Dagobert Cuvier, who studied the vertebrate fossil bones found in the sedimentary beds of the Paris basin. He was an antievolutionist and attacked Lamarck on two grounds. Lamarck believed that organisms became more complex through evolution, but Cuvier noted that the fish that he found in ancient rocks were already very complex animals. He also could see no evidence of gradual evolution in the fossils that he studied. The bones in each successive bed were different from those in the underlying bed, and there was no trace of any of the intermediate steps necessary in evolution. For these reasons, he believed in a separate creation for each bed and that the change in rock type from bed to bed was caused by the revolution that killed the organisms and so set the stage for the next creation. Cuvier did some excellent work in his studies of fossil vertebrates. He founded modern comparative anatomy, and it is ironic that comparative anatomy is one of the more important lines of evidence for evolution. If the skeletons of vertebrates are compared, similar bones are found in each, although they may serve a somewhat different function in each. In some cases, the correspondence is very close (Fig. 4.1). It is highly unlikely that most vertebrates would have the same bones unless all evolved from a single common ancestor. Surely the most efficient design of a whale fin, a bird wing, and a human hand would not all use the same bones for such different functions.

Darwin

The stage is now set for Charles Robert Darwin, who was born on the same day as Abraham Lincoln. In 1831, when he was 22, he sailed on the *Beagle* as naturalist on a five-year voyage around the world. During the early days of the voyage, while fighting seasickness, he read Charles Lyell's *Principles of Geology*, which had been published the preceding year. In his book, Lyell had gathered much evidence for uniformitarianism and had shown that the earth is much older than was then

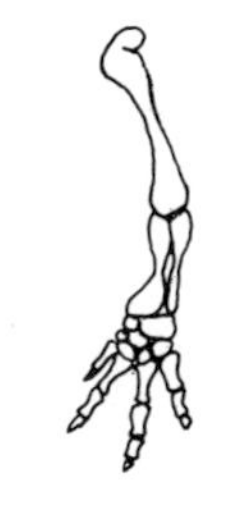

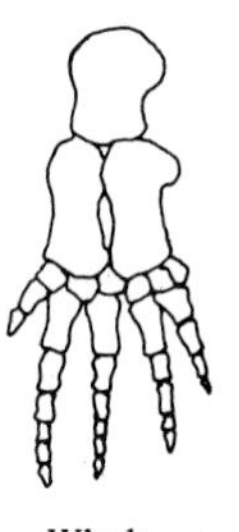

FIGURE 4.1
Similarity of bones in forelimbs of vertebrates. Similar bones serve different purposes in each of the examples. It seems very unlikely that such different limbs should have similar bones without evolution.

currently thought. This book had a great influence on Darwin, and it started him thinking of slow biologic changes over an immense length of time.

During the voyage, Darwin made many observations, both biologic and geologic, but it is his study of finches in the Galápagos Islands that has become best known. The Galápagos Islands are 800 to 970 km (500 to 600 mi) off the coast of Ecuador, too far for these birds to fly. Darwin found 13 species of finches that are unknown elsewhere in the world, although they are closely related to South American finches. The 13 species vary mainly in the size and shape of their beaks, and this reflects differences in their food. (See Fig. 4.2.) Darwin recognized the importance of his discovery at the time, but he did not develop his theory until

FIGURE 4.2
Darwin's finches of the Galápagos Islands. These finches apparently evolved from a common ancestor. Notice the differences in the shapes of their beaks and in the size of the birds.

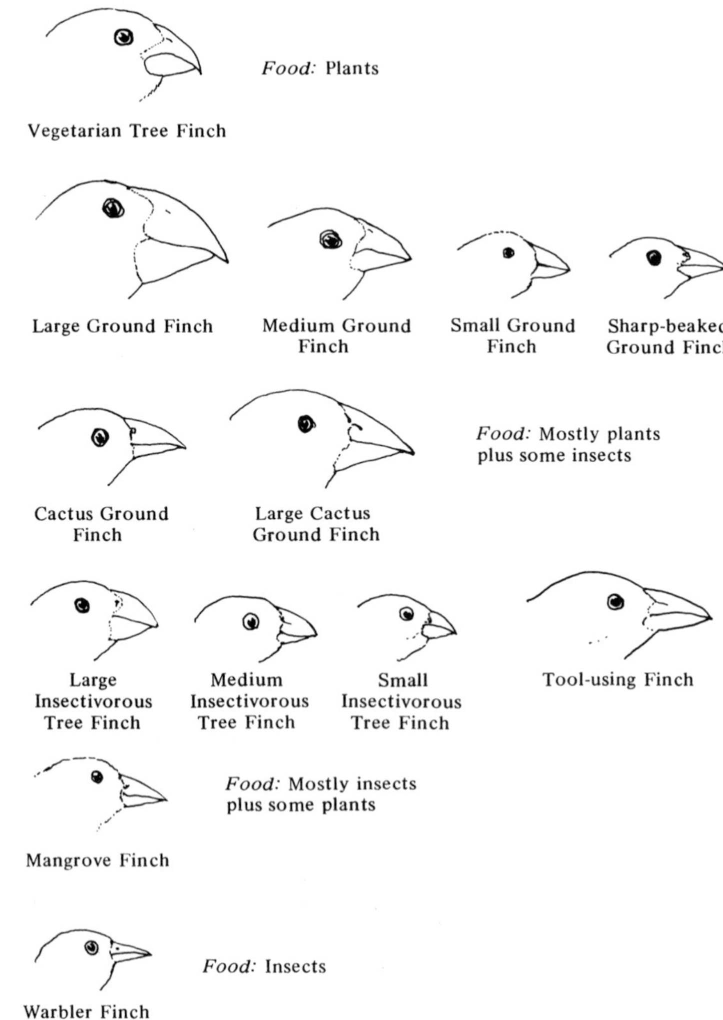

later. Darwin believed that in some accidental way, such as being blown by a storm, a few mainland finches reached the islands, and from these birds the various species evolved, each species being adapted to a different food supply. The finches were able to take over the various food sources because no other birds except ocean birds are found in the Galápagos Islands; thus there was no competition.

In 1838, long after the voyage, Darwin read Thomas Malthus's essay on population that had been published forty years before. Malthus's main point was that populations grow faster than their food supplies, and so food supply will ultimately control the world's population, an idea that is causing much concern today. Malthus used the phrase "struggle for existence" and noted that most organisms produce more young than can easily survive. Darwin reasoned that those young with any slight advantage over the others would be the most likely to survive and produce offspring. He called this "natural selection" by survival of "the fittest."

As soon as Darwin realized that species are not fixed, he knew that he must answer all possible questions and produce an airtight case because he was sure to be attacked. He began to write, documenting the evidence for evolution. He produced a 35-page abstract in 1842, and in 1844 he allowed a few friends to read a 230-page essay. By 1858, after finishing his studies of the materials collected on the voyage of the *Beagle,* he had eleven chapters written of what would be *The Origin of Species* and was proceeding at a leisurely rate. At that point, he received an essay from a much younger man, Alfred Russel Wallace, that contained the main points of Darwin's work. Darwin was ready to abandon his work in favor of Wallace, but his friends urged a joint presentation. Darwin finally published *The Origin of Species* late the next year. The attack that Darwin had expected came; it was bitter and still, in a few places, continues today. Until Darwin's time, most geologists were trying to fit their geological history into the then-current interpretation of biblical history. Darwin finally broke this tradition. He was attacked because he did not believe that all species came from the Ark and because many interpreted his work as meaning that humans evolved from apes. Darwin's ideas on humans were finally published in 1871 in the book *The Descent of Man, and Selection in Relation to Sex.*

Mendel—The Laws of Inheritance

One reason for the attacks on Darwin was that, though he had documented the evidence for evolution, he had no process or mechanism to cause evolution to occur. He leaned toward Lamarckism, but he also recognized its weakness. Had Darwin known it, the mechanism had been discovered in 1865 and published in 1866 by Gregor Johann Mendel, an obscure Austrian monk. Mendel experimented over an eight-year period with garden peas. He used several characteristics, such as whether the peas were round or wrinkled. He discovered that in cross-pollinating round and wrinkled peas, the first generation are all round peas. These hybrid round peas are not all the same, however, because the following generation is in the ratio of three round peas to one wrinkled pea. (See Fig. 4.3.) In succeeding generations, the wrinkled pea produces only wrinkled peas. In the same way, one-third of the round peas produce only round peas in succeeding generations. The remaining two-thirds of the round peas produce in the next generation round and wrinkled peas in the ratio of three round to each wrinkled one. Mendel called characteristics such as roundness "dominant" and wrinkledness "recessive." He continued his experiments, using two characteristics such as color and roundness, and discovered again that the offspring occur in fixed ratios. He showed that traits are inherited as discrete units.

Mendel's genius was that he could apply mathematics to breeding. He was ahead of his time because no one else understood the importance of his discoveries. Darwin himself experimented with cross-breeding and noticed the same three-to-one ratio, but the meaning and importance escaped him. Mendel's work lay buried

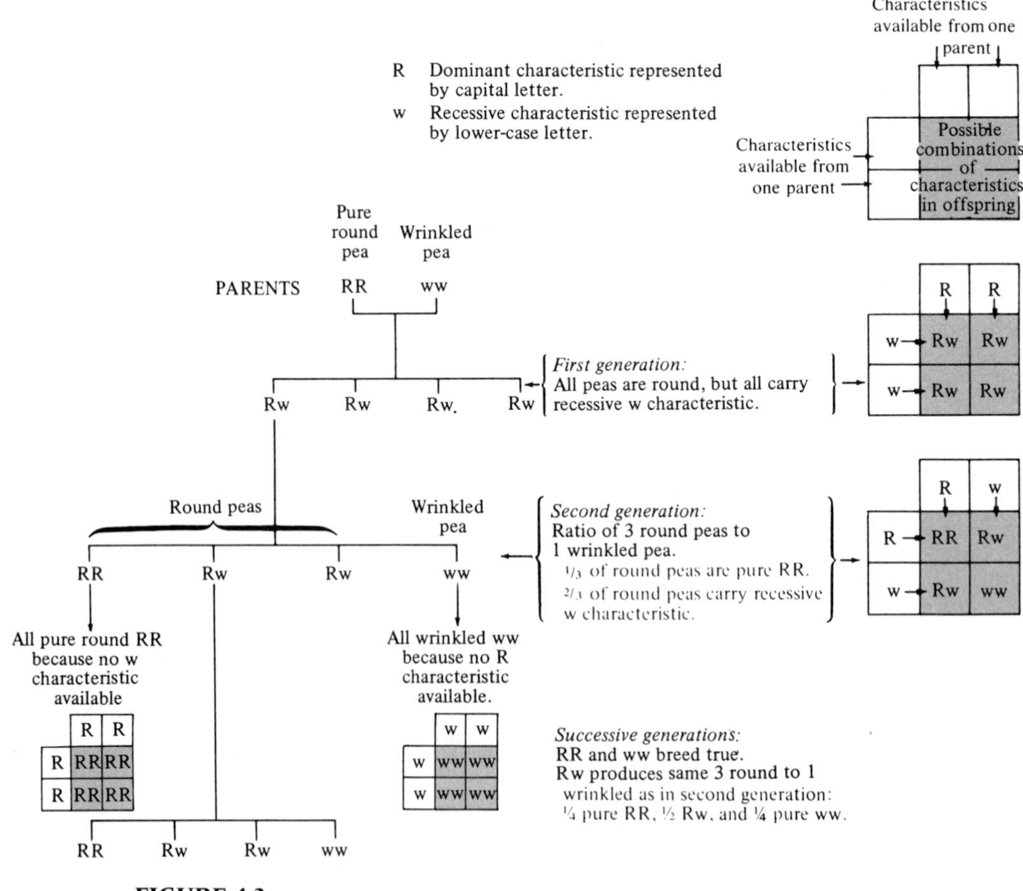

FIGURE 4.3
Mendel's inheritance law.

until 1900, when three separate workers rediscovered both his findings and then his published work. The Darwin-Wallace simultaneous discovery and the triple rediscovery of Mendel's work are two examples of parallel discoveries that occur from time to time in science.

Mendel discovered that characteristics are passed to offspring by discrete units that are now called **genes.** At first, the mechanism was believed to be mutations that caused big steps. Darwin had referred to these as "sports." Further study and reflection show that most big changes are detrimental and the mutant generally dies before producing any offspring. Evolution is caused by mutation; but the changes are typically very small steps, and natural selection of these small changes is the mechanism of evolution. Genes change spontaneously, but, of course, at a very low rate. Changes are also caused by radiation. Natural radioactivity and cosmic rays can cause changes in genes, as can radiation such as X rays. Many feel that such changes are the real danger from atomic bombs, and this is why atomic power plants are so carefully designed and monitored.

Evolution can occur because in any species, especially if there are a large number of individuals, there is a large pool of genes. These genes are varied enough that favorable characteristics are available to meet almost any change in environment. Even if the mutation is only slightly favorable, it will, over a number of generations, replace the original characteristic and become the normal characteristic of the population. Thus it is the population that undergoes evolution.

Punctuated Equilibrium vs. Gradualism

The geologic record is important in all theories of evolution, but that record is not complete. The remains of most organisms are not preserved by fossilization, and even the few that are preserved may be destroyed later by erosion, metamorphism, or rock deformation. In spite of these problems, at a few places the fossil record is reasonably complete. At those places we find one species being replaced by another species, and that species replaced by still another, and so forth. The problem is the almost complete lack of the intermediate populations that must have existed if one species evolved into another species. The geologic record shows long intervals of nonchanging species, or, in other terms, successful species do not change.

Most theories of evolution assume that changes in species occur slowly, but that external changes, such as climatic change, may at times cause more rapid evolutionary changes. *Slow evolution by small steps is called* **gradualism.** Gradualism is not supported by the geological record because of the lack of intermediate stages. The question is whether the lack of intermediate stages is real or whether it is caused by a lack of data. Gradualists believe that lack of data is the problem, but others believe that the intermediate stages are not found for other reasons.

The other possibility is that *at times evolution is rapid, but most of the time very little change occurs. This is called* **punctuated equilibrium.** In this case, evolution occurs in small, stressed populations, generally at the fringes of the habitat. The changes occur rapidly in small populations and so are not apt to be preserved. After such a short burst of evolutionary change, perhaps caused by a climatic or other environmental change, the new species is better adapted and so challenges and rapidly replaces the old species.

PATTERNS OF EVOLUTION

Divergence (Radiation)

As the history of the earth is outlined, it will be clear that some organisms have changed and others have not. Sponges, for instance, have not changed much during geologic history. They are very well suited to their form of life. Other organisms have changed, or adapted **(adaptation)**, to take advantage of new or changing conditions. These changes are generally a *spreading out, or* **divergence (radiation)**, *to different habitats.* Radiation occurs again and again in the geologic record. Darwin's finches (Fig. 4.2) are an example of divergence.

Convergence

A striking feature of evolution is the way that *different organisms adapting to the same habitat develop similar characteristics. This is called* **convergence**, and an example is the similarities between the present placental mammals and the marsupials of Australia. There is a marsupial in Australia to match almost every placental mammal habitat. For instance, there is a marsupial "mouse," a marsupial "mole," and the Tasmanian "wolf." Another example is the similar body shapes of the shark, a fish; the dolphin, a mammal; and the ichthyosaur, an extinct reptile (Fig. 4.4).

Extinction

Some of the most pronounced changes in life occurred at times when the environments changed, such as when the seas retreated from the continents. This could be more apparent than real, however, because of the unrecorded time represented by the resulting unconformity. When the environment changes, a species can do one of three things: it can migrate, it can adapt, or it can die out. The reasons for natural extinction are seldom simple and, generally, must be speculations because of the lack of evidence.

Disease epidemics have been suggested as the cause of the mass extinctions of the geologic past. The pathogenic organisms that cause disease generally attack only a single species, so this cannot explain the extinction of many species, such as the dying out of all the dinosaurs at the end of the Cretaceous.

Climatic change has probably been the most popular explanation for mass extinctions. At many of the times that the mass extinctions of the geologic past occurred, mountain building also occurred, and seas withdrew. These changes would, of course, drastically change the climate. Movements of continents on

FIGURE 4.4
Convergence is shown by the similar body shapes of the shark, a fish; the dolphin, a mammal; and the ichthyosaur, an extinct reptile.

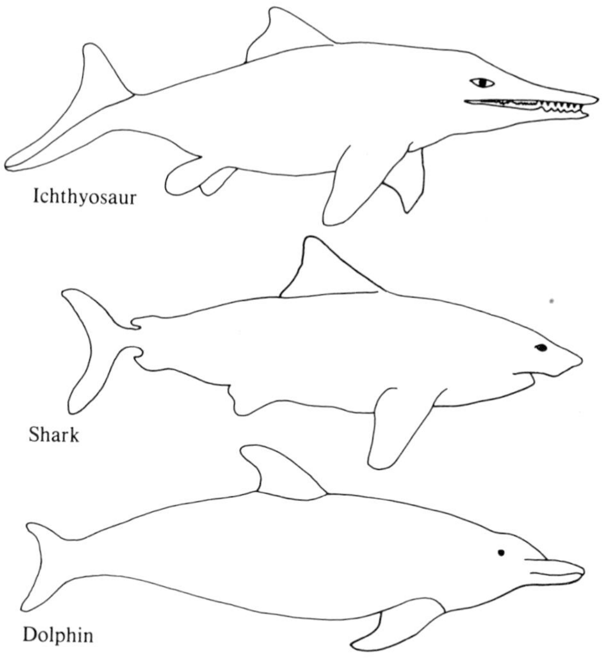

tectonic plates can also change climates. Many species would be unable to adapt or migrate and would die out, and new species, better adapted to the environment, would evolve. Close examination reveals flaws in this theory. At the times of great animal extinctions, the animal population changes drastically but not the plants, and plants seem to be at least as sensitive to climate as animals, probably more so. The last glacial events record a climate change, but the associated extinctions occurred long after the main glacial advances.

Another possibility is that many species must adapt or die if, in some way, an element on which they depend for food is destroyed or consumed by a more efficient competitor. This reduces the problem to finding a reason for the death of only one species instead of whole populations. Many of the theories just offered could explain the death of a single species. Many of the mass extinctions are associated with withdrawal of the sea, suggesting that such an event might shrink the environment of a key species, causing the demise of many, perhaps somewhat marginal, species. New species would develop when the seas again advanced, expanding the environment.

A variant of this theory is that most of the life of the ocean would be affected if the plankton died because the plankton, directly or indirectly, are the food for most forms of life in the ocean. The record suggests such a demise of at least some of the plankton. Chalk deposits of late Cretaceous age, composed almost entirely of the shells of plankton, are common in many parts of the world. These and other

FIGURE 4.5
Types of fossilization. A. Mold and cast. Left is cast, right is mold of *Maclurites,* an Ordovician snail from New York [7.6 cm (3 in.)]. B. Internal cast of *Turritella,* a Cretaceous snail [length, 5.1 cm (2 in.)]. C. A leaf preserved as a carbonaceous film. *Liquidambar* from Miocene of Oregon. D. Insect preserved in amber, Oligocene, East Prussia; insect preserved as a carbonaceous film. Parts E, F, G, H, I, and J on the following page.

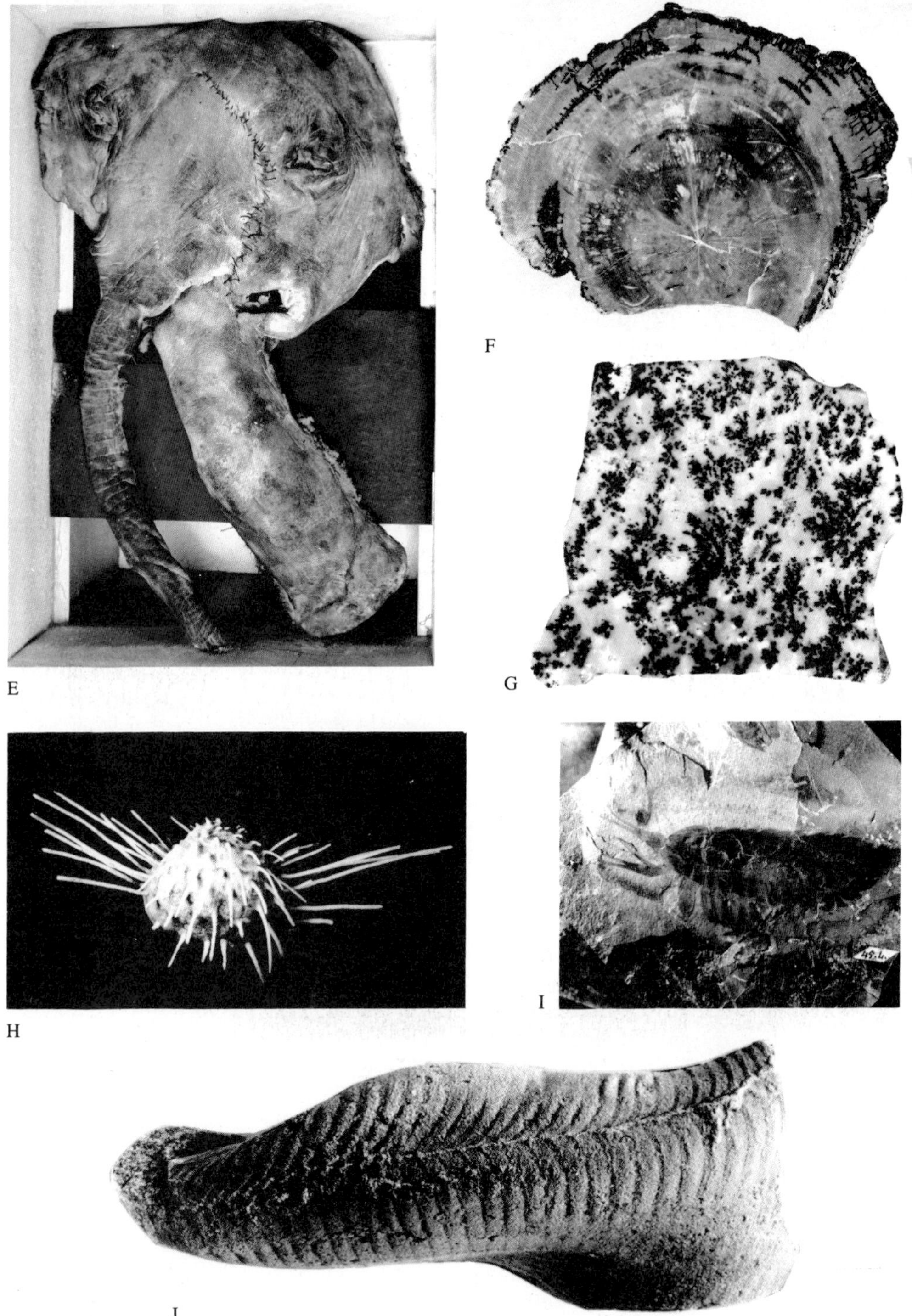

FIGURE 4.5 (cont.)
E. Baby woolly mammoth found in frozen ground in Alaska. F. Petrified wood. Silica has filled the voids in the wood, thus preserving it. G. False fossil. The structures in this specimen of moss agate are inorganic. H. Spiny brachiopod recovered by dissolving enclosing limestone with acid. I. Delicate features are preserved in this wormlike animal by carbonization. J. Crawl marks. Parts A and B courtesy Ward's Natural Science Est., Inc. Parts C, D, F, G, H, and I courtesy of the Smithsonian Institution. Part E Neg. # 320496 courtesy of the American Museum of Natural History. Part J courtesy of P. E. Cloud, Jr.

late Cretaceous deposits suggest that at many places this was a time of little erosion on the continents. It has been suggested that this lack of erosion reduced the amounts of inorganic nutrients in the ocean and so caused the dying of the plankton. Changes in the amount of oxygen may have caused some extinctions. It was recently shown that the present-day representatives of organisms that underwent extinction need more oxygen than those that did not undergo extinction. The correlation is very high. Near the end of the Cretaceous Period, when a great many organisms died out, a new group of photosynthesizers expanded greatly. They are the coccoliths, and their remains form great chalk deposits, such as the White Cliffs of Dover, England.

The most recent theory is that comets, or perhaps meteorites, crashing into the earth have thrown up clouds of dust that changed the climate worldwide. This idea was first proposed to explain the extinction of the dinosaurs at the end of the Cretaceous. It was proposed to account for traces of rare elements found at many places all over the world at that horizon. The evidence is outlined in Chapter 9. This theory was later extended to account for many other extinctions when it was recognized that many extinctions seem to occur at about 26-million-year intervals. Astronomers believe that this is the interval between times that the earth encounters the part of space in which comets are abundant and so is showered with comets.

In every case of extinction of a large group, some other group expands and so takes over the habitat. In most cases it is not possible to say why some groups died and others flourished.

FOSSILIZATION

Fossils are any evidence of past life, and this evidence may be preserved in many different ways. When most animals die, the body is rapidly destroyed; scavengers consume the body or bacteria cause it to decay. Only a very few escape this destruction, possibly to become fossils. Generally, if the body has hard parts, those parts are most likely to be preserved; if the body is rapidly buried, its preservation is more likely. Burial may also bring on physical and chemical changes that may cause destruction. Later uplift is necessary to expose the fossil, and weathering and erosion may destroy it then. Thus fossilization is a relatively rare event, and this is why the record of fossil life is so incomplete. For these reasons, the most common fossils are the shells of shallow-water marine animals such as clams.

The incompleteness of the fossil record makes reconstructions of past life very difficult. It has been estimated that only about 1 out of 44 species of the fossilizable marine invertebrates of the past is known. It is estimated also that of the 10,000 or so species found on a tropical river bank, only 10 to 15 are likely to be preserved.

Fossils form in the following ways:

Preservation of part of the animal, such as a clam shell, can occur.

Petrifaction takes place by replacement, recrystallization, or permineralization of the original shell or other part of the animal. This is a very common type of fossilization, and most fossil shells are either recrystallized or replaced. Ground water is the agent that causes the replacement in most cases. Silica and calcite are the main replacing minerals, but many other minerals are found as replacements. In some cases, very delicate internal structures may be preserved. Some of the best fossils with internal structures and surface ornamentations preserved are replacements by silica in limestone. In such cases, the fossils can be very easily recovered by dissolving the limestone in acid. (See Fig. 4.5H.) In permineralization the permineralizing mineral fills the voids in the fossil, as silica does in petrified wood.

Molds of fossils form if the fossil is dissolved, leaving an empty space. (See Fig. 4.5A.)

Casts form if molds are filled with minerals or sediments. (See Figs. 4.5A and B.)

Carbonization occurs if a fossil such as a leaf is squeezed, generally on a bedding plane, and the fluids are removed, leaving a thin carbon film. Worms and other animals without hard parts can be preserved in this way. Very tiny, delicate structures can be preserved by carbonization. (See Figs. 4.5C and I.)

Tracks and *burrows* are also evidence of life. (See Fig. 4.5J.)

Coprolites, or fossil excrement, can tell much about the eating habits of fossil life.

Frozen mammoths have been uncovered with every part of the animal intact. Such occurrences are, of course, very rare. (See Fig. 4.5E.)

Mummified animals are also quite rare. Mummification could occur through the trapping and sinking of an animal in a tar seep; the tar would inhibit decay of the body. Peat bogs have yielded specimens with soft parts well preserved. Mummification can occur also in very dry climates.

SUMMARY

Biological evolution is the continuing change in populations of organisms that is now occurring and has occurred in the geologic past.

A species is a group of similar individuals that can interbreed to produce fertile offspring.

A genus is a group of closely related species.

Species of fossil organisms must be based on form and structure of a specimen.

A type specimen is used by geologists to define a species.

Linnaeus was first to classify organisms, using species and genus to show the degree of relationship.

Lamarck was an early evolutionist. Lamarckism has come to mean that offspring inherit the acquired traits of parents.

Cuvier did not believe in evolution, but his comparative anatomy is important evidence for evolution. That is, it is highly unlikely that most vertebrates would have equivalent bones unless all evolved from a common ancestor.

Darwin marshaled the evidence for evolution. He believed in "natural selection" by survival of "the fittest," but he could find no mechanism of inheritance.

Mendel, experimenting with cross-pollinated peas, discovered how traits are inherited.

Genes are the discrete units by which characteristics are passed to offspring. They change spontaneously at a very slow rate and more rapidly when exposed to radioactivity.

Evolution occurs because in any large population there are genes that favor almost any change in environment; the mutation, even if it is only slightly favorable, will, over a number of generations, replace the original characteristic and become the normal characteristic of the population.

The lack of intermediate stages in evolution has led to the concept of punctuated equilibrium, the concept that at times evolution is rapid, but most of the time very little change occurs.

Adaptation is a change by an organism to take advantage of new or changing conditions.

Radiation is a spreading out into different habitats.

Convergence is the development of similar characteristics by different organisms in the same habitat.

When the environment changes, a species can migrate, adapt, or die out.

It is not yet possible to explain the large extinctions of the geologic past.

Fossils are evidence of past life.

Fossilization is a rare event, so the record of fossil life is incomplete.

Most fossils are hard parts that have been preserved by petrifaction, generally by recrystallization or replacement by silica or calcite.

Fossilization may also occur by preservation, molds, casts, carbonization, freezing, and mummification.

Tracks, burrows, and coprolites (fossil excrement) are also fossils.

QUESTIONS

1. List the evidence for biological evolution.
2. Who first used genus and species in naming organisms? When?
3. What is a species?
4. What is a genus?
5. How does a geologist's use of species differ from a biologist's use?
6. What is a type specimen?
7. What were Lamarck's ideas about evolution?
8. What was Cuvier's evidence for separate creations? Is it good evidence?
9. What led Darwin to his ideas on evolution?

10. What was the chief weakness in Darwin's *Origin of Species?*
11. What is Mendel's law of inheritance?
12. How do genes control evolution?
13. What is biologic radiation?
14. List and describe the types of fossilization.
15. Discuss the relationship between geologic changes and evolution of new species.
16. Describe several possible explanations for the extinction of species.
17. What is meant by punctuated equilibrium?

SUGGESTED READINGS

Clarke, B., "Causes of Biological Diversity," *Scientific American* (August 1975), Vol. 233, No. 2, pp. 50–61.

Ehrlich, Paul, and Anne Ehrlich, *Extinction. The Causes and Consequences of the Disappearance of Species.* New York: Random House, 1981, 305 pp.

Fridriksson, Sturla, *Surtsey: Evolution of Life on a Volcanic Island.* New York: John Wiley and Sons, Inc., 1975, 198 pp. Describes the development of life on a newly formed island.

Kimura, Motoo, "The Neutral Theory of Molecular Evolution," *Scientific American* (November 1979), Vol. 241, No. 5, pp. 98–126. This paper presents an alternative to Darwinism.

Kolata, G. B., "Paleobiology: Random Events over Geologic Time," *Science* (August 22, 1975), Vol. 189, No. 4203, pp. 625–626, 660. Describes a different view of evolution.

Raup, D. M., "Biological Extinction in Earth History," *Science* (March 28, 1986), Vol. 231, No. 4745, pp. 1528–1533.

Russell, D. A., "The Mass Extinctions of the Late Mesozoic," *Scientific American* (January 1982), Vol. 246, No. 1, pp. 58–65.

Scientific American (September 1978), Vol. 239, No. 3. Entire issue devoted to evolution.

Stanley, S. M., *Macroevolution. Pattern and Process.* San Francisco: Freeman, 1979, 332 pp.

Stone, Irving, *The Origin. A Biographical Novel of Charles Darwin.* Garden City, N.Y.: Doubleday, 1980, 744 pp.

Walliser, O. H., *Global Bio-Events. A Critical Approach.* New York,: Springer-Verlag, 1986, 442 pp.

Walter, M. R., "Interpreting Stromatolites," *American Scientist* (September-October 1977), Vol. 65, No. 5, pp. 563–571.

5

Early History of the Earth—Astronomical Aspects*

PRIVATE LIVES OF STARS—WE ARE CHILDREN OF THE STARS

Astronomers believe that the universe began with a huge explosion (the *Big Bang*) that occurred sometime between 10 and 19 billion years ago. The time range results from the assumptions made in their calculations. In the first few minutes after the explosion, hydrogen, the simplest element, was created; then some of it was transformed into helium, the next heaviest element. The fusion of hydrogen to helium (the hydrogen bomb reaction) takes place at a very high temperature (10 million degrees), but the expansion, caused by the atoms moving in all directions away from the site of the explosion, quickly cooled the gases and stopped the conversion. The hydrogen and helium atoms were not uniformly distributed in space but formed great cosmic gas clouds. Within these clouds smaller more dense clouds developed. These smaller clouds condensed to form stars, and the bigger clouds defined galaxies. The condensation was caused by gravity. The gravitational compaction of the hydrogen and helium atoms caused the temperature to increase to the point that fusion of hydrogen to helium began in the cores of the new stars. A star is born when it begins to radiate energy. The origin of our sun is discussed in more detail in the next section.

*In this chapter the astronomical convention of capitalizing the names of all planets will be followed.

Note that up to here the whole universe is made of hydrogen and helium and that none of the elements that form our bodies or Earth itself are present. The other elements were made in stars as they reached the ends of their lives. Because stars create energy by atomic reactions that turn one element into another, it is clear that eventually the supply of an element will be used up. Their lives are long even in terms of geologic time, but eventually a star must die. The lifetime of a star and its death process depend on many factors, especially its mass. A small star may simply collapse and fade away when its core ceases to produce energy. The collapse of a larger star may produce enough compression to cause the core temperature to increase so that a different atomic reaction can occur. If the new reaction produces enough energy it may cause expansion of the star and turn it into a giant star. If the expansion is rapid enough the star may explode as a supernova. One result of the last two cases is the formation of heavier elements. The explosions associated with *novas* and *supernovas* create new dust clouds that contain heavier elements as well as hydrogen and helium. These clouds, too, condense to form new stars that in turn may produce even heavier elements. Our solar system condensed from such a cloud. Most of the atoms that form our bodies and the Earth we walk on were created in old distant stars. We are truly children of the stars.

ORIGIN OF THE SOLAR SYSTEM

The planets and most of the other objects that orbit the sun are believed to have formed at the same time the sun formed. The sun is believed to have originated by the gravitational condensation of a cosmic dust or gas cloud called a *nebula* (cloud). Condensation causes the temperature to rise; when the temperature rises high enough, nuclear reactions begin and the new star begins to radiate energy.

The planets are believed to have formed in a similar way at the same time. During condensation, such a gas cloud would almost certainly rotate because of the initial velocities of the particles brought together. As condensation continued, the rotation became more uniform as the random velocities canceled out. Ultimately, the cloud flattened and became disk-shaped, with most of the mass concentrated near the center. Further contraction and flattening made the disk unstable, and it broke into a number of smaller clouds. These smaller clouds were all in the same plane. Condensation occurred in each of these clouds, concentrating the heavier elements at the center. Because the nebula had about the same composition as the sun, the ancestral planets were surrounded by envelopes of hydrogen and helium. At about this stage, condensation had heated the ancestral sun to the point that nuclear reactions began. The sun's radiation ionized the remaining part of its nebula; these ionized particles interacted with the sun's magnetic field, reducing the sun's angular momentum (spin). Thus, the sun was slowed to its present slow rate of rotation, and the particles in the dust cloud gained in energy. Ionized particles radiated from the sun swept away the remaining gaseous dust surrounding the sun and the planets in the same way that the solar wind and the sun's radiation make the tail of a comet point away from the sun, regardless of the direction of motion of the comet. In this process the planets nearest the sun lost the most material. They are much smaller and denser than the more distant planets. As each planet condensed, it rotated faster, just as spinning skaters spin faster when they bring their arms in to their bodies. This theory explains the size and density of the planets, the distribution of momentum in our solar system, and why most of the planets revolve in the same direction.

The asteroid belt between Mars and Jupiter may have originated from a number of small centers of accumulation rather than from a single center, as did the other planets. Collisions among these bodies increased their total number, and it is the resulting fragments broken from larger bodies that became most of the meteorites.

Earth-Moon System

Interaction between Earth and the moon is believed to be slowing Earth's initially rapid rotation. This process is a continuing one, and the length of day is increasing. The year, the period of Earth's trip around the sun, is not affected because it is determined by the distance between the sun and Earth. Thus the number of days in the year is decreasing. The present average rate of slowing of Earth's rotation is estimated at between 1 second per day per 120,000 years to 2 seconds per day per 100,000 years. This is enough to produce apparent errors if the times of eclipses computed back in terms of our present day are compared with the ancient records. An ingenious check on this lengthening of the day in the geologic past has been devised. Some corals grow by adding layers much like tree rings. Thinner layers, apparently daily layers, occur within the annual layers. These daily layers are commonly not well preserved, but the few that can be counted suggest that in Middle Devonian time, about 350,000,000 years ago, the year had 400–410 days and in Pennsylvanian time, about 280,000,000 years ago, 390 days. (See Fig. 5.1.) The figures are in good agreement with the estimated slowing of Earth's rotation. Other types of fossils also show daily, monthly, tidal, and annual cycles, and it may soon be possible to determine the early history of the Earth-moon system from such studies.

The slowing of Earth's rotation is caused by the tidal drag of the moon. The total rotational energy of the Earth-moon system is conserved, so the energy lost by Earth is gained by the moon. The moon moves farther from Earth, and the length of the month, the moon's period around Earth, increases. Calculations show that ultimately Earth's rotation (spin) and the moon's orbital period will be the same, so that one part of Earth will always face the moon just as only one part of the moon now faces Earth. Similar calculations can be made to find the moon's past position and orbit. The results of such calculations depend to a large extent on the assumptions made. This is unfortunate, because otherwise the problem of the moon's origin and early history probably could be solved.

The origin of the moon cannot be determined, and there are serious objections to all theories so far proposed. The theory that Earth and the moon

FIGURE 5.1
Growth lines in coral. The fine bands are daily growth layers. By counting the number of such bands within the yearly growth bands, it is possible to determine the number of days in the year at the time the coral lived. Photo courtesy of S. K. Runcorn.

formed at about the same time at different centers of accumulations in a dust cloud was just described. The main objection to this is that both Earth and moon should have the same composition if they formed in this way, but their average specific gravities are different. The average specific gravity of Earth is 5.5, and that of the moon is 3.3. Some of the recent calculations of the moon's early orbit strongly suggest a capture origin. However, if Earth captured the moon, the orbital energy (kinetic energy) of the moon would have heated Earth, probably to near its melting point, an event not suggested by Earth's thermal history. The currently favored hypothesis is that the moon may have originated from the collision of a Mars-sized object with the protoearth. The material ejected from Earth would then condense to form the moon.

SOLAR SYSTEM

Our **solar system** *is composed of the sun and the planets that orbit around the sun, the satellites of the planets, and comets, meteors, and asteroids.* All of the orbital motions are explained by the law of gravity.

The sun is an ordinary, average-size star, composed mainly of hydrogen and helium, but containing most of the elements found on Earth. The sun has 99.86 percent of the mass of the solar system, and Earth has about 1 percent of the remaining 0.14 percent.

The nine planets all revolve in the same direction around the sun. The orbits are all elliptical, very close to circular, and lie close to a single plane, except Pluto, whose orbit is inclined and whose orbital motion is somewhat irregular. Seen from Earth, the sun and the planets follow the same path across the sky. The planetary data are summarized in Table 5.1. The inner planets, Mercury, Venus, Earth, and Mars, form a group in which the individual members are more or less similar in size. The next four planets, Jupiter, Saturn, Uranus, and Neptune, form a similar grouping of much larger planets. Pluto is much smaller than the other outer planets and has a tilted, eccentric orbit. Between Mars and Jupiter are several thousand small bodies, the asteroids, that are probably the origin of most meteorites.

Terrestrial Planets

We will consider the moon with the terrestrial planets and proceed in order of their sizes, beginning with the smallest, our moon.

Moon Just as the planets orbit the sun, some planets have moons, or satellites, that revolve around them. Most of the moons revolve in the same direction as the planets, but some move in the opposite direction. Most of the moons are much smaller than their planets. Ours is much closer in size (1738 km [1078 mi] radius) to Earth so that the Earth-moon system should probably be considered a double planet.

The moon orbits Earth and rotates on its axis at the same rate and so only one side of the moon is visible from Earth. Our only knowledge of the far side of the moon comes from space probes that have photographed the far side. They have revealed a cratered surface with almost no smooth areas.

The visible face of the moon has large, dark, smooth areas and lighter-colored, more rugged areas. The early astronomers called the dark areas seas (*maria* [pl.] , *mare* [sing.]) and the lighter areas lands, in a fanciful analogy to Earth. Both have craters but the lighter uplands have more craters than the lower maria. The craters range from hundreds of miles across down to a few inches in diameter. Most of these craters probably were formed by meteorite impacts. The moon, unlike Earth, is not protected by an atmosphere, so more meteorites hit the moon than Earth. The relative ages of various parts of the moon's surface can be determined by the density of the craters, the older surfaces having more craters. Some of the craters

TABLE 5.1
Planet data.

Body	Distance from Sun (average)		Radius		Specific Gravity	Atmospheric Pressure at Surface (Earth = 1)	Composition of Atmosphere
	Millions of Kilometers	Millions of Miles	Kilometers	Miles			
SUN	—	—	695,088	432,000	1.4	—	—
Mercury	58	36	2414	1500	5.4	—	None
Venus	108	67	6114	3800	5.2	90	Mainly carbon dioxide
Earth	150	93	6376	3963	5.5	1	78% nitrogen 21% oxygen
Mars	227	141	3379	2100	4.0	0.007	Mainly carbon dioxide, possibly minor water
Jupiter	777	483	69,992	43,500	1.3	200,000	Mainly hydrogen and helium; some methane and ammonia
Saturn	1426	886	59,533	37,000	0.7	200,000	Same as Jupiter
Uranus	2869	1783	25,342	15,750	1.2	8	Similar to Jupiter, with more methane, no ammonia
Neptune	4494	2793	26,549	16,500	1.7	?	Same as Uranus
Pluto	5899	3666	1500 (?)	900 (?)	?	?	?

have rays. (See Fig. 5.2.) The rays are believed to be material splashed from the crater by impact.

Evidence of volcanic activity on the moon is abundant. Some of the craters may be volcanic in origin, and volcanic activity probably modifies many of the craters. Figure 5.2 shows a number of domes with pits or craters at the top, strongly suggesting volcanoes.

Some very slow erosion does occur because the older craters are more rounded than younger ones. This rounding is believed to be caused mainly by the impact of micrometeorites.

FIGURE 5.2
Oblique view of the crater Kepler taken by *Orbiter III* from a height of 52.6 km (32.7 mi). Kepler has rays extending radially in all directions. The rays are believed to be the material splashed out by the impact and are visible from Earth only near full moon. The light color streaks in the lower right may be ray material, especially as the associated smaller craters form lines radial from Kepler. Many domes are visible; and one in the foreground has a crater at its summit, suggesting that the domes may be volcanic. The line in the foreground may be a fault. Photo from NASA.

FIGURE 5.3
The surface of the moon. In the foreground is the orange "soil" found by the crew of *Apollo 17*. The tripodlike object is used as a photographic reference and to establish sun angle. Note the footprints in the fine surface material. Photo from NASA.

The moon has a rugged surface with many mountains; however, these mountains have apparently been formed by impacts and volcanic activity. Thus, they are very different from the compressive folded and faulted mountains of Earth.

Several types of rocks have been returned from the moon. Those from the maria resemble Earth basalt, but have much more titanium and much less of the more volatile elements such as sodium and potassium. Gas holes are common. Many of these rocks are breccias and were formed by meteorite impact. The fragments in these breccias are basalt, gabbro, rock (breccia), glass, anorthosite (feldspar rock), and peridotite. Much of this material is dust size. (See Fig. 5.3.) Melting and fracturing from impact shocks are visible in most samples, and spherical glass particles and iron-nickel spheres of impact origin are common. (Figs. 5.4, 5.5.) The mare basaltic rocks all have radiometric dates between about 3900 and 3100 million years and are the youngest rocks so far found on the moon.

The rocks from the rugged uplands of the moon are different. In composition many are anorthosite, a rock composed of between 70 and 90 percent plagioclase,

FIGURE 5.4
Glass spheres from the moon's surface. The largest is 0.5 mm in diameter. Colors are clear, yellow, red, and black. Their origin is probably from melting during meteorite impact. Photo from NASA.

and in texture they are breccias. This suggests that the uplands may be underlain by anorthositic (plagioclase-rich) gabbro that has been brecciated by meteorite impacts. Many of these breccias were partially melted. Some of the highland rocks are lavas. The highland rocks all have radiometric ages of about 4000 million years.

Carbon compounds have been found in the moon rocks, but no trace of life has been found.

Seismic studies reveal that the moon is layered. The surface to about 19 km (12 mi) is apparently rubble. The base of the crust is at about 64 km (40 mi), and there the velocity increases. The mantle extends to about 966 km (600 mi). At that depth the shear waves are somewhat reduced in amplitude, suggesting that the core may be partly melted.

The history of the moon begins with its formation, probably about 4550 million years ago. This is the same date as the origin of Earth and the solar system. Much of the outer part of the moon probably melted at this time. The infall of material and the resulting compaction would cause heating. This melting would produce the plagioclase-rich crust of the moon, which crystallized about 4000 million years ago. After the crust formed, the interior of the moon would begin to heat. This heating would be caused by radioactive elements. It would first melt the layer below the crust and could be the origin of the mare basalt that formed between 3900 and 3100 million years ago. Melting would then progress downward as the upper layers cooled. This process could produce the partially melted core

FIGURE 5.5
Moon rocks. A. This rock, returned by *Apollo 12,* was collected from the bottom of a one-meter-deep crater. The bottom part is covered by a thin layer of glass that was probably produced by melting caused by impact. The rest of the rock also appears to be shattered by impact. B. Rock returned by *Apollo 16*. This fragment was broken from a larger rock. Gas holes are prominent. Photos from NASA.

A

B

predicted by the seismic data. After 3100 million years ago, the main process acting on the surface has been meteorite impacts.

Mercury Mercury is too small and too close to the sun for detailed telescope study even though it has no atmosphere to obscure observation. A space probe has revealed cratered highland areas and presumably younger, uncratered basin areas. In this, Mercury is similar to Mars and the moon. The basins are much like those on the moon but have many more cracks and wrinkle ridges. (See Fig. 5.6.) The rocks filling these basins appear to be volcanic, like the similar rocks on the moon; however, unlike Mars and the moon, none of the photos of Mercury show any volcanoes. As on Mars, the distribution of cratered and basin areas is not uniform.

The big basins on Mercury are apparently impact craters caused by huge meteorites. At points on the opposite side of the planet from such basins are what is called weird terrain. (See Fig. 5.7.) It is believed that the shock effects of great impacts are concentrated at such places. The moon has similar big basins, and somewhat similar weird terrain is found there also.

Mercury has many scarps over 483 km (300 mi) long and 3 km (1.75 mi) high. (See Fig. 5.8.) These scarps cut both smooth and cratered areas and may have been formed by faulting. They may be compressional features, and, if so, they are the only compressional features seen on any planet except Earth.

Mercury's magnetic field is about one percent that of Earth's, but even this weak field is much greater than that of Venus or Mars. This suggests that Mercury may have or may once have had a liquid iron core or that the present field is caused by remanent magnetism in the rocks. Mercury's specific gravity is almost as high as Earth's, further suggesting that it may have an iron core. All of this suggests that Mercury melted, at least partially, early in its history.

Mars Mars has a reddish color and so was named for a god of war. The surface is covered with fine material and boulders that appear to be weathered basalt. It is farther from the sun than is Earth and so has lower temperatures. Its rotational axis is tilted with respect to its orbital plane, and so it has seasons like Earth. Telescopic observation reveals that polar ice caps form and disappear each year, and the color of the planet changes from bluish-green in spring to brown in the fall, which once suggested seasonal growth to some observers. *Mariner 9* photographs reveal dust clouds, so the observed color changes may be caused by dust storms and clouds. Mars' atmosphere is composed mainly of carbon dioxide and a little water vapor. *Mariner* spacecraft revealed that this inhospitable atmosphere is less than one percent as dense as Earth's. The average temperature at the equator at noon is probably 4°C (40°F), and the temperature may fall to −130°C (−200°F) at night. Thus the conditions are not very favorable for life on Mars. The pictures returned by *Mariner* space probes show evidence of both internal and surface activity.

Dust storms have been observed on Mars. As a storm subsides, first the higher elevations come into view and finally, the lowlands. In spite of the thin atmosphere, winds up to 483 km (300 mi) per hour are possible and 275-km-per-hour (170-mi-per-hour) winds are not uncommon. The dust may result from meteorite impacts or other processes. The wind may move the dust to lower elevations, and, in this way, create the featureless plains of Mars. The pits and basins of Figure 5.9 could be caused by wind erosion. The cliff, over a kilometer

FIGURE 5.6
Part of Caloris Basin is shown on the left side of this mosaic of *Mariner 10* images. Caloris Basin is the largest feature so far discovered on Mercury and is 1300 km (800 mi) in diameter. The floor of the basin has many cracks and ridges. Photo courtesy Jet Propulsion Laboratory.

FIGURE 5.7
This unusual terrain, seen only on Mercury, is termed *weird terrain.* The hills and ridges cut across both smooth and cratered areas. This image is on the opposite side of Mercury from Caloris Basin shown in Figure 5.6. Photo courtesy Jet Propulsion Laboratory.

high, that surrounds Olympus Mons (see Fig. 5.10) may also have been produced by wind erosion.

In the volcanic area is Olympus Mons (see Fig. 5.10), the largest volcano ever seen, although Venus may have larger ones. It is about 24 km (15 mi) high and 483 km (300 mi) across at the base; it has a broad complex crater, and its flanks are covered with what appears to be flow material. Nearby are three big volcanoes in a row, one of which is over 24 km high.

The equatorial region near the volcanic area is a plateau cut by what appear to be faults. (See Fig. 5.11.) The valleys shown in Figure 5.11A are part of a parallel system that extends more than 4830 km (3000 mi). They are interpreted as tensional features, perhaps associated with uplift. Figure 5.11B shows another valley that also may be of fault origin. Note the lines of craters that parallel the main valley.

Mars also has features that suggest erosion by running water, although very little water is present in the atmosphere. Figure 5.12 shows an example of a long, sinuous gully that resembles a river course. The water might have come from melting part of the polar caps. It is also possible that water in the form of ice is present in the "soil." The pits and basins shown in Figure 5.9 are near the south pole and so could have formed by the melting of subsurface ice rather than by wind erosion. The chaotic terrain of short ridges and small valleys shown in Figure 5.13

FIGURE 5.8
The scarps running from upper left to lower right in this image of Mercury are about 300 km (185 mi) long. They may be compressional features. Photo courtesy Jet Propulsion Laboratory.

is similar to landslide and collapsed areas on Earth caused by melting subsurface ice. Clouds on some of the big volcanoes suggest the possibility that they may be emitting gases, possibly water vapor, that might at times be abundant enough to cause rain and so erosion.

Craters are common in many areas, and, although some are of volcanic origin, many are caused by meteorite impact. This is not unexpected because Mars is near the asteroid belt and has a thin atmosphere. The effects of meteorite impacts are clearly shown on Mars' moon Phobos. (See Fig. 5.14.) Mars' equator almost separates cratered from uncratered areas. Why such a hemispherical division occurs is not known.

The polar ice caps appear to be dual ice caps. Much of the ice is frozen carbon dioxide, and this is the part that disappears and reforms seasonally. The rest is composed of water ice. The laminated structure near the ice caps suggests that

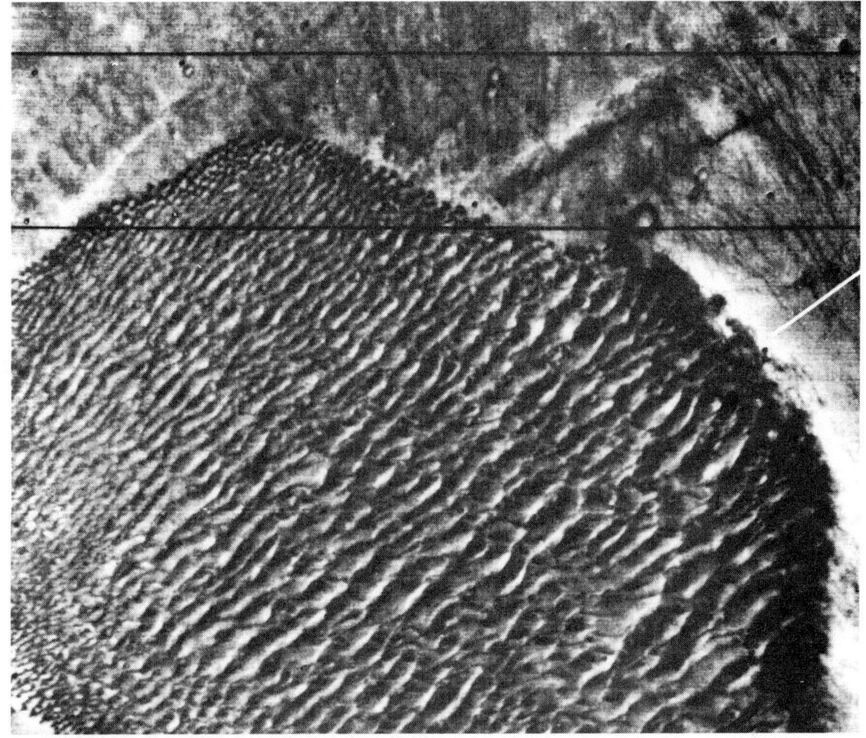

FIGURE 5.9
A possible dune field within a 150-km (93-mi) wide crater on Mars. Photo from NASA.

windblown dust accumulates in the ice at times. These deposits seem to be at least 100 m (300 ft) thick.

Venus Venus is named for the goddess of love because it is visible only near sunrise or sunset. It has a thick atmosphere, composed almost entirely of carbon dioxide, and the carbon dioxide causes a greenhouse effect, which accounts for Venus' high temperature. Also present in minor amounts are hydrogen, water vapor, carbon monoxide, and hydrofluoric and hydrochloric acids.

The surface temperature is near 465°C (870°F), making life a very remote possibility. The atmosphere has a solid cloud cover 72 to 97 km (45 to 60 mi) thick, so the surface is not visible from Earth.

Venus is the only planet that rotates on its axis in a clockwise direction (see discussion of Uranus below). No magnetic field has been detected, suggesting that it, too, lacks a liquid core, although its rate of rotation may be too slow to develop

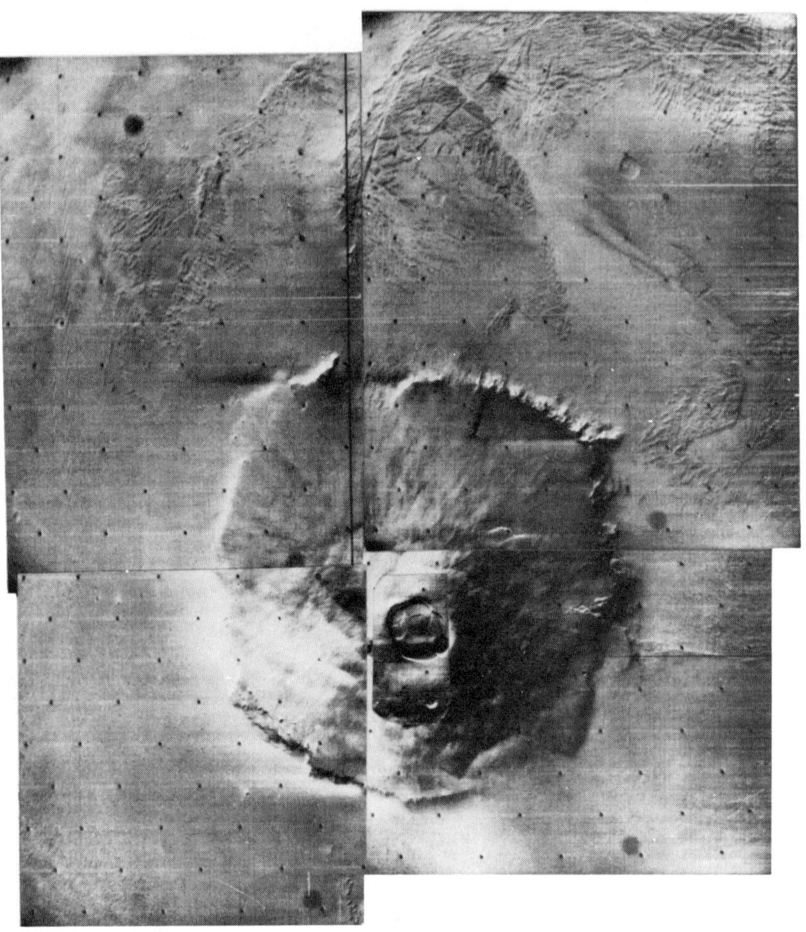

FIGURE 5.10
Olympus Mons, a gigantic volcanic mountain on Mars. The mountain is 500 km (310 mi) in diameter across the base. The complex crater is 64 km (40 mi) in diameter. Photo from NASA.

a magnetic field. The photographs sent back to Earth by soft-landers that operated for about an hour before the high surface temperatures incapacitated them showed angular rocks and no sand or dust.

Radar observation of Venus has revealed some features like Earth's. Mountains that appear to have lava flows may be the largest volcanoes in the solar system. Fault zones with lines of volcanoes and rift zones have also been detected. What may be folded rocks have also been seen and, if they are, Venus is the only other planet with compressive features. Uplands and lowlands are both cratered.

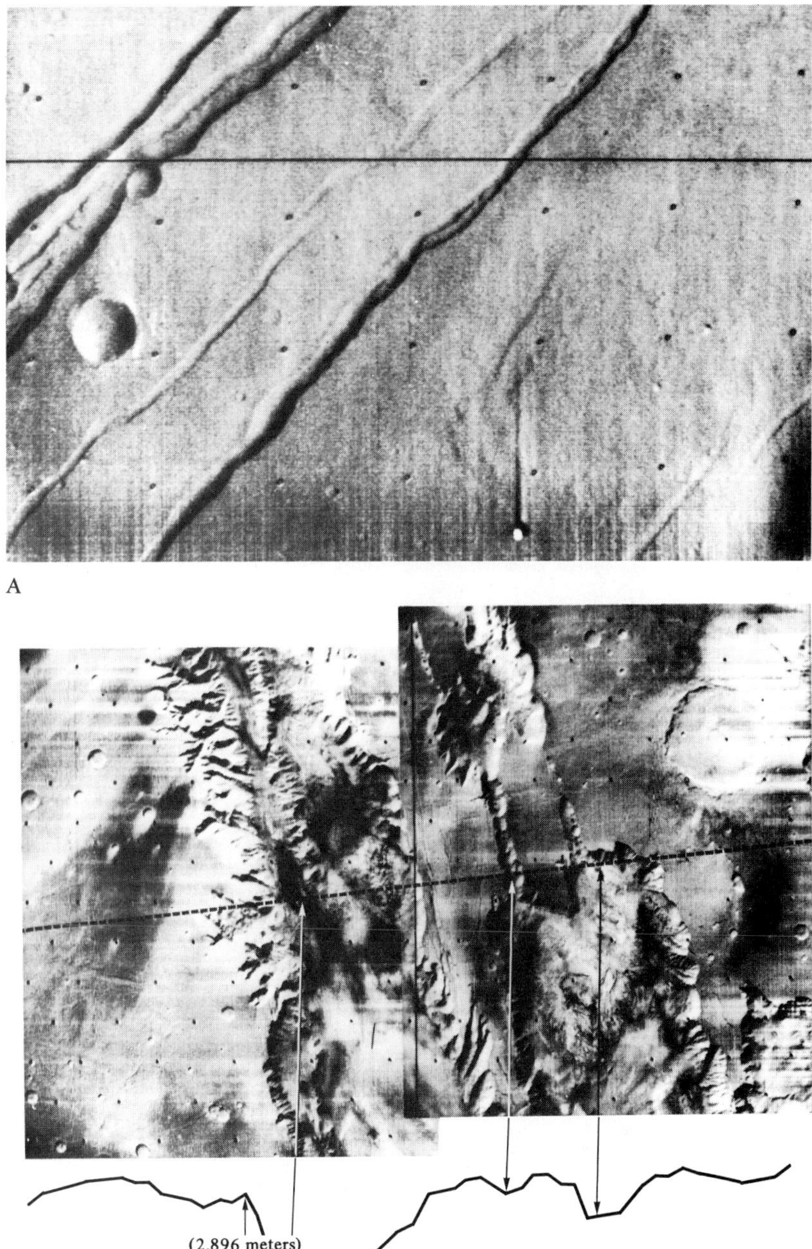

FIGURE 5.11
Valleys of probable fault origin on Mars. A. These structures are part of a system of parallel fissures extending more than 4830 km (3000 mi). The widest valley is about 1.6 km (1 mi) across and has a smaller valley inside it. The photo covers an area 34 by 42 km (21 by 26 mi). B. A valley on Mars much deeper than the Grand Canyon. This valley is 120 km (75 mi) across. Photos from NASA.

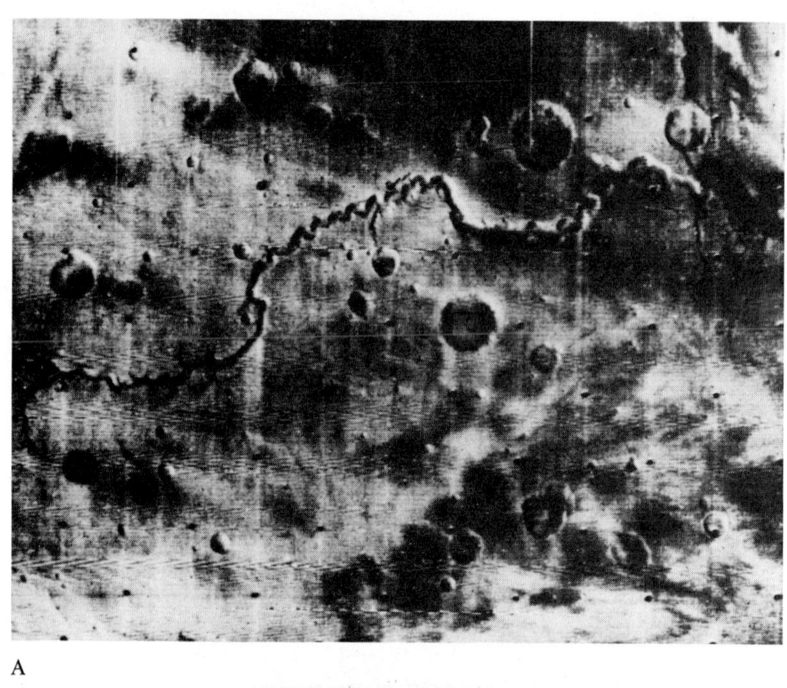

A

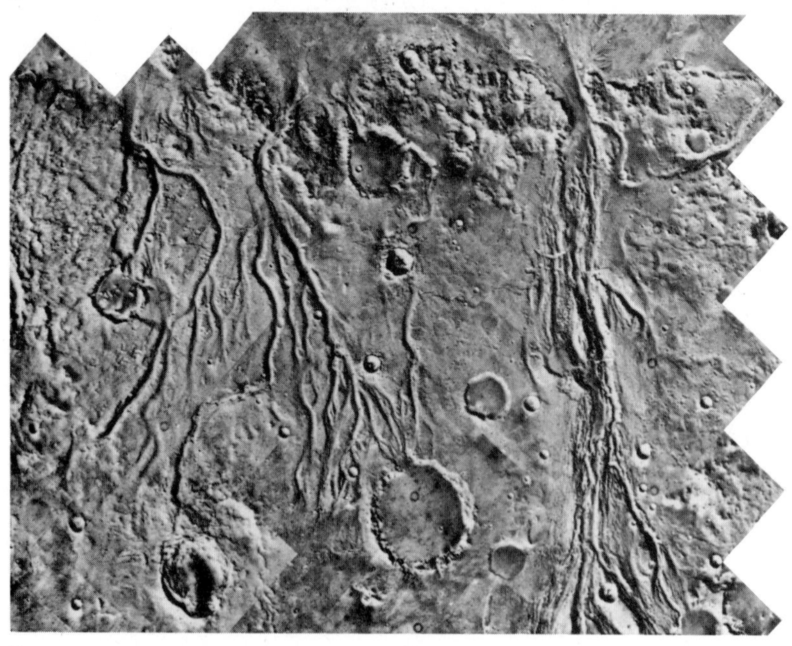

B

FIGURE 5.12
Erosional features on Mars. A. Riverlike gully suggests erosion by running water. B. The pattern of these gullies suggests erosion by floodwater.

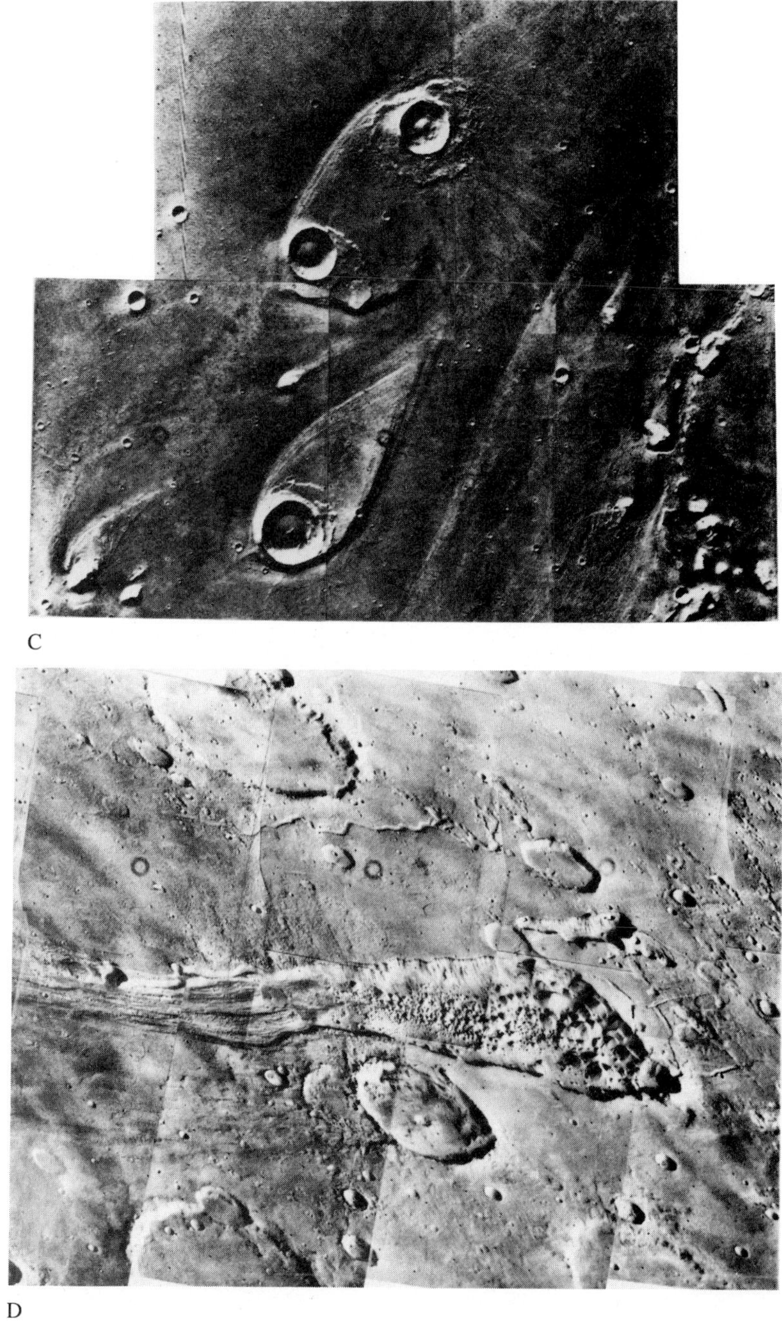

FIGURE 5.12 (cont.)
Erosional features of Mars. C. These teardrop-shaped islands are also suggestive of erosion by floodwater. D. The chaotic terrain at the head of this gully suggests that melting of subsurface ice may have been the source of the fluid that eroded the gully. Photos from NASA.

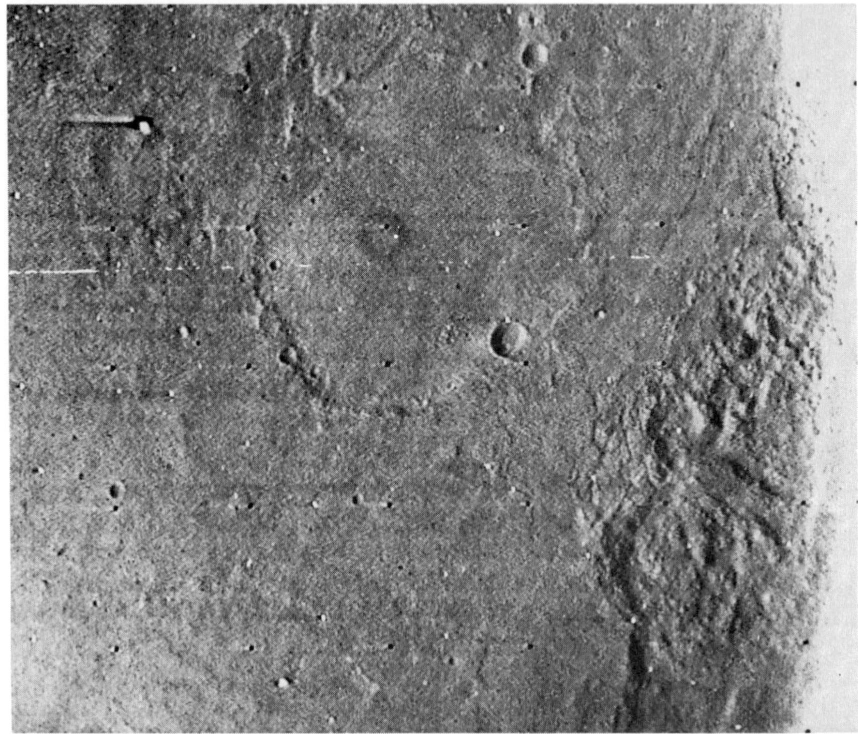

FIGURE 5.13
Chaotic terrain on Mars is shown on the right. The chaotic terrain is a jumble of ridges, each of which is 2.4 to 3.2 km (1.5 to 2 mi) wide and 1.6 to 8 km (1 to 5 mi) long. *Mariner 6* photo from NASA.

Great Planets

The great planets, Jupiter, Saturn, Uranus, and Neptune, are so called because they are much larger than the other planets. Although larger in diameter and mass, they are much less dense than the other planets. They all have thick atmospheres, which are, of course, all that we can see. Their rapid rotation causes pronounced bulges of the gaseous exteriors. The upper surfaces of their atmospheres are very cold because of their distances from the sun. They all have methane in their atmospheres, with ammonia, also, in Jupiter and Saturn; most of their atmospheres, however, are hydrogen and helium. Their low overall densities suggest that they, like the sun, are composed mainly of hydrogen and helium. Saturn, Jupiter, and Uranus have rings; Neptune has partial rings. Saturn's rings are best known and probably are dust and ice. This material is too close to the planet to consolidate and form a moon. The rings may be left over from the formation of the planet, or they might be satellites that approached the planet too closely and were

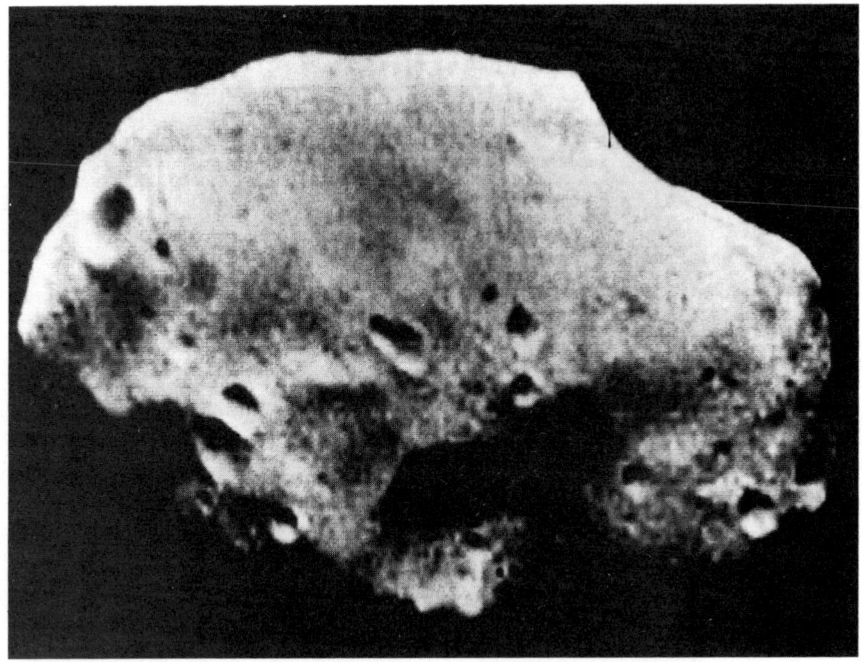

FIGURE 5.14
Mariner 9 photograph of Mars' moon Phobos. Phobos is about 21 by 26 km (13 by 16 mi). The age and strength of Mars' innermost moon is shown by the impact crater. Photo from NASA.

broken up when the gravitational attraction overcame their strength. Uranus is unique in having its rotational axis tilted 98 degrees to its orbital plane; alternately, it could be considered tilted 82 degrees and revolving in the opposite direction to the majority of the planets. Its moons orbit in the same direction as its rotation.

Larger Moons of Jupiter and Saturn

The larger moons of Jupiter and Saturn range in size from slightly smaller than our moon to somewhat larger than Mercury. Jupiter and Saturn have at least 16 satellites each, but only four of Jupiter's and one of Saturn's are similar in size to the inner planets. Only one other moon, Neptune's Triton, is of similar size, but very little is known about it.

The most interesting of these bodies is Jupiter's Io, which is slightly larger, 3640 km (2260 mi) in diameter, than our moon and has about the same density. Spacecraft images have revealed eight active volcanoes (Fig. 5.15). The volcanic eruptions are caused by tidal flexing by Jupiter and the other nearby large satellites, especially Europa. The gravitational attraction of the other satellites deforms Io and

FIGURE 5.15
Io is the only object, except Earth, on which volcanic eruptions have been observed. *Voyager 2* photograph provided by the National Space Science Data Center.

so heats it by friction. The material erupted by the volcanoes is mainly sulfur and sulfur dioxide; these materials are responsible for Io's colors—red, orange, black, and white. Some silicate material may also be erupted. Io and the other large satellites are obviously much different from the inner planets.

The other large satellites of Jupiter are composed of rock and water. Europa (3050 km [1890 mi] diameter) has a density of 3.03 g/cm^3, which is somewhat less than our moon's. It is believed to be composed of rock with a layer of ice about 100 km (60 mi) thick at the surface. The surface shows a tangle of lines, perhaps cracks in the ice (Fig. 5.16). Ganymede (5270 km [3270 mi] diameter) has a density of 1.93 g/cm^3 and is believed to be about half rock and half water. Callisto (5000 km [3000 mi] diameter) is less dense and contains more water than rock. Ganymede's surface has dark cratered areas and lighter grooved, perhaps faulted, areas (Fig. 5.17). Callisto's surface is covered with craters (Fig. 5.18).

Saturn has only one large moon, Titan, 5000 km (3000 mi) in diameter. Titan

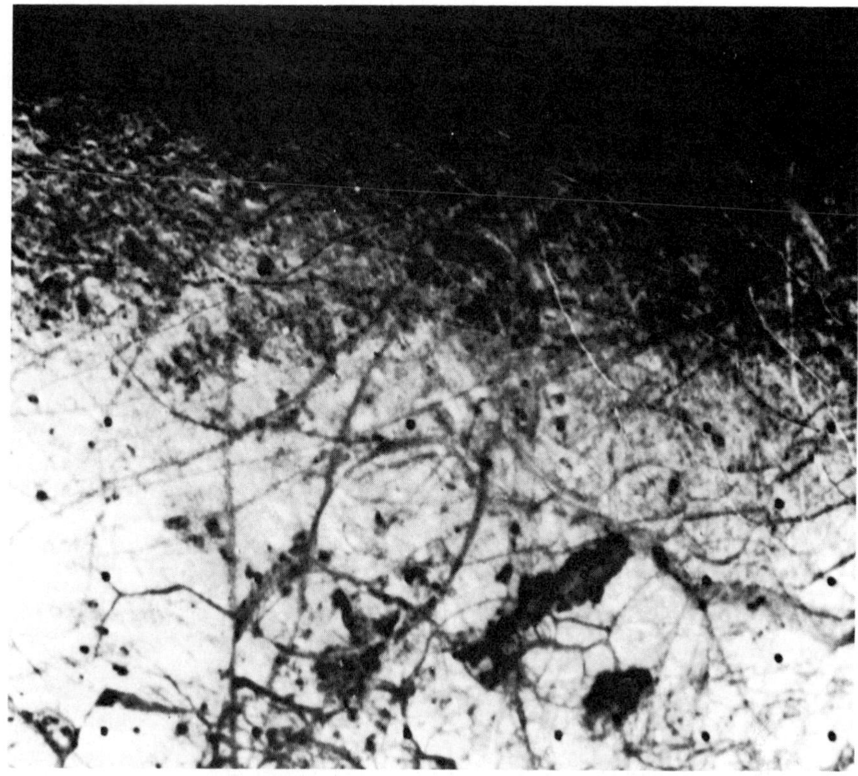

FIGURE 5.16
Europa shows a complex tangle of lines. *Voyager 2* photograph provided by National Space Science Data Center.

has a thick atmosphere, so little is known about its surface. Although it is only slightly larger than Mercury, Titan is able to retain an atmosphere because it is so cold. The atmosphere is, like Earth's, composed largely of nitrogen. The other constituents of Titan's atmosphere are hydrocarbons, especially methane. The surface temperature is estimated to be about −184°C (−300°F), and liquid nitrogen may be present on the surface. The action of the sun's energy on this atmosphere has produced brown gases that are probably similar to photochemical smog on Earth, which forms as the result of the sun's actions on hydrocarbons in our atmosphere.

Images of some of Saturn's smaller inner satellites have also been returned (Fig. 5.19), showing them all to be much smaller, ranging in size from Mimas, 390 km (240 mi) in diameter, to Rhea, 1530 km (950 mi) in diameter. They are probably all composed largely of water ice, and all except Enceladus are cratered. Mimas has a crater with a diameter about one-quarter of the diameter of the

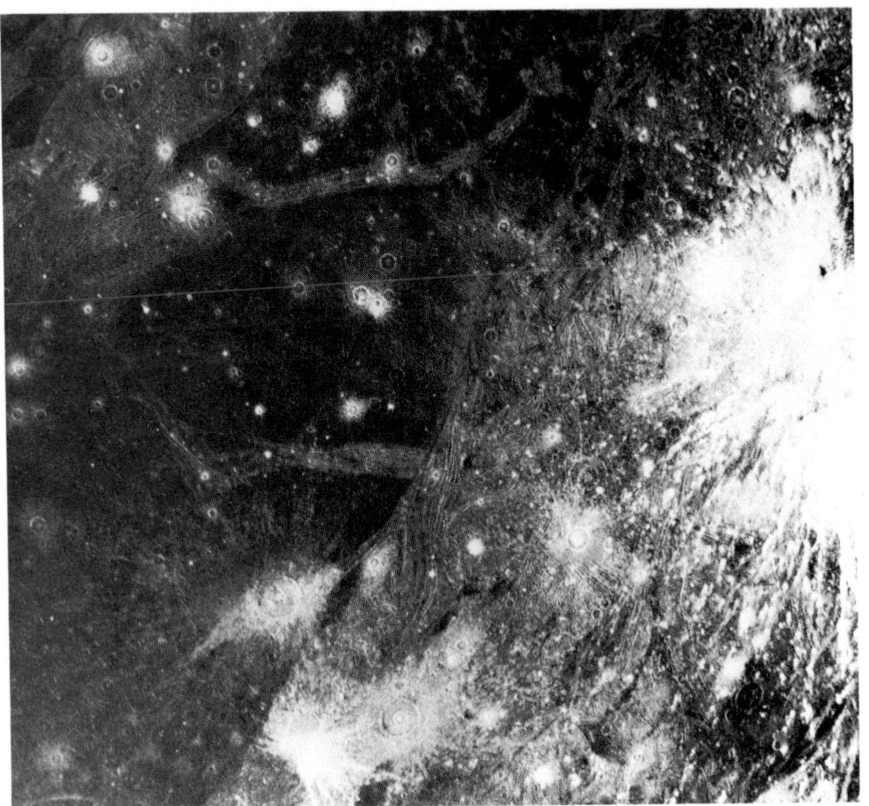

FIGURE 5.17
Ganymede is cratered. It has light-colored grooved terrain and dark, heavily cratered areas. *Voyager 2* photograph provided by National Space Science Data Center.

satellite. Tethys is fractured with a valley 70 km (40 mi) wide and 800 km (500 mi) long. Smaller sinuous, branching valleys are seen on Dione.

The puzzling features of all of the icy satellites may, at least in part, be caused by ice volcanism. This has been suggested for some of the moons of Uranus.

Meteorites

Meteorites are objects from space that hit Earth. The iron-nickel meteorites are believed to be similar in composition to the core of Earth, and the stony meteorites similar to the mantle. One interesting aspect of meteorite recognition is that over 90 percent of those seen to fall are stony meteorites, but only a third of the finds not seen to fall are stony.

Meteorite specimens range from dust to 264,000 kg (290 tons). Those that have produced the large impact craters on Earth were much larger. (See Fig. 5.20.)

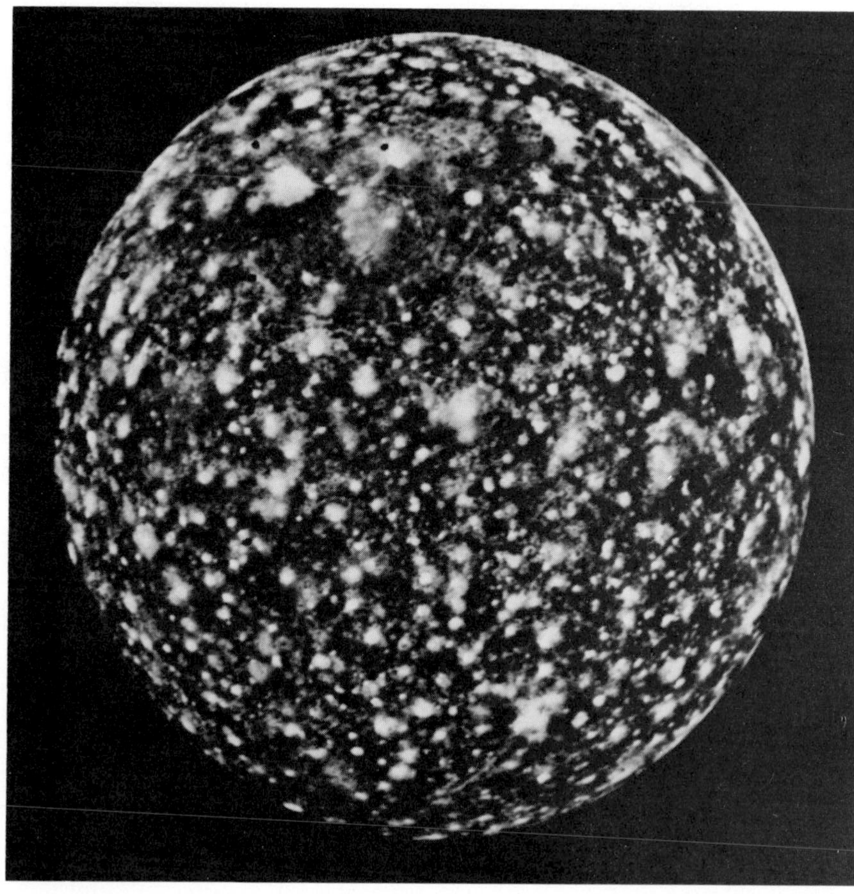

FIGURE 5.18
Callisto has many bright craters. *Voyager 2* photograph provided by National Space Science Data Center.

Many of the smaller meteorites do not survive the heat generated by friction in passing through Earth's atmosphere, producing meteors, or shooting or falling stars. It is estimated that between 1000 and 1,000,000 tons of meteorites fall on Earth each year.

Meteorites are believed to have formed at the same time as the rest of the solar system. Thus, they are used to date the formation of the solar system and as models of Earth's structure. Meteorites are fragments of much larger bodies. This is shown by the coarse crystalline nature of many, suggesting slow cooling that is possible only in a large body. Many, however, have complex crystallization histories. Some meteorites are also differentiated bodies; this melting can occur only in bodies large enough to generate the necessary heat, again suggesting that meteorites are

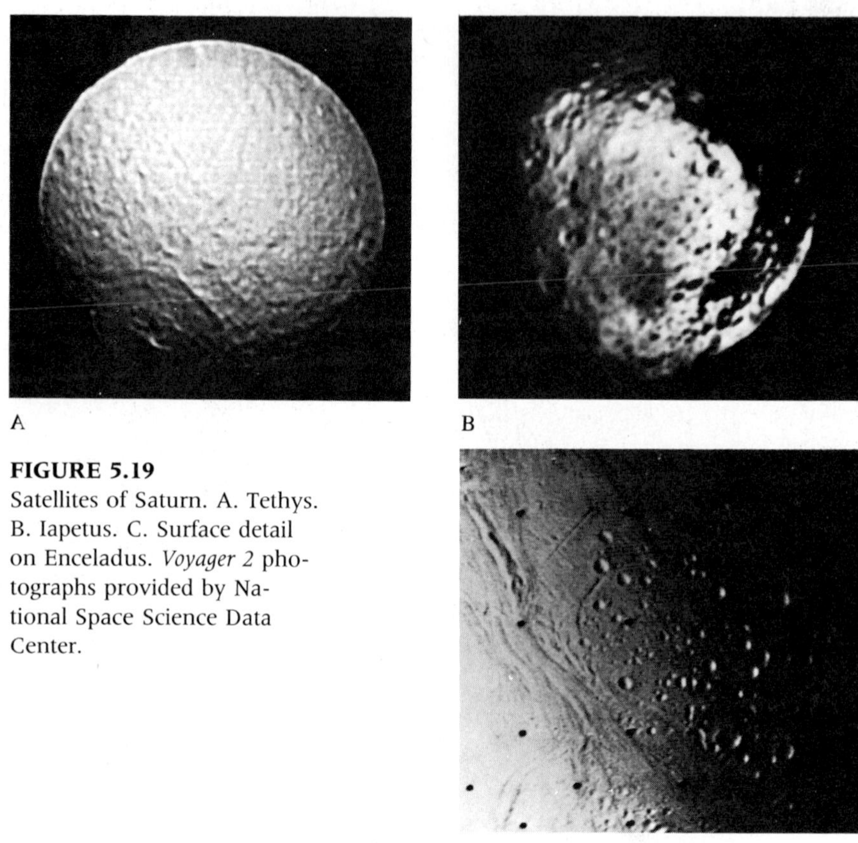

FIGURE 5.19
Satellites of Saturn. A. Tethys. B. Iapetus. C. Surface detail on Enceladus. *Voyager 2* photographs provided by National Space Science Data Center.

fragments of large bodies. Meteorites are generally believed to be fragments from the asteroid belt. A few are known to have originated from the moon and Mars. The asteroids are a group of about 30,000 small bodies ranging from about 805 km (500 mi) to about 1.6 km (1 mi) in diameter. Thousands of millions more, down to dust size, are thought to exist.

Comets

Comets are among the most awe-inspiring spectacles in nature. (See Fig. 5.21.) In the past they were considered harbingers of evil times. We now know their makeup in general terms and can predict their reappearance. Comets have very small masses. One passed through Jupiter's moons without noticeable effect on the moons, and Earth has passed through the tail of a comet in historic time. **Comets** *are believed to be masses of frozen gas with some very small particles similar in composition to meteorites.* This is sometimes called the "dirty snowball" theory.

FIGURE 5.20
Meteor Crater (Barringer Crater), Arizona. A similar crater would not be as well preserved in a more humid climate. Photo from Yerkes Observatory.

When distant from the sun, a comet is very small, perhaps 0.8 km (0.5 mi) in diameter. At this distance it is seen only by the sunlight that it reflects. When closer to the sun, the sun's radiation melts some of the ice and ionizes some of the resulting gases, making the comet more visible. Thus, comets gradually disintegrate by passing near the sun and die after a number of passes. During the process of disintegration, the spectra of water, carbon dioxide, methane, and ammonia can be seen. At this stage an average comet may have a head 128,800 km (80,000 mi) in diameter, made mostly of very thin gas. The solar wind pushes this gas outward from the head to form a tail up to 80 million km (50 million mi) long. This is the spectacular stage of a comet and occurs only when the comet is near the sun. One interesting point is that the solar wind pushes the tail away from the sun without regard to the direction in which the comet is moving. Most comets have very elliptical orbits, which explains why they reappear at intervals. As with the planets, the sun is at one focus of the ellipse; however, when distant from the sun, comets are too small to see. Some comets have orbits that bring them back only after thousands of years; some may be parabolic and so never return. (See Fig. 5.22.)

The origin of comets is not known; they may form at the fringe of the solar system or may come from outside the solar system. Their composition is not too unlike the material from which the solar system is believed to have formed. Thus they may be fragments "left over" from the origin of the solar system.

INFERENCES AND SPECULATIONS

The terrestrial planets may give us a view of what Earth was like in the past. The most obvious difference between Earth and the moon and inner planets is the absence of

FIGURE 5.21
Halley's comet. Photo from Mount Wilson and Palomar Observatories.

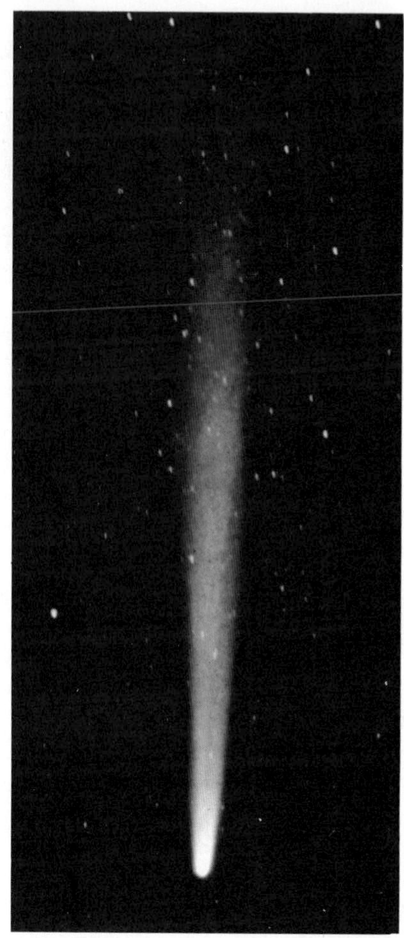

FIGURE 5.22
The solar system showing some possible comet orbits. The comet is shown on one orbit. Note that the tail always points away from the sun and that the comet gets larger where it is heated by the sun. The other orbit shows possible elliptical and parabolic orbits.

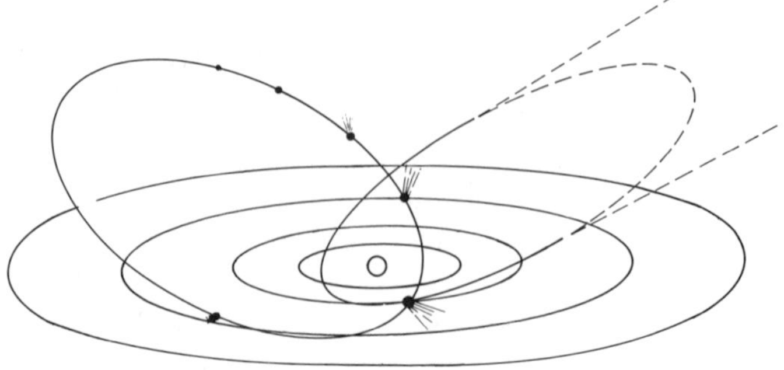

or greatly reduced amount of erosion on the latter because, with the exception of Venus, they have very tenuous or no atmospheres. This aids our interpretation of their internal processes because surface forms are not destroyed by erosion.

Plate tectonics is the main internal process on Earth, and a somewhat similar process appears to operate on some of the other planets. The energy for these internal processes must come from within the planet. *The amount of internal energy apparently is proportional to the size of the planet.* The other planets are smaller than Earth, and they show only some aspects of plate tectonics.

The similarities among the moon and the inner planets so far studied by space probe are striking. The similarity of surface features strongly suggests similar histories for the moon and these planets—Mars and Mercury. Rocks have been returned only from the moon, so the dating of events on Mars and Mercury depends on correlation with the moon on degree of cratering. Their history begins with their formation about 4550 million years ago. From then until about 4000 million years ago, cratering occurred at a rapid rate, forming the cratered highlands. Apparently the inner planets experienced a high influx of meteorites during this interval. Erosion has largely obliterated evidence of this event on Earth. From 4000 million to about 3100 million years ago, the number of impacts diminished rapidly before the less cratered volcanic lowlands formed.

The moon's cratered uplands differ somewhat in composition from the volcanic maria. This is seen from Earth as a difference in reflectivity. Similar differences in reflectivity are apparent on Mars and Mercury. This suggests that they, like Earth, underwent some melting early in their histories. The resulting layering is revealed by seismic studies on Earth and the moon. The energy released by crystallization of Earth's fluid core may be the energy that causes deformation of the crust. The early crystallization of the cores of the smaller planets may account for their lack of compressional features and for the great age of their surface features. Mars and Mercury have some surface deformation. Mars has great fractures, and Mercury has long scarps, suggesting again that size and internal energy are related.

The smaller planets also differ from Earth in that their surface features are not evenly distributed. Mars' equator is almost the dividing line between a cratered hemisphere and a smooth one. Mercury also has cratered and smooth areas. The side of the moon opposite Earth is more cratered than the side facing us. Why such a distribution of upland and lowland exists is not clear, although at present there is more continent in Earth's northern hemisphere than in the southern. At times in the geologic past, Earth's continents are believed to have been joined into a single large continent. Thus, had Earth run out of internal energy earlier, as the smaller planets did, a land hemisphere and an ocean hemisphere might have resulted.

Venus is nearly the same size as Earth and has many similar features. That planet has two levels like ocean and continent, volcanoes, and both tensional and compressive features and is the only other planet with possible folded mountains.

In the past the composition of Earth's atmosphere may have been like that of one of the great planets or like that of Mars and Venus. This aspect of early Earth is discussed in the next chapter.

SUMMARY

The solar system is composed of the sun and the planets, comets, meteors, asteroids, and the satellites of the planets.

The inner planets—Mercury, Venus, Earth, and Mars—are smaller and denser than the more distant great planets.

The moon orbits Earth and rotates on its axis at the same rate, so only one side is visible from Earth. The dark, smooth areas are called maria (mare, sing.) and the lighter areas are uplands.

Most of the moon rocks are breccias formed by the meteorite impacts that also cause the many craters. The mare rocks are basaltic in composition, and those of the uplands are feldspar-rich rocks.

Most moonquakes are low magnitude and are caused by tidal effects. They originate from a single zone and reveal a layered moon with about 19 km (12 mi) of surface rubble with velocity increases at about 19 and 64 km (12 and 40 mi) deep. The core, starting at about 966 km (600 mi), may be partially melted because seismic waves are found to be somewhat attenuated there.

The moon probably formed about 4550 million years ago and probably melted soon after this. The oldest rocks so far found are from the highlands and are about 4000 million years old. The maria basalts were extruded between 3900 and 3100 million years ago. Since then the moon has been quiet, with only meteorite impacts and perhaps some limited local volcanic activity.

Mars has dust storms, seasonal ice caps, huge volcanoes, evidence of faulting, apparent erosion by running water, and impact craters, as well as featureless and chaotic terrains. It is the other planet most likely to have life.

Venus has a thick atmosphere and is very hot.

None of the inner planets except Venus has compressive features like Earth's.

The great planets—Jupiter, Saturn, Uranus, and Neptune—are composed mainly of hydrogen and helium.

Meteorites are objects from space that hit Earth. Iron-nickel meteorites are believed to be similar in composition to Earth's core. Over 90 percent of the meteorites seen to fall are stony. Meteorites range from dust to 264,000 kg (60 tons). Large meteorites cause craters. Their origin is probably the asteroids.

Comets are masses of frozen gas with some small particles. They orbit the sun and are visible when they are close enough to the sun that their gases are ionized by the sun's energy.

Growth rings on corals and other marine animals show that Earth's rotation has been slowing. This is caused by the tidal drag of the moon.

QUESTIONS

1. List the components of the solar system.
2. How do the inner planets differ from the great planets?
3. How can the different numbers of moons that each planet has be explained?
4. Which planets might contain life? Why?
5. What is the origin of the moon's surface features?
6. How do comets differ from meteorites?
7. How long would it take for meteorite falls to produce Earth? Is this a reasonable theory for the origin of Earth?
8. The uranium-lead and thorium-lead methods are used to date meteorites in spite of the problems in correcting for original lead. Why are not other radiometric methods used?

SUGGESTED READINGS

Planets—General

Baker, V. R., "Planetary Geomorphology," *Journal of Geological Education* (September 1984), Vol. 32, No. 4, pp. 236–246.

Head, J. W., and others, "Geologic Evolution of the Terrestrial Planets," *American Scientist* (January-February 1977), Vol. 65, No. 1, pp. 21–29.

Murray, Bruce, M. C. Malin, and Ronald Greeley, *Earthlike Planets. Surfaces of Mercury, Venus, Earth, Moon, Mars.* San Francisco: Freeman, 1981, 388 pp.

Sagan, Carl, "The Solar System," *Scientific American* (September 1975), Vol. 233, No. 3, pp. 22–31. This entire issue is on the solar system.

Mars

Arvidson, R. E., A. B. Binder, and K. L. Jones, "The Surface of Mars," *Scientific American* (March 1978), Vol. 238, No. 3, pp. 76–89.

Carr, M. H., "The Volcanoes of Mars," *Scientific American* (January 1976), Vol. 234, No. 1, pp. 32–43.

Horowitz, N. H., "The Search for Life on Mars," *Scientific American* (November 1977), Vol. 237, No. 5, pp. 52–61.

Moon

Boss, A. P., "The Origin of the Moon," *Science* (January 24, 1986), Vol. 231, No. 4736, pp. 341–345.

Hartmann, W. K., "Cratering in the Solar System," *Scientific American* (January 1977), Vol. 236, No. 1, pp. 84–99.

Mason, Brian, "The Lunar Rocks," *Scientific American* (October 1971), Vol. 225, No. 4, pp. 48–58.

Great Planets

Hubbard, W. B., "Interiors of the Giant Planets," *Science* (October 9, 1981), Vol. 214, No. 4517, pp. 145–149.

Ingersoll, A. P., "Jupiter and Saturn," *Scientific American* (December 1981), Vol. 245, No. 6, pp. 90–108.

Owen, Tobias, "Titan," *Scientific American* (February 1982), Vol. 246, No. 2, pp. 98–109.

Soderblom, L. A., "The Galilean Moons of Jupiter," *Scientific American* (January 1980), Vol. 242, No. 1, pp. 88–100.

Soderblom, L. A., and T. V. Johnson, "The Moons of Saturn," *Scientific American* (January 1982), Vol. 246, No. 1, pp. 100–116.

Waldrop, M. M., "The Puzzle That Is Saturn," *Science* (September 18, 1981), Vol. 213, No. 4514, pp. 1347–1351.

6

The Precambrian—The Oldest Rocks and the First Life

One of the most important horizons in geologic history is the base of the Cambrian System, where the oldest easily recognized fossils became abundant. This event occurred about 570 million years ago; all of the time prior to that event, almost 90 percent of geologic time, is called the Precambrian Era. All interpretation in geology is at least somewhat speculative, but Precambrian history is more speculative than that of younger parts of the geologic column. The reasons for this are many. The rocks are mainly metamorphic, making their origin difficult to decipher. They have been intruded by batholiths. They are generally deformed and may have been involved in more than one period of mountain making. (See Fig. 6.1.) At many places, only fragments of once-extensive rock units are preserved. Perhaps most important, dating is difficult, generally requiring extensive laboratory work. Thus Precambrian history is even more sketchy, and more subject to major revision as new data become available, than the geologic history of later times.

PREGEOLOGIC HISTORY

The pregeologic history of the earth begins with the origin of the earth and ends with the oldest rocks, at which horizon conventional geologic history begins. *The oldest crustal rocks so far dated by radiometric methods are about 3962 million years old* and are in the Northwest Territories of Canada. The earth is appreciably older than this. Meteorites yield dates of 4550 million years; this is the time that they

FIGURE 6.1
Deformed Precambrian metamorphic rocks in Northwest Territories, Canada. Such rocks are much harder to map than most sedimentary rocks. Photo by J. A. Donaldson, Energy, Mines and Resources. Reproduced with the permission of the Minister of Supply and Services Canada, 1990.

crystallized. As noted earlier, meteorites are dated using the uranium-lead and thorium-lead methods, and the correction for original lead is a problem. Meteorites are believed to have the same origin as the earth and to have the same composition as the interior of the earth. *Thus the 4550-million-year date is perhaps the time that the interior of the earth crystallized.* This event occurred sometime after the formation of the probably initially homogeneous earth. How much after is the problem because the size of the planet probably is important in determining the rate at which recrystallization occurred. At any rate, 4550 million years is the time at which the bodies that became meteorites crystallized; the solar system, including the earth, formed sometime earlier.

The earth probably formed from a nearly homogeneous and relatively cool dust cloud. Contraction caused some heating, and radioactivity also must have caused heating. The radioactive heating must have been much greater than at present because much more radioactive material was present on the earth. This follows because after each half-life, half of the original radioactive element is gone. Meteoritelike material hitting the earth probably also caused local heating. *The heating of the earth caused melting, and, under the influence of gravity, differentiation of the earth into core and mantle began.* The heavier materials such as iron and nickel melted and moved toward the center, and the lighter silicates moved outward. The silicate mantle probably solidified from the bottom outward. The core, unable to lose its heat, remained liquid. The inner core may be crystallized. The radioactive elements that caused most of the heating—uranium, thorium, and potassium—all moved outward. These are all large atoms that do not fit into the iron-magnesium silicates, so they stayed with the melt as crystallization progressed from the core outward.

The actual condensation of the earth is believed to have occurred 5000 to 6000 million years ago, perhaps closer to 6000 million years. This estimate is reached from at least two lines of reasoning, but neither method gives more precise figures. The age of the sun is estimated to be about 6000 million years. This figure is based on knowledge of stellar structure and rate of energy production. This does put an upper limit on the age of the earth, however, because the whole solar system is believed to have originated at the same time. The second line of reasoning is based on estimates of the time necessary for melting and recrystallization to occur. About 500 to 1000 million years is a reasonable time for these processes, so this figure should be added to the 4550-million-year date of meteorites.

The early mantle was at the surface and probably was cooling by convection. Magnesium-rich lava formed at the surface and most probably sank because of its density and convection. Partial melting of this lava after it sank to deeper, hotter levels would have produced basaltic and granitic rocks, the earliest crust. The first of this crust that was preserved marks the end of pregeologic history. The oldest crustal rocks appear to be metamorphosed sediments, suggesting that the early ocean and atmosphere also formed in pregeologic time.

ORIGIN OF OCEANS AND ATMOSPHERE

The present atmosphere and oceans have developed since the earth solidified, but the development is known only in general terms. Although the earth and the sun formed from the same dust clouds, the earth's overall composition now differs from that of the sun. Therefore, it is clear that sometime after the initial accumulation from a dust cloud, the earth lost much of its original thick gaseous envelope of mainly hydrogen and helium. We must account for the loss of these gases by processes not now operating because the earth is at present losing very little of its atmosphere. Much of the original atmosphere was lost when it was ionized by the sun's radiation and then swept away by moving ionized particles from the sun. The remaining gas nearer the earth was probably lost as a result of the rapid rotation of the earth and as a result of the heating that melted the earth and formed the core and mantle.

It is difficult to reconstruct the next stage in the development of the early atmosphere; it may have consisted of such gases as ammonia, methane, hydrogen, and steam. This is the composition of the atmospheres of the great planets such as Jupiter (where the water is ice and helium is also present). In the case of the great planets, the source is believed to be the initial dust cloud. The earth, being nearer the sun, had lost by this stage the gases of the original dust cloud; however, frozen chunks of these gases left in the earth would have melted and been released during the later remelting of the earth to form this atmosphere. Evidence for the existence of this atmosphere is very weak. If such an atmosphere ever existed on earth, it probably was lost quickly because of the prevailing high temperature. Ammonia would be quickly decomposed by the sun's ultraviolet energy. Methane would cause fixation of carbon in the early sedimentary rocks, but there is no evidence that this occurred. *The present atmosphere developed mainly from volcanic emanations,* and the cooler, more slowly rotating earth retained these gases. Methane, ammonia, hydrogen, and water would form an atmosphere in which life could have begun. However, as outlined here, the temperature was probably too high for life at the time the atmosphere had this composition. Life could also form in a volcanic-derived atmosphere. The origin of life will be discussed in more detail later. Returning to our main theme, volcanoes today emit mainly steam, carbon dioxide, and nitrogen, together with some sulfur gases; there is no reason to believe that earlier volcanoes were different. Thus the atmosphere became steam, carbon dioxide, and nitrogen. With the cooling of the earth some of the steam condensed. The oxygen now in the atmosphere came from photosynthesis by plants after life began, and life probably could not form until the temperature was below 93°C (200°F). At the present time the atmosphere is 78 percent nitrogen and 21 percent oxygen, with all other gases forming about 1 percent. The disposition of the volcanic gases probably follows this pattern: the steam has become water and is mainly in the oceans; the nitrogen is in the atmosphere; and the carbon dioxide has been partly broken down by photosynthesis to carbon in life and to oxygen in the atmosphere and is partly in carbonate rocks.

This development of the atmosphere seems reasonable, and it is possible to work out a crude timetable. By about 4000 million years ago, the crust had formed. By 3800 million years ago, the ocean and atmosphere probably were present because water-deposited sedimentary rocks of this age have been found. Prior to that time, anaerobic life may have existed. Oxygen, however, was not present in the atmosphere because up to about 1800 million years ago, clastic fragments of pyrite and uraninite are found in sedimentary rocks. If oxygen had been present, these minerals probably would have been oxidized. Another point suggesting that the early atmosphere had no oxygen is that organisms had to develop oxygen-mediating enzymes because, without them, oxygen would destroy the organisms. Life developed, probably in the ocean, and oxygen was being released by photosynthesis by about 3200 million years ago. This oxygen stayed in the ocean; not much of it reached the atmosphere.

As long as there was no oxygen, the iron in the ocean was in solution. This iron came from weathering of rocks and submarine volcanic activity. As oxygen was released into the ocean by photosynthesis by primitive organisms in the ocean, it combined with the iron (and silica) to form banded iron formation. **Banded iron formation** *is a chemically precipitated rock with alternate thin layers of chert and iron oxide.* It was formed only in the time from about 3200 million to 2000 million years ago, with most forming between 2500 million and 2000 million years ago. *After about 2000 million years ago, apparently all of the iron in the oceans was precipitated, and oxygen could escape from the oceans and begin to accumulate in the atmosphere.*

Free oxygen appeared in the atmosphere at about 2000 million years ago because from that time on, **red beds**—*that is, clastic rocks in which the iron is oxidized—are common.* Thick, extensive red beds appear in the range 1800 to 2000 million years ago and so slightly overlap the last of the banded iron formation. Apparently at this time, oxygen-mediating enzymes formed, allowing life to expand rapidly. The life, probably algae, produced much oxygen; as the oxygen accumulated in the atmosphere, iron was oxidized in weathering, preventing further development of banded iron formation.

Banded iron formation is found on every continent and is a valuable mineral resource. This unique ore could form only during the time that iron was in solution in the oceans. Only in the absence of oxygen can the iron released by weathering be transported in solution and exist in the oceans in solution. As soon as oxygen was present in the oceans, the iron there was oxidized and precipitated as banded iron formation; later, when oxygen appeared in the atmosphere, iron was oxidized in the weathering process, cutting off the supply of iron to the oceans.

Up to this time, ultraviolet from the sun was lethal to life at the surface, so the organisms must have lived in water deep enough to shield them.

Some of the oxygen in the atmosphere was changed into ozone by ultraviolet radiation from the sun. Ordinary atmospheric oxygen is in the diatomic form, O_2, in which two oxygen atoms share electrons to form stable outer electron shells. In the ozone molecule, O_3, three oxygen atoms share electrons. Ozone is not stable and ultimately returns to ordinary diatomic oxygen. The formation of ozone absorbs much ultraviolet radiation that would otherwise reach the surface of the earth. Higher forms of life could not originate until the ozone layer had formed.

PRECAMBRIAN ROCKS AND HISTORY

Precambrian time includes almost 90 percent of geologic time. This section is about the rocks and the physical history that can be determined from them. This should be the longest and most interesting chapter in geologic history. However, the difficulties of dating these rocks, the repeated deformation many have had, and their fragmentary nature due to long erosion or younger covering rocks make their study one of the most difficult and demanding aspects of geology. The story begins after the formation of the crust and extends to the base of the Cambrian System, where extensive fossils begin and dating rocks becomes easier. All time correlation in Precambrian rocks must be done with radiometric dating. Locally, lithologic similarities or similar methods can be used to establish rock correlations.

From earlier chapters, it is obvious that geologic history, viewed very broadly, is a sequence of plate tectonic events and the associated geosynclines. Such events have shaped the earth's surface. It seems likely that such events occurred in Precambrian time as well. However, the beginning had to be different. Somehow the first continent had to form. The record of these events is not clear, and much speculation is involved in the following story.

Because Precambrian time is so long, some subdivisions are desirable; but

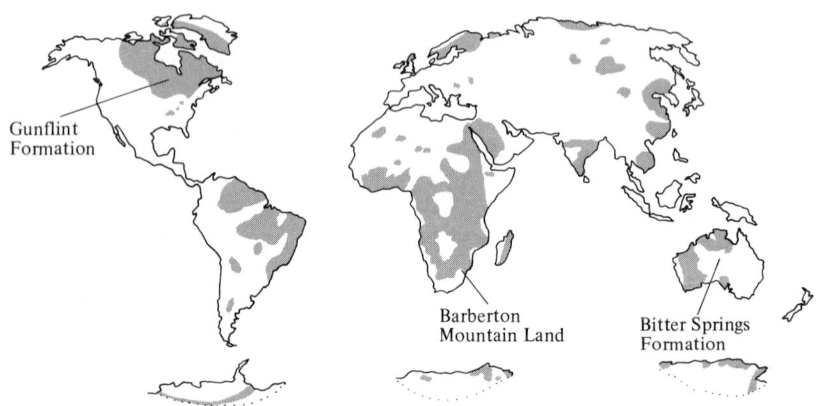

FIGURE 6.2
Areas where Precambrian metamorphic rocks are exposed.

subdivision was not easy before radiometric dating became available. The traditional subdivision is into an older Archean and a younger Proterozoic Era. When these terms were first introduced, workers thought that the metamorphic rocks were older than the sedimentary rocks, so Archean and Proterozoic were used in that sense. Radiometric dating showed that many metamorphic rocks are younger than some sedimentary rocks. This led to much confusion and the use of other terms in local areas. The twofold subdivision is useful and is now again being used, but the division now is made on the basis of radiometric dating. *Rocks older than 2500 million years are assigned to the* **Archean,** and *the younger Precambrian rocks are assigned to the* **Proterozoic.**

Extensive exposures of Precambrian rocks are found in central, southern, and eastern Canada. (See Fig. 6.2.) Glaciation has exposed many of these rocks, and their mineral wealth has encouraged study. Precambrian rocks underlie the central interior of North America, where they are covered by younger rocks. This is shown by the Precambrian rocks exposed in the uplifted Rocky Mountains. Precambrian rocks also appear to underlie parts of the Appalachian Mountains.

Archean History

Archean history begins with the oldest rocks and, at least in North America, ends with the **Kenoran orogeny,** *in which at least the core of the North American continent was first assembled.* The oldest rocks so far found are 3962 million years old and are in Canada's Northwest Territories. At Isua, West Greenland the rocks are 3800-million-year-old high-metamorphic-grade granitic gneiss and appear to have formed from sedimentary rocks. The implication is clear that the continental crust and the ocean and atmosphere had begun to form by that time. Some Australian rocks have sand grains composed of zircon that have a radiometric age of 4100 million years, suggesting that the first crust may be even older.

Typical Archean rocks are **high-grade granitic gneiss** *with belts of* **greenstone.** At the places where the relationships are clear, the high-grade gneiss is older than the greenstone. The greenstone belts are composed of deformed and metamorphosed volcanic and sedimentary rocks. They are called greenstones because basalt, the dominant rock, develops green-colored minerals when metamorphosed. The greenstones are lower-grade metamorphic rocks than are the high-grade gneisses. Batholithic intrusive rocks and metamorphism permeate the Archean terrane, so relationships among the rocks are not clear. The greenstone belts range from small to up to 100,000 km^2 (38,500 mi^2) in area. They range in age from 3500 to 1800 million years, with many at 2700 million years.

The origin of these Archean greenstone and high-grade gneiss terranes is not clear. The occurrence of high-grade and lower-grade metamorphic rocks in close contact suggests vertical movements of the crust. One possibility is that the greenstones originated at something like volcanic island arcs. At the time that these terranes formed, the earth must have been more active because it had much more radioactive energy than at present. This suggests the possibility of active convection currents in a warmer mantle.

At places convection in the mantle would cause volcanic eruption of mantle rocks through the thin crust. The weight of the resulting volcanic pile would cause the area to sink into the hot mantle, resulting in metamorphism and perhaps melting of the deeply buried rocks. The top of such a pile might extend above sea level, and the resulting erosion and deposition would produce sedimentary rocks at places above the volcanic rocks. Whatever processes acted, the Archean must have been a time of many active areas of accumulation of volcanic and sedimentary rocks.

These processes were the beginning of plate tectonics and must have built many microcontinents and volcanic arcs. It is estimated that about half of the area of the present continents formed during the Archean. Convection in the mantle probably swept many of them together, forming larger continents. *Plate tectonics was established by at least 2650 million years ago because this is the age of the oldest ophiolite yet found.* **Ophiolites** are bits of oceanic crust thrust onto continents at the suture marking the point of collision of two continents. *The Kenoran orogeny lasted from about 2700 to 2500 million years ago, and at that time part of the core of the North American continent was assembled by continental collisions. This event marked the end of the Archean Era.* (See Fig. 6.3.)

Early Proterozoic History

In North America the early Proterozoic is from the Kenoran orogeny to the **Hudsonian orogeny,** *or from roughly 2500 to 1700 million years ago.* The Kenoran orogeny changed the nature of the crust profoundly by making it much more stable. This process probably began in the late Archean with the emplacement, starting about

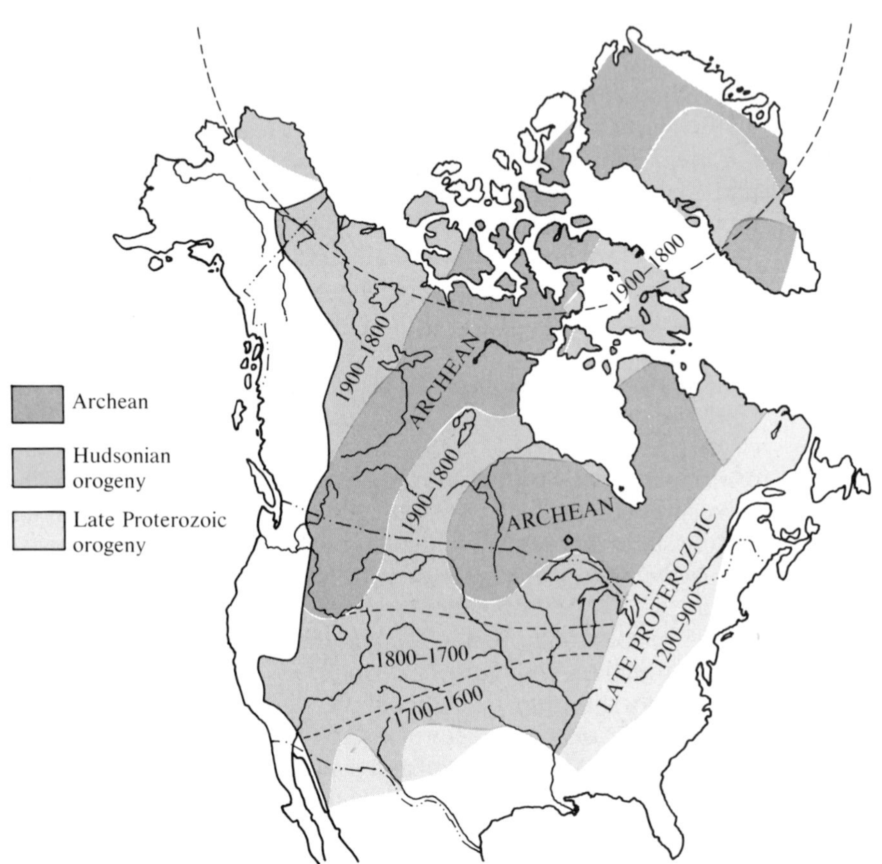

FIGURE 6.3
Precambrian development of North America. The core is made of Archean rocks (over 2500 million years old) that were assembled from smaller crustal pieces during the Kenoran orogeny. The Archean continents were brought together in the Hudsonian orogeny, which lasted from 1900 to 1600 million years ago. At that time a single supercontinent may have formed. Three pulses of that collisional orogeny have been recognized in North America. Later rifting separated the present continents. The belt of late Proterozoic orogeny in the east records rifting and later collision at that time.

3000 million years ago, of many batholiths. In any case the early Proterozoic was a time of much intrusive activity. One theory is that the larger continents that started to assemble in the late Archean may have insulated the mantle and so caused the temperature there to rise, promoting intrusive activity.

At about 1900 million years ago, very large intrusive bodies of gabbroic composition were emplaced at many places around the earth. These bodies crystallized slowly, with the crystals falling to the bottom and forming layers. These layered complexes are the sites of many platinum and chromium mines. Examples are the Bushveld complex in South Africa, the Great Dyke in Zimbabwe, and the older Stillwater Complex in Montana. Around 2200 million years ago, huge basaltic dike swarms were emplaced in the Lake Superior region, Scotland, India, Australia, and Africa. A major effect of the new crustal stability was the development of large thick sedimentary basins. Some of the best studied of these are in Africa, where the stabilization began earlier than in North America. Beginning about 3000 million years ago, a series of basins formed on granitic crust in Swaziland in southern Africa. The oldest rocks were basaltic volcanics with younger sedimentary rocks. They were followed by conglomerates, sandstones, and volcanic rocks about 2800 million years old. These rocks are covered unconformably by the Witwatersrand Supergroup, 2700 to 2500 million years old. The Witwatersrand is composed of shale, quartz sandstone, banded iron formation, volcanic rocks, and conglomerate. It is famous as the earth's most productive gold source, and it also contains uranium mines. The sedimentation in this area continued until about 2300 million years ago; thus it continued into the Proterozoic. Apparently, by 3000 million years ago, the continents had become big enough for large basins to develop, and from then on they did on all continents.

The Proterozoic rocks are generally similar to those just described—large basins with thick accumulations of sedimentary and volcanic rocks.

In North America the early Proterozoic ended in the Hudsonian orogeny, which lasted from about 1900 to 1700 million years ago. Evidence of orogeny at that time is found on all of the present continents, suggesting that all of the continents came together then to form a single supercontinent. At any rate, the core of the present North American continent was assembled at that time. (See Fig. 6.3.)

Late Proterozoic History

In the late Proterozoic, conditions became more like those of the Paleozoic. The major events were the rifting of the supercontinent into smaller continents. Conditions were not completely like those of later times because in the period between 1600 and 1200 million years ago, about 75 percent of the earth's anorthosite bodies were emplaced. Anorthosite is a coarse-grained rock composed almost completely of plagioclase. The Adirondack Mountains in New York and many areas in eastern Canada are examples.

In western North America, rifting occurred soon after the Hudsonian orogeny, forming a continental shelf–slope type of geosyncline. The edge of the continent at that time was well to the east of the present shore. Most of the rocks deposited in the geosyncline have not been metamorphosed. In the northern Rocky Mountains, over 21,350 m (70,000 ft) of sedimentary rocks were deposited in Idaho, Montana, and Alberta. These rocks have been dated at between 850 and 1450 million years old. (See Fig. 6.4.) Unconformably above these is another sedimentary sequence over 1830 m (6000 ft) thick. The unconformity may indicate another rifting event. The second sequence is covered without notable unconformity by Cambrian rocks

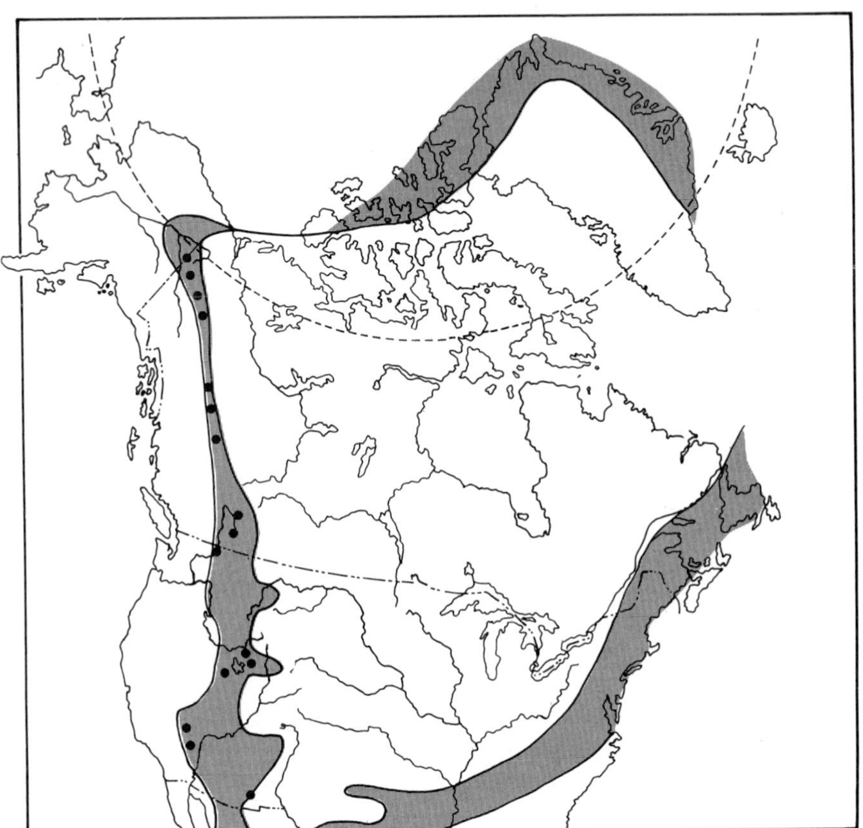

FIGURE 6.4
Late Proterozoic sedimentary rocks were deposited on continental shelf–slopes on the rifted margins of both coasts. On the west coast, sedimentation began much earlier and two sequences formed. Glaciation occurred between them. Reconstructions like this are made from many scattered outcrops because later erosion and deformation have removed many of the rocks.

at some places and unconformably at other places by rocks as young as Silurian. Other thick sequences of Precambrian sediments occur in northern Utah in the Grand Canyon and south of there in Arizona. At other places, Precambrian sediments grade upward to Cambrian rocks without apparent unconformity; they are described in the next chapter with the Cambrian rocks with which they are more closely allied.

In eastern North America the story is not as clear. Rifting may have occurred earlier, but in any case collision with Europe happened between 1200 and 900 million years ago. Sometime after that, rifting did occur and a shelf-slope type of geosyncline formed in the area of the present Appalachian Mountains. Sediment accumulation began in the Precambrian and continued into the Cambrian so we will consider them in the next chapter.

On all continents except Antarctica, glaciation may have occurred in very late Precambrian time (750 to 850 million years). The evidence for this is rocks that resemble glacial deposits in that they contain conglomerates and overlie striated rocks. Distinguishing between glacial deposits and other conglomerates is difficult. Similar rocks 950 and 2000 to 2400 million years old suggest much earlier Precambrian glaciations.

ORIGIN OF LIFE

All discussions on the origin of life are necessarily speculations. Life must have begun very early in the history of the earth, probably almost as soon as it became cool enough. The role of early life in the development of the atmosphere was described earlier.

One of the first people seriously to consider the problem of the origin of life was A. I. Oparin, a Russian biochemist. He speculated in the 1920s and 1930s that life could originate only in an oxygen-free environment containing the elements needed to form amino acids and larger organic molecules. He suggested an atmosphere containing ammonia, methane, hydrogen, and water vapor. At that time, this was the composition thought to have existed on the early earth. In 1953, Stanley L. Miller, an American graduate student, put these materials in a closed container in which an electric spark could be caused. The spark acted like lightning in the atmosphere. He obtained a number of amino acids, suggesting that life could have originated in this way. This experiment was repeated in 1957 by Philip Abelson, who used mainly carbon monoxide, carbon dioxide, and nitrogen. This mixture is much closer to the probable atmosphere of the early earth. He obtained most of the amino acids necessary for life.

The steps involved in the formation of life are believed to start with the formation of amino acids. Next, these amino acids are brought together to form proteins, nucleic acids, and other compounds. The most difficult step is the transformation into living, self-replicating structures.

The amino acids and other organic molecules probably collected in the oceans, forming what has been called a "thin soup." The early life probably got its food directly from the ocean, using material dissolved in the ocean water. The next advance was probably the development of chlorophyll by some of the organisms. This enabled them to use the sun's energy and carbon dioxide in their life cycles and produced oxygen as a byproduct. From that point on, oxygen became available in the oceans and atmosphere.

Life could have formed only in an environment without free oxygen. If oxygen had been present, it would have destroyed the amino acids by oxidation. After life formed and oxygen was produced by photosynthesis, no new life could be formed. Thus, all of the present life on the earth must have evolved from that first population of living organisms.

Clearly, if life formed in the manner described, it could form anywhere where these conditions exist. That is, it could have formed on another planet in the solar system or elsewhere in the universe. The conditions needed are the elements carbon, nitrogen, and hydrogen, plus water in the liquid state. Such conditions must occur at many places in the vast universe, and so there must be other life in the universe. Amino acids and hydrocarbons were found in small amounts on moon rocks. They are believed to be of inorganic origin. On September 28, 1969, a carbonaceous meteorite fell at Murchison, Victoria, Australia. This meteorite, which was dated as 4500 million years old, contains amino acids. These discoveries suggest that the conditions necessary to form life are not uncommon and that amino acids may have formed in pregeologic time in the solar system.

THE OLDEST FOSSILS

Until the late 1950s, very few Precambrian fossils were found. Many of the reported fossils found before that time are probably of inorganic origin. Microfossils have been found at many localities around the world, and larger fossils have been found in upper Precambrian rocks in a dozen or so places. The record is complete enough to be able to continue the story that began in the last section.

The first living organisms probably could not produce their own food but lived on material dissolved in the ocean. This situation could exist for only a short time; otherwise, the life would use all of the available nutrients. The development of photosynthesis probably made the first self-sustaining organism.

The oldest fossils are found in the 3500-million-year-old Warrawoona Group of Western Australia. The cherts in these rocks contain well-preserved microfossils. The slightly younger (3400 million years old) Swaziland Supergroup in eastern South Africa also contains microfossils. The microfossils in these old rocks are similar to modern cyanobacteria (blue-green algae) and bacteria and are accepted as evidence of life by most who have studied them. They could be nonbiologic because somewhat similar structures were produced in experiments, like those discussed in the last section, that synthesized amino acids and other organic molecules. The next youngest rocks with microfossils are found in Australia, Canada, and Rhodesia. These rocks are 2700 to 2800 million years old, and these microfossils are generally accepted as genuine.

Abundant Precambrian fossils from the Gunflint Chert on the north shore of Lake Superior in Canada have been studied. (See Fig. 6.5.) Again, cyanobacteria (blue-green algae) and bacteria are found; these organisms were clearly photosynthetic. The rocks are about 2000 million years old.

The organisms found at these localities are primitive. Cyanobacteria (blue-green algae) and bacteria (prokaryotes) are the only living things whose reproduction never involves genetic material from two parents, and so their form does not change. Thus these organisms in rocks over 3000 million years old are much like living species. Because redistribution of parental genes is not involved, any mutation that occurs, in general, dies out in a few generations. This is not at all like the evolution described in Chapter 4.

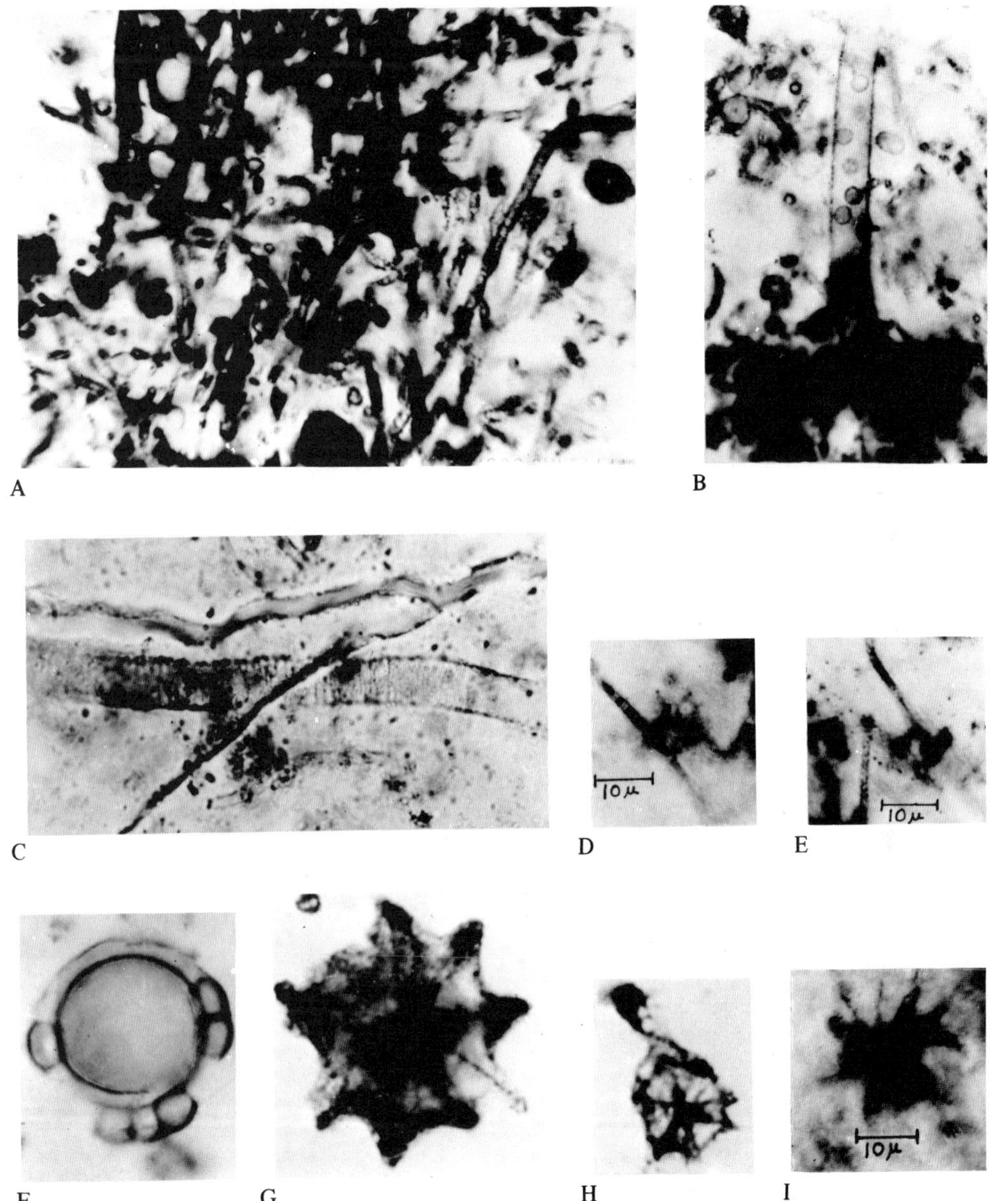

FIGURE 6.5
Microfossils from the Precambrian Gunflint Chert. A. Tangled filaments, mainly *Gunflintia minuta;* and sporelike bodies, *Huroniospora.* Diameter of field 0.1 mm. B. *Entosphaeroides amplus* with sporelike bodies within the filament. Diameter of field 40μm. C. *Animikiea septata* probable algal filament with transverse septae. Diameter of field 50 μm. D. and E. Filaments with well-defined septae similar to iron-precipitating bacteria and cyanobacteria (blue-green algae). F. *Eosphaera tyleri.* Diameter of field 32 μm. G. and H. *Kakabekia umbellata.* Diameter of field of G is 20 μm and H is 12 μm. I. Radiating structure similar to manganese- and iron-oxidizing colonial bacteria. (One micrometer [μm or μ] is one-millionth of a meter.) Photos A, B, C, F, G, and H courtesy of E. S. Barghoorn; D, E, and I courtesy of P. E. Cloud, Jr.

The first organisms that can combine genetic material from two parents (eukaryotes) are found in the 1300-million-year-old Pahrump Group in California and in the 900–1000-million-year-old rocks at Bitter Springs in the Northern Territory of Australia. The organisms are green algae, and all green algae differ from cyanobacteria (blue-green algae) in being capable of sexual reproduction. Some of the fossils recovered from this chert show cell division. The development of these early forms of life is summarized in Figure 6.6. Younger Precambrian fossils are shown in Figure 6.7.

These discoveries suggest that true evolution could not begin until about 1400 million years ago. Up to that time, the organisms were too primitive and evolutionally too conservative to change very much. After that time, evolution could proceed, and apparently did, because about 600 million years ago, life became abundant. Thus, in less than 800 million years, all of the many types of life found in Cambrian rocks may have evolved from green algae. Representatives of every phylum capable of fossilization are found in Cambrian or early Ordovician rocks.

Very little evidence of this evolution has yet been found. Perhaps the reason is that the organisms had few hard parts capable of fossilization. At the three localities just described, the algae were preserved in chert beds and were found only after searching with both optical and electron microscopes. Thus intensive search may unearth more evidence. Another possibility is that, although the number of organisms was increasing rapidly during this time, only by the start of the Cambrian Period was life abundant enough to form abundant fossils. This is suggested by the

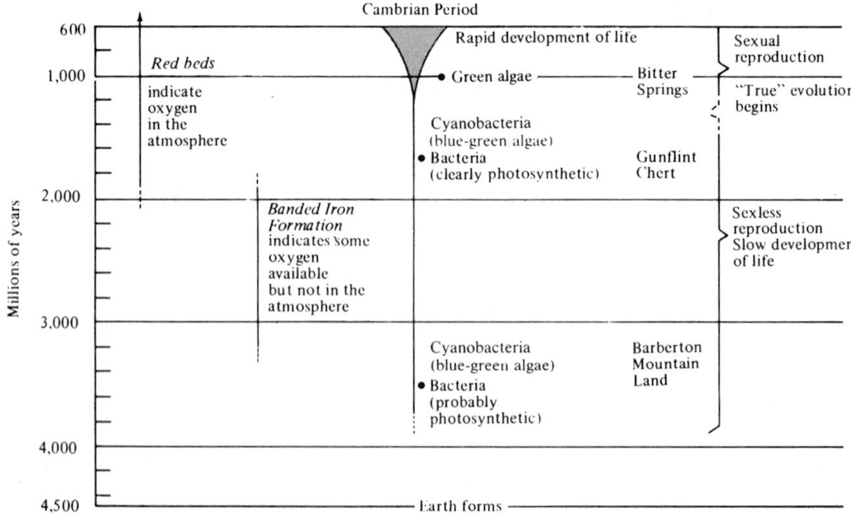

FIGURE 6.6
Development of Precambrian life and oxygen in the atmosphere.

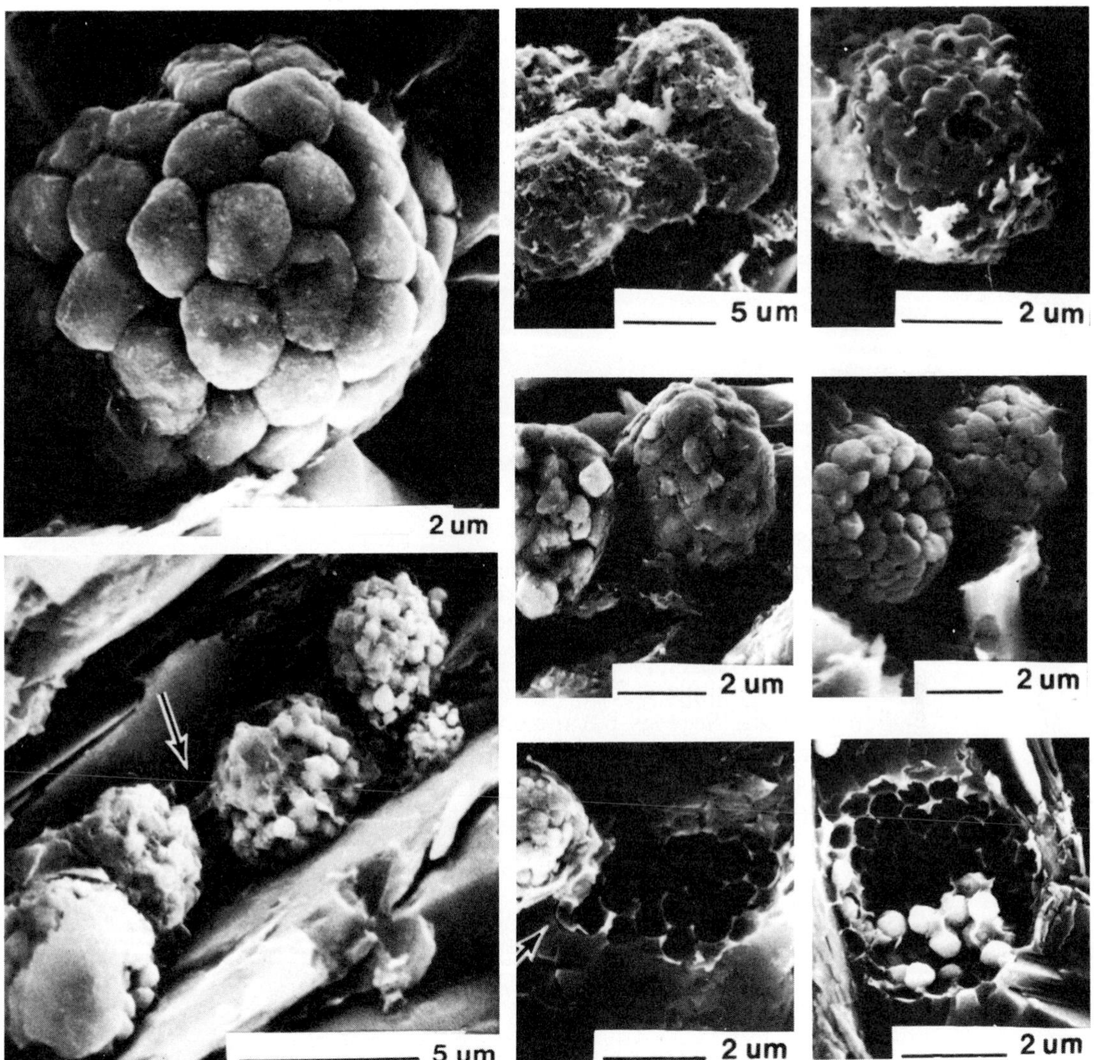

FIGURE 6.7
Late Precambrian microflora from southern British Columbia. These microphotographs were made with a scanning electron microscope and reveal more surface detail than is possible with a light microscope. These fossils were found in hard black shale. One micrometer (μm) is one-millionth of a meter. Photos courtesy of B. J. Javor and E. W. Mountjoy, and the Geological Society of America.

FIGURE 6.8
Map showing locations where the unusual late Precambrian fauna shown in Figure 6.9 have been found. Other Precambrian faunas have been found elsewhere. After M. Glaessner.

discovery at several places of a Precambrian fauna in rocks between 600 and 700 million years old. (See Figs. 6.8 and 6.9.) This suggests that by late Precambrian, life was becoming abundant but had few preservable hard parts.

This late Precambrian fauna (Fig. 6.9) is important also because the organisms are not microscopic. They probably had genes and so may signal more rapid

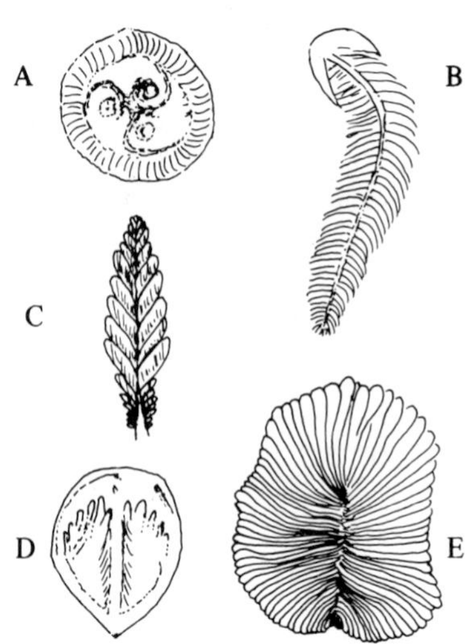

FIGURE 6.9
This unusual Precambrian fauna is found at the places shown in Figure 6.8. These organisms are found as impressions in sandstone.
A. Circular form of unknown affinities, possibly with coiled arms. About 1.9 cm (0.75 in.). B. Probable worm. About 2.5 cm (1 in.) long. C. Leaflike form of unknown affinities. About 17.8 cm (7 in.). D. Oval form of unknown affinities. About 1.9 cm (0.75 in.). E. Wormlike form. About 4.5 cm (1.75 in.).

A

B

FIGURE 6.10
Algal structures in rocks about 1000 million years old in Glacier National Park, Montana. A. Top view, and B. side view of *Collenia undosa*. Photos by R. Rezak. U.S. Geological Survey.

FIGURE 6.11
Possible marks or tubes of a bottom-dwelling or burrowing organism. Some worms make similar markings. These markings are in sedimentary rocks 2000 to 2500 million years old. Scale in centimeters. Photo by Energy, Mines and Resources. Reproduced with the permission of the Minister of Supply and Services Canada, 1990, through the courtesy of H. J. Hoffman, who first described these occurrences.

FIGURE 6.12
Brooksella canyonensis, a supposed jellyfish from Precambrian rocks in the Grand Canyon. Diameter of field is 8.9 cm (3.5 in.). Photo from Smithsonian Institution.

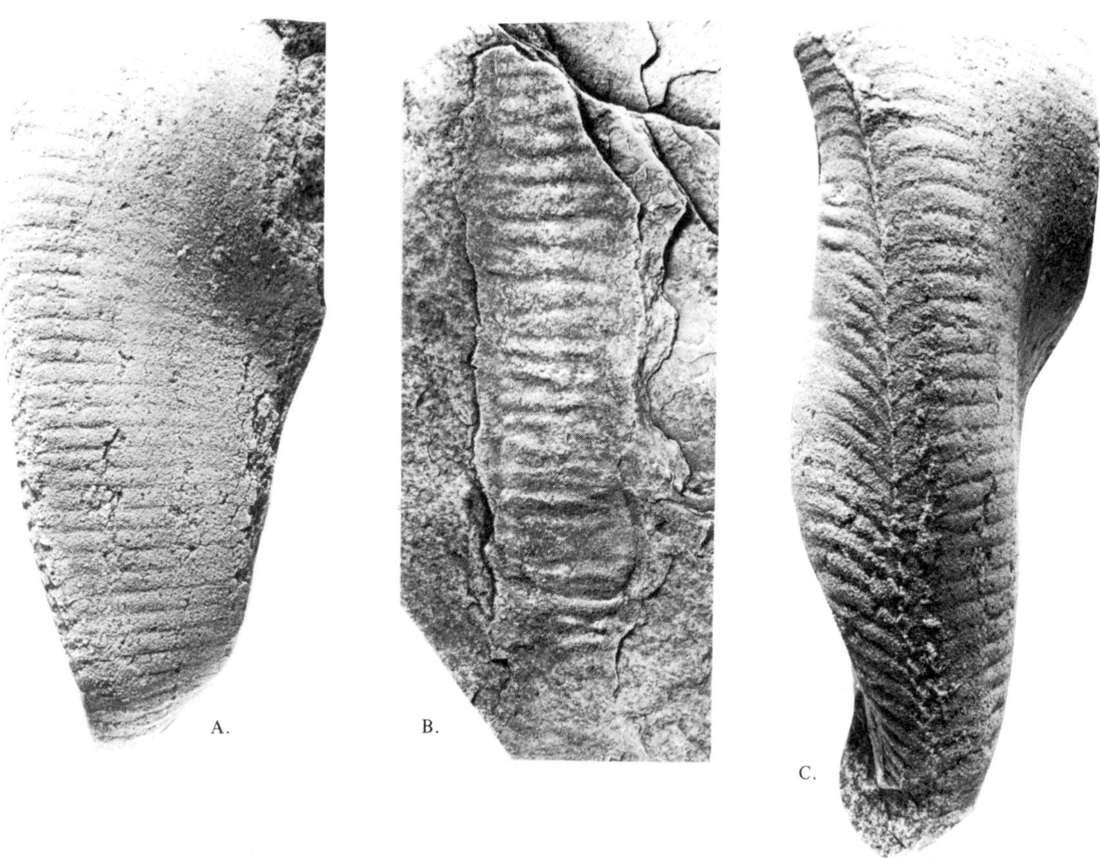

FIGURE 6.13
Crawl and scratch marks, probably of trilobites. Marks such as these are commonly found both in Cambrian and younger rocks and in rocks well below (older than) the first known trilobites. Parts A and C are from rocks of the Nama System, South West Africa, from well below the first trilobites. Part B, from the Deep Spring Formation, White Mountains, California, well below the first trilobites. Photos courtesy of P. E. Cloud, Jr.

evolution. Bigger organisms may have become possible about 700 million years ago because by then oxygen in the atmosphere had increased to about 10 percent of its present amount. This may have aided respiration, but, more important, the ozone layer may have formed and protected the organisms from lethal ultraviolet in the sunlight.

The most common Precambrian fossils are algal mounds or columns (stromatolites) similar to those made by living cyanobacteria (blue-green algae). These algal structures can be used in a general way to make broad correlations of Precambrian rocks. (See Fig. 6.10.) Figure 6.11 shows some unusual markings that may be worm tubes, or they may be of inorganic origin. Figure 6.12 shows another type of marking found in Precambrian rocks. The example shown is thought by some to be a jellyfish. Crawl and scratch marks are found at many places in both young and old rocks. Figure 6.13 shows both Precambrian and Cambrian crawl marks, and the similarities suggest that trilobites were present in some late Precambrian rocks.

SUMMARY

The earth's crust formed about 3800 million years ago.

Meteorites crystallized about 4550 million years ago.

After formation, the earth melted about 4550 million years ago to form its layered structure. Compaction (compression) and radioactivity were the probable energy sources.

The earth formed about 5000 to 6000 million years ago.

The oceans and the atmosphere formed in the interval 4500 to 4000 million years ago.

The early atmosphere may have been ammonia, methane, hydrogen, and steam, like the present great planets.

The present atmosphere and the oceans came from volcanic gases.

Oxygen came from carbon dioxide, the process being photosynthesis by plants after life had formed.

The oldest sedimentary rocks found so far are about 3800 million years old, so water was present by then.

Oxygen was not present in the atmosphere until about 1800 million years ago because fragments of pyrite and uraninite are found in sedimentary rocks.

Oxygen formed between 3200 and 2000 million years ago produced banded iron formation.

Red beds became common between 1800 and 2000 million years ago, suggesting that oxygen was in the atmosphere.

As soon as oxygen was present, the ozone layer formed, shielding lethal ultraviolet and so allowing higher forms of life to evolve.

Time correlation of Precambrian rocks must be done radiometrically.

The Precambrian is subdivided into the Archean Era, more than 2500 million years, and the younger Proterozoic Era.

The oldest rocks yet found are 3962 million years old and are in the Northwest Territories, Canada.

Most Archean areas are made of high-grade granitic gneiss, with lower-grade greenstone belts. The greenstones are made of metamorphosed volcanics with some sediments. The greenstone belts may be early volcanic arcs.

Plate tectonics began in the Archean and the continents became larger.

In the early Proterozoic many batholiths were emplaced, making the continents more stable. Large sedimentary and volcanic basins formed on the continents.

In the late Proterozoic, North America was assembled and then rifted. Geosynclines formed on both sides of the continent.

During the Proterozoic, huge gabbroic intrusive bodies were emplaced at many places.

Much of the interior of North America is underlain by Precambrian igneous and metamorphic rocks.

Precambrian sedimentary rocks, some very thick, are found in the Rocky Mountains and the Appalachian Mountains.

Glacial deposits occur in rocks 2000 to 2400 million years old, about 950 million years old, and between 850 and 750 million years ago.

The formation of life probably began with formation of amino acids.

Amino acids can be produced by an electrical discharge in a mixture of water, methane, ammonia, and hydrogen, or of carbon monoxide, carbon dioxide, and nitrogen.

To form life, amino acids must join to form proteins and nucleic acids, and these must be transformed into living, self-replicating structures.

Life could form only in an oxygen-free environment.

Life probably originated in the oceans.

Life could form on any planet, given the same conditions.

Fossil bacteria and cyanobacteria (blue-green algae) are found in rocks as old as 3400 million years.

Cyanobacteria (blue-green algae) and bacteria do not evolve rapidly.

The first organisms with sexual reproduction are green algae 1300 million years old in California and about 1000 million years old at Bitter Springs, Australia.

True evolution began about 1400 million years ago, and all phyla probably had evolved by the start of the Cambrian about 600 million years ago. Very little record of the evolution has been found, perhaps because the early organisms had no hard parts capable of fossilization.

A few fossils of soft-bodied late Precambrian animals have been found, but algal columns, crawl marks, and burrows are the only nonmicroscopic Precambrian fossils.

QUESTIONS

1. Explain why the initial melting of the earth occurred sometime after the formation of the earth.
2. How did the earth's layered structure form?

3. When did the first life form?
4. Outline the steps by which the atmosphere is believed to have formed.
5. What is the significance of banded iron formation?
6. Can the beginning of Cambrian time be recognized at all places where sedimentary rocks of this age occur?
7. Where are the oldest rocks in North America found? What is their age?
8. What is the age of the oldest rocks so far found? Where are those rocks found?
9. Where are Precambrian metamorphic rocks exposed in North America?
10. Where are Precambrian sedimentary rocks exposed in North America? How thick are they?
11. How is life believed to have originated on the earth? Can this process account for the development of new types of life?
12. List the types and ages of the Precambrian fossils so far found.
13. What are some of the explanations for the sudden appearance of abundant fossils at the base of the Cambrian?
14. Why is Precambrian history so difficult to interpret?
15. How is the Precambrian subdivided?
16. Describe the Precambrian greenstone belts. How did they form?
17. How do Proterozoic rocks differ in general from those of the Archean?

SUGGESTED READINGS

Bickford, M. E., "The Formation of Continental Crust: Part 1. A Review of some Principles; Part 2, An Application to the Proterozoic Evolution of Southern North America," *Bulletin, Geological Society of America* (September 1988). Vol. 100, No. 9, pp. 1375–1391.

Calvin, Melvin, "Chemical Evolution," *American Scientist* (March-April 1975), Vol. 63, No. 2, pp. 169–177.

Cloud, Preston, "Evolution of Ecosystems," *American Scientist* (January-February 1974), Vol. 62, No. 1, pp. 54–66.

Cloud, Preston, James Wright, and Lynn Gover, III, "Traces of Animal Life from 620-million-year-old Rocks in North Carolina," *American Scientist* (July-August 1976), Vol. 64, No. 4, pp. 396–406.

Conway, Morris S., "The Search for the Precambrian-Cambrian Boundary," *American Scientist* (March-April 1987), Vol. 75, No. 2, pp. 156–167.

Cowie, J. W., and M. D. Brasier, Eds. *The Precambrian-Cambrian Boundary:* Clarendon (Oxford University Press), New York. 1989, 213 pp.

Frieden, E., "Chemical Elements of Life," *Scientific American* (July 1972), Vol. 227, No. 1, pp. 52–64.

Glaessner, M. F., *The Dawn of Animal Life.* Cambridge: Cambridge University Press, 1984, 244 pp.

Goodwin, A. M., "Precambrian Perspectives," *Science* (July 3, 1981), Vol. 213, No. 4503, pp. 55–61.

Groves, D. I., J. S. R. Dunlop, and Roger Buick, "An Early Habitat of Life," *Scientific American* (October 1981), Vol. 245, No. 4, pp. 64–73.

Hargraves, R. B., "Precambrian Geologic History," *Science* (July 30, 1976), Vol. 193, No. 4251, pp. 363–371.

Kerr, R. A., "Origin of Life: New Ingredients Suggested," *Science* (October 3, 1980), Vol. 210, No. 4465, pp. 42–43.

King, P. B., *Precambrian Geology of the United States: An Explanatory Text to Accompany the Geologic Map of the United States.* U.S. Geological Survey Professional Paper 902, Washington, D.C.: U.S. Government Printing Office, 1977, 85 pp.

Moorbath, Stephen, "The Oldest Rocks and the Growth of Continents," *Scientific American* (March 1977), Vol. 236, No. 3, pp. 92–104.

7

Early Paleozoic—Cambrian, Ordovician, Silurian

PATTERNS IN EARTH HISTORY

The first step in the search for patterns was the establishment of the geologic time scale. The first systems to be defined were natural groupings; that is, they were rocks bounded by unconformities or other distinct horizons. It soon became apparent that the systems were, in general, natural subdivisions only at the places where they had been defined, and their boundaries transgress time as do all unconformities. At other places no natural boundary existed, and disputed zones were found where rocks had been deposited during the time of the unconformity at the type section. The type sections were not all in the same area, and this also led to correlation problems and other disputed zones. As a result of these problems, the type sections and extent of each system were defined arbitrarily. This established a workable time scale on which geology developed, but it is not in any sense a natural time scale. Because the time scale developed in Europe, it is even more unnatural in North America.

Another development has been the recognition of broad patterns of deposition and unconformity in the stable continental interior of North America. (See Fig. 7.1.) It has been possible to recognize these sequences in the continental interior because there the rocks are relatively undeformed, are fossiliferous, and have been extensively studied. Such a pattern requires a relative change in sea level. Deposition occurs when the continent is below sea level, and nondeposition and erosion occur when the continent is above sea level. Either the continents move up and down, or sea level changes, or both occur. If the continent rises and

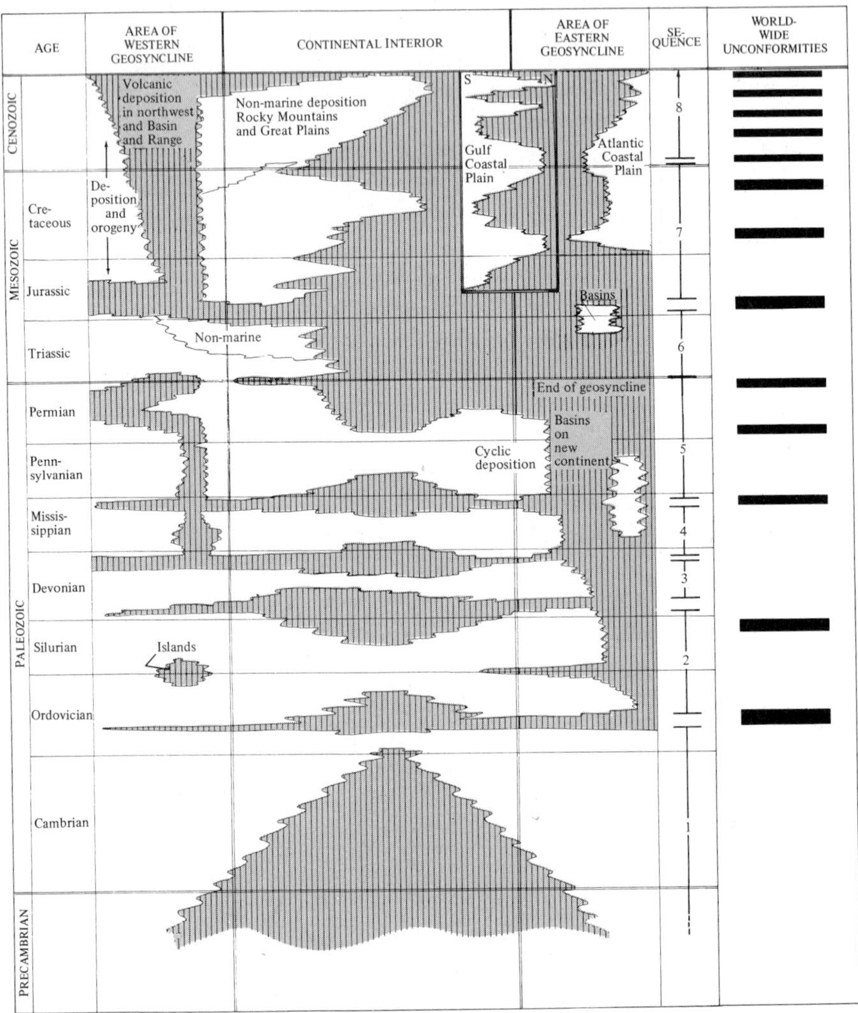

FIGURE 7.1
The geologic sequences of deposition in central North America. This diagram shows in a general way where and when deposition occurred. Geosynclinal deposition in Texas and Oklahoma is omitted. Erosion and deformation cause many difficulties in constructing such a diagram. To gain a real understanding of geologic history, the student should add the rock types and their sources to the depositional sequences in the diagram. In general, the times of nondeposition in the geosynclines correspond to orogenic periods. The vertical lines indicate nondeposition.

falls, the resulting sequences of erosion and deposition would be local to that continent, although at least minor changes in sea level would occur worldwide. The flat-topped seamounts and atolls record changes in sea level in parts of the oceans, suggesting that the shape of the ocean basins may change, causing a change in volume and thus a change in sea level. Such changes are probably associated with plate tectonics. During a time of rapid plate motion, the mid-ocean ridges or spreading centers would be active and hot; therefore, because of thermal expansion, the ridges would have a larger volume. This would cause sea level to rise worldwide, thus causing flooding of the continents. Conversely, at times of slow plate movement, the spreading centers would be much smaller and so the ocean basins would have a larger volume.

The pattern of geologic history of the interior of North America, then, is the advance and retreat of seas, causing sequences of deposition followed by unconformity in the interior. Had the geologic time scale been defined in the interior of the North American continent, these sequences probably would have been the geologic systems.

Fossils also reveal changes in life that are milestones in geologic history. The eras were defined by marked changes in life that were obvious to the early geologists.

Orogenic or mountain-building events are also easily recognized and provide a framework for geologic history. Most of these events are caused by plate collisions and so are best recorded near continental margins. Orogenies will be important in the geologic history of North America that follows.

WORLD GEOGRAPHY IN THE EARLY PALEOZOIC

It is not yet possible to reconstruct the positions and movements of the continents in detail in the early Paleozoic. When those data are available, it will be possible to describe the history of the whole earth in a unified way. From Permian onward, the reconstructions are somewhat easier, although disagreements do exist among those who have studied the problems closely. Much of what is said in these sections will require revision as more data are gathered.

Very little is known of the positions of the continents throughout most of the Precambrian. Paleomagnetic data suggest that the continents came together in the late Precambrian. The evidence that parts of this supercontinent were together in the Precambrian is the matching of ages and structures. The fold belt of Precambrian age is an excellent match between South America and Africa. (See Fig. 1.8.) Late Precambrian glacial deposits have been reported. Many of these reported glacial tills may be merely conglomerates and not of glacial origin. Those near the pole may be glacial, but those near the equator most likely are not.

In the Cambrian there were five large continents: North America, Europe, Siberia, Asia, and Gondwanaland (Fig. 7.2). *The southern continent, called* **Gondwanaland,** *consisted of South America, Africa, Antarctica, Australia, and India.* Gondwanaland remained a single continent throughout the Paleozoic. The North

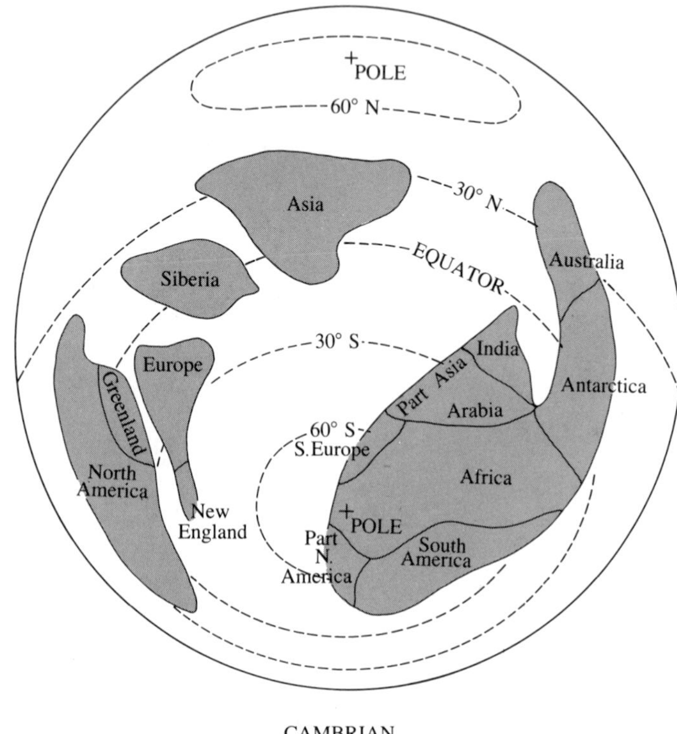

FIGURE 7.2
The continents of the Cambrian Period. This map and the similar ones in this and later chapters are modified from Mintz, 1981.

American continent, which we will follow most closely, probably contained parts of the present Europe, and the Europe of the Cambrian Period probably included parts of the present eastern North America. Much of southeastern United States and probably Mexico were part of Gondwanaland, as was part of present-day southern Europe and Asia.

In the Cambrian Period, the North American continent had continental shelf–slope geosynclines on all sides. On the west coast this persisted until the Devonian. Europe was close by, probably separated by a mid-ocean ridge. In the Ordovician, a subduction zone with a volcanic arc formed along the east coast of North America (Fig. 7.3). In the Silurian Period, Europe collided with North America (Fig. 7.4). The effects of these events on North America are discussed in the next section. In the Silurian Period, Siberia and Asia also collided (Fig. 7.4).

From late Precambrian through Ordovician, the south pole remained in North Africa, although it did move somewhat. Glacial deposits of Ordovician age have been found in the Sahara, confirming the location of the paleomagnetic pole. In the Silurian Period, Gondwanaland moved so that the south pole was in southern Africa. The equator passed through North America in a north-south direction in the Cambrian and in a southwest-northeast direction in the Silurian.

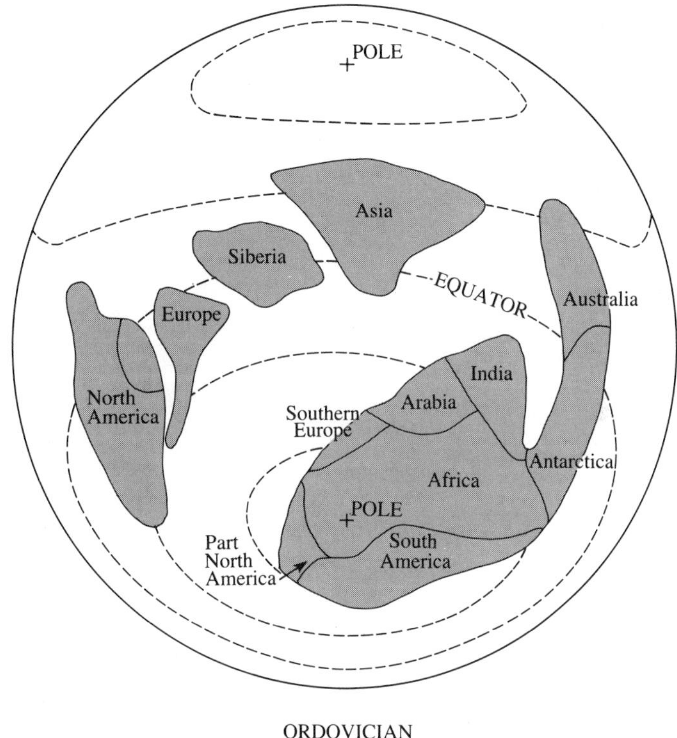

FIGURE 7.3
The continents in the Ordovician Period. A subduction zone has formed between Europe and North America.

LATE PRECAMBRIAN TO EARLY ORDOVICIAN

The base of the Cambrian rocks, with their abundant fossils, is one of the most important points in geologic history. The base of the Cambrian System is generally defined as the first occurrence of certain fossils (trilobites); however, some place it at the first occurrences of easily recognized fossils. Thus recognition of this important time line depends on chance preservation of fossils. At many places the first Cambrian fossils occur in the midst of thick sedimentary sections.

A general reconstruction of early Paleozoic geography is shown in Figure 7.5. On this map and the other paleogeographic maps, the present outline of the continent and some of the present rivers are shown for reference. This does not imply that they existed at the times depicted.

At the beginning of the Cambrian Period, the main framework of North American geology was established. Geosynclines of the continental shelf–slope type occupied both sides of the continent as well as the far north. During the Cambrian, the seas spread over more and more of the continental interior until by Early* Ordovician time, most of the interior was awash. The seas then withdrew,

*The terms Early, Middle, and Late are used when referring to time. Lower, Middle, and Upper are used when referring to rocks. All of these terms are capitalized when used in the formal sense.

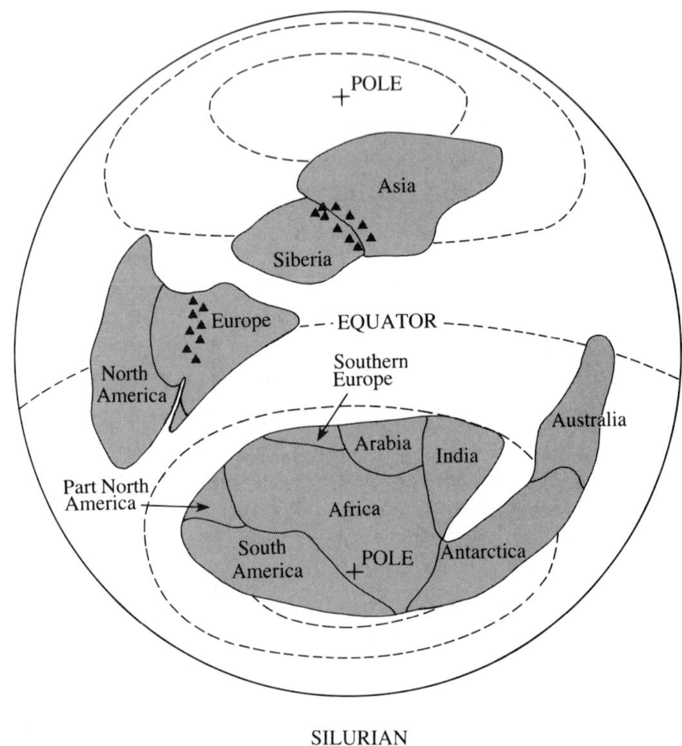

FIGURE 7.4
The continents in the Silurian Period. Gondwanaland has moved so that a pole is in southern Africa; Europe and North America, and Siberia and Asia collided.

first from the interior, and finally from the geosynclines, ending this simple episode of geologic history.

Interior

In late Precambrian time, the interior of North America was dry but was surrounded by geosynclines. The seas gradually expanded over much of the southern, western, and far northern parts of the interior. (See Fig. 7.5.) Much of northeastern Canada and most of Greenland remained dry land, as did a broad arch from roughly Lake Superior to the Gulf of California. This pattern of land and sea was to be repeated but with a variety of minor variations throughout the Paleozoic. The land areas were undergoing erosion and provided the clastic sediments deposited in the seas. This is indicated by the fact that the sediments become coarser as these land areas are approached. The present continental interior is underlain by Precambrian metamorphic and igneous rocks, such as those presently exposed north of the Great Lakes. Erosion reduced these rocks to relatively smooth surfaces that underlie the Cambrian rocks. At some places, such as the Ozark Mountains of Missouri, Upper Cambrian rocks lap up against the flanks of the ancestral Ozark Mountains, showing that highlands did exist in the interior. The movement of the

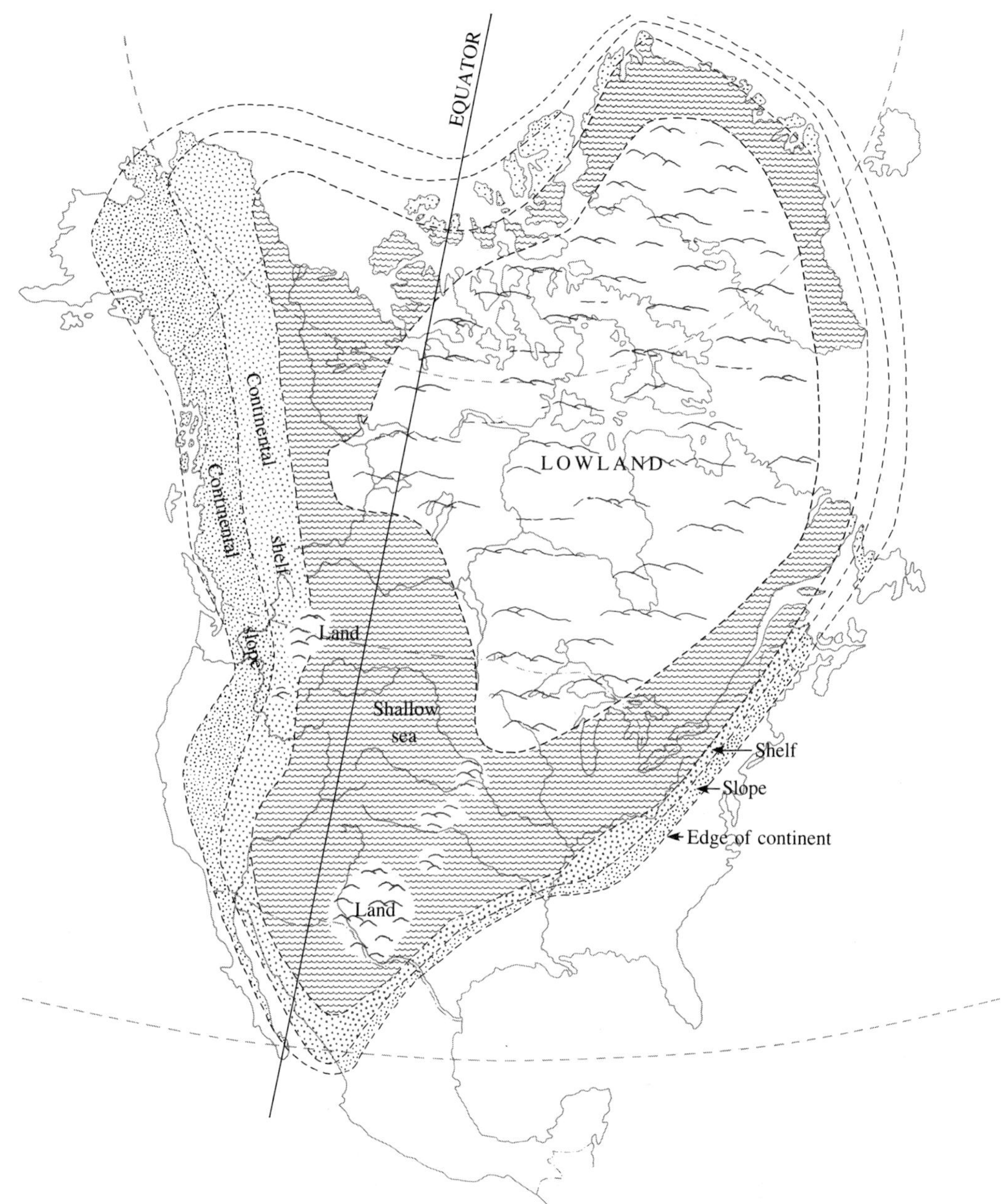

FIGURE 7.5
Generalized map of North America in the Cambrian Period. Continental shelf–slope geosynclines surround the continent. The extent of the shallow sea in the interior changed during the period. This general paleogeography of marginal geosynclines and shallow interior seas persisted through most of the Paleozoic. In eastern Canada and New England, the edge of the continent shown on this map is the dividing line between two different trilobite faunas of the same age.

seas into the interior was not a uniform advance, but probably had a number of short retreats and readvances. This is suggested by both rock types and changes in fossils. At most places in the interior, the oldest rocks are sandstones. These rocks are generally less than three hundred meters thick, although they are much thicker at places. Except at the margins of the interior platform, the oldest rocks are Upper Cambrian in age. Lower and Middle Cambrian rocks are, for the most part, confined to the margins of the interior. The Lower Ordovician rocks conformably overlying these rocks are largely limestones with some shale. This change from clastic rocks to carbonates suggests that the land areas had been reduced to low relief in early Ordovician time.

East

In the late Precambrian, a geosyncline formed in the east. The oldest rocks are in the southern Appalachian Mountains. They consist of a sequence of 9150 m (30,000 ft) of clastic rocks. Overlying them is an 1830-m (6000-ft) quartz-sandstone and shale unit, in which the first trilobites are found near the top. The two units are more or less conformable, although some workers consider the contact unconformable. At any rate, it seems clear that sedimentation began in Precambrian time. The source of these rocks was the interior of the continent. Overlying them are about 2135 m (7000 ft) of Middle and Upper Cambrian shales and limestone. The amount of carbonate increased upward as the seas covered more and more of the interior, moving the source of the clastics farther away. Above these are about a thousand meters of Ordovician carbonates. The transition from Cambrian to Ordovician is in this conformable sequence of carbonates.

Cambrian shale and sandstone up to 2440 m (8000 ft) thick are found in Vermont. These are overlain by thick Lower Ordovician clastic rocks. Later mountain building has produced complex structures in New England and to the north, making interpretation difficult.

The trilobites in eastern New England and adjacent Canada are different from those in the rest of the eastern geosyncline. The boundary separating the two trilobite faunas is the edge of the Cambrian continent shown in Figure 7.5. The eastern trilobites are similar to those found in Europe. In the past, this distribution of fossils was interpreted as indicating that a barrier existed in the geosyncline and separated the faunas. We now believe that the rocks containing the European trilobites originated in Europe and were welded to North America when Europe collided with North America in the middle and late Paleozoic.

West

The early Paleozoic also began in the west with the establishment of a geosyncline. The seas moved into the area from both north and south. A barrier existed between these two seas in Idaho until Middle Cambrian time. Although a much younger batholith obscures the evidence for the barrier, its presence is indicated by a difference in fauna between the two seas and by coarser clastic rocks near the old shoreline.

A very thick section of Precambrian and lower Paleozoic rocks is found in the southwestern ranges of the Great Basin near the California-Nevada border. At that time this was near the edge of North America. Here sedimentation began in Precambrian time. The oldest sedimentary rocks are a 2440-m (8000-ft) sequence of clastics with some carbonates. Unconformably above those rocks are later Precambrian and Cambrian rocks. They begin with 460 m (1500 ft) of dolomite, followed by up to 3050 m (10,000 ft) of clastic sediments, largely quartz sandstones. Conformably above these rocks is another unit of similar clastic rocks about 1525 m (5000 ft) thick. Near the top of this unit, the first Cambrian fossils are found. The Lower Cambrian rocks here are about 3050 m (10,000 ft) thick and are mainly sandstone, grading into shale near the top. The Middle and Upper Cambrian rocks are close to 2440 m (8000 ft) thick and are mainly carbonates. They are overlain by similar thick Lower Ordovician carbonates.

The Cambrian clastic rocks had their source in the continental interior, as was the case with the other areas described. During the Cambrian Period, the seas moved eastward onto the interior. As the seas transgressed onto the continent, the basal Cambrian deposits were a distinctive beach sand. This unit can be traced from southern Nevada through Utah and Montana to the Midwest. This sandstone is progressively younger to the northeast, ranging from Precambrian to middle Cambrian in age. (See Fig. 7.6.)

Farther north in the western geosyncline in Canada, similar rock accumulated in great thicknesses. The history is similar, except that at most places, the Lower Cambrian rocks appear to lie unconformably on the late Precambrian sediments. The Precambrian sedimentary rocks are about ten thousand meters thick here.

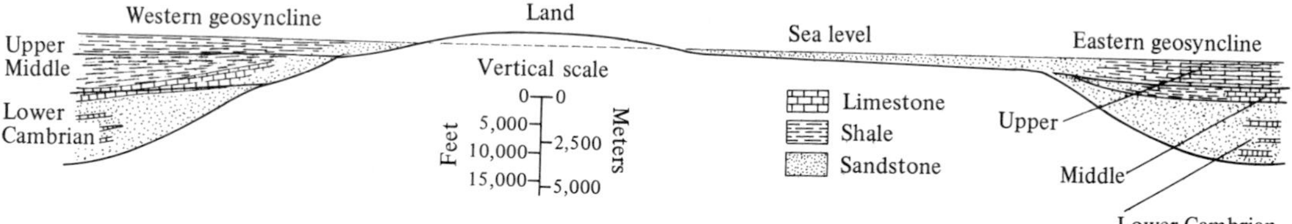

FIGURE 7.6

Generalized east-west restored section of Cambrian rocks. The section is drawn from the California-Nevada border to Virginia. Note that the basal Cambrian sandstone was deposited in a sea that gradually encroached on the continent, so the sandstone becomes younger as the interior of the continent is approached. In northern United States the late Cambrian sea was continuous across the continent. A section such as this is an interpretation from many fragmentary occurrences of deformed rocks.

Far North

Geosynclinal rocks are found in the Arctic islands of Canada. They apparently connect with similar rocks in Greenland and in northern Alaska. This is a very difficult area in which to study geology. Very few Cambrian rocks have been found. Ordovician rocks are widespread. To the north in the geosyncline, they are up to 3050 m (10,000 ft) thick and consist of volcanic and clastic rocks. Farther south, they are thick shales; still farther to the south, on the edge of the Canadian Shield, they grade into carbonates up to 915 m (3000 ft) thick. In the geosyncline in East Greenland, about 12,200 m (40,000 ft) of clastic rocks of Precambrian age are found. Overlying these are up to 2440 m (8000 ft) of Cambrian and Ordovician carbonates. The Ordovician rocks are unconformably overlain by Devonian clastics.

MIDDLE ORDOVICIAN TO EARLY DEVONIAN

A volcanic arc, or subduction zone, had formed near the east coast, and it became a new and important aspect of the environment.

Interior

In Middle Ordovician time the seas returned, first to the geosynclines, and then across the interior. The pattern of land and sea was much like that of the Cambrian. The new sequence began with the deposition of a remarkably widespread, thin quartz sandstone over most of the interior. This sandstone is up to 92 m (300 ft) thick at places and covers almost two million square kilometers. Its source was apparently the Canadian Shield. Overlying this sandstone are carbonates west of the Mississippi Valley and shale to the east. The carbonates are mainly dolomite, only about a hundred meters thick, and extend from western United States to arctic Canada. The shale in the eastern interior grades into carbonate in the Silurian. The source of shale was a highland to the east in the area of the present Appalachian Mountains. Mountain building had begun in the geosyncline. The seas withdrew from the interior in late Silurian time and from the geosynclines in early Devonian time. In Late Silurian time, a remnant of the sea was isolated in one of the basins in Michigan, Pennsylvania, and New York. Here a thick section of evaporites, especially rock salt (halite) and gypsum, was deposited.

During the Ordovician, the interior changed slowly from a flat lowland on which uniformly thick sediments were deposited to a number of very broad domes and basins. (See Fig. 7.7.) The domes and basins were pronounced by Devonian time, when they dominated the depositional pattern.

East

The eastern geosyncline was an area of active mountain building throughout this sequence. This was the beginning of a new pattern here that culminated in the

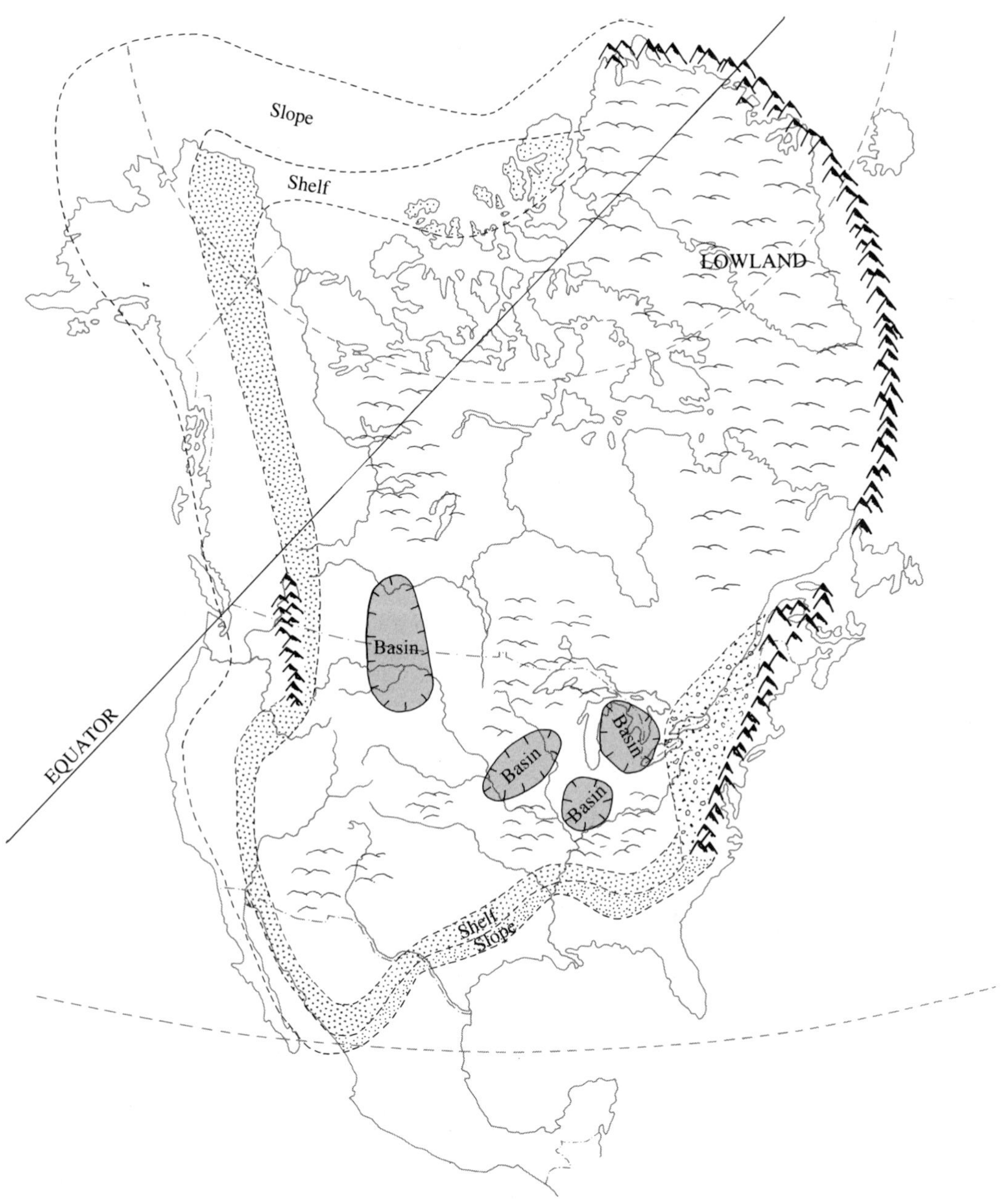

FIGURE 7.7
Ordovician paleogeography. A subduction on the eastern side of the continent caused mountains that shed clastics. Basins and domes (hilly areas) began to form in the interior and were pronounced by Devonian time.

formation of the Appalachian Mountains. During the Ordovician and Silurian, mountain building took place at different times and places throughout the geosyncline. None of the events described occurred simultaneously through the whole geosyncline. These and later orogenies and their associated igneous and metamorphic events have destroyed much of the evidence, so reconstruction of the history is difficult.

In the Ordovician, a volcanic arc and subduction zone developed along northeastern North America, causing deformation and mountain building there. Clastic rocks of Middle Ordovician age are the first indication of the change. The source of these rocks was clearly the eastern part of the geosyncline because they thicken and become coarser to the east. Large fragments of Cambrian and Lower Ordovician limestones are found in thick clastic sections in the northern Appalachian Mountains, suggesting that uplifted and probably deformed lower Ordovician and older rocks were eroded to form Middle Ordovician deposits. This is shown diagrammatically in Figure 7.8B. Volcanic rocks in eastern New England suggest volcanic island arcs there.

Mountain building continued throughout the Ordovician, filling the geosyncline with clastics and spreading shale far to the west. From West Virginia to New York, Late Ordovician clastics unconformably overlie Middle Ordovician rocks. These clastics were shed so rapidly that the geosyncline was filled and the sediments accumulated above sea level. Some of these rocks are red beds, so called because their iron has been oxidized to hematite, which is believed to happen when the rocks are subjected to seasonal wetting and drying. A similar middle Ordovician clastic fan had its source in eastern Tennessee.

This orogeny reached a peak near the end of the Ordovician. Folding, faulting, and uplift were widespread. The Taconic Mountains in Massachusetts and Vermont were formed at this time. This range is a block of Cambrian and Ordovician shales 242 by 48 km (150 by 30 mi) that were thrust at least 48 km (30 mi) to the west. Middle and Upper Ordovician rocks were folded at this time. This event is called the **Taconic orogeny.**

In the Silurian, Europe collided with North America, causing the **Caledonian orogeny** and forming mountain ranges on both continents. The Silurian record begins with a widespread quartz sandstone that is unconformable on Upper Ordovician rocks in the Appalachian Mountains. (See Figs. 7.8C and 7.9.) Farther to the west, Silurian carbonates overlie Upper Ordovician carbonates conformably. Uplifts in the highlands to the east occurred throughout the Silurian Period, but the resulting clastic sediments did not spread west of the geosyncline. Thus the mountains to the east were lower than during the Ordovician. In parts of New England and in maritime Canada, mountain building continued, resulting in thick sections of volcanic and clastic rocks. These rocks were folded and metamorphosed, and batholiths were emplaced. Elsewhere, the Silurian rocks were not affected by this mountain building.

The Lower Devonian rocks are a continuation of the Silurian patterns. Thick clastic deposits are found to the east, and thin limestones to the west. In the

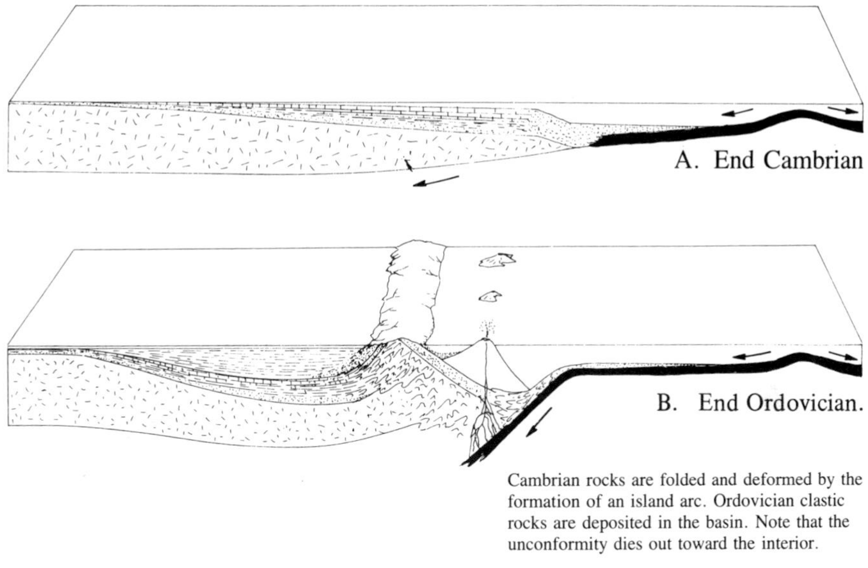

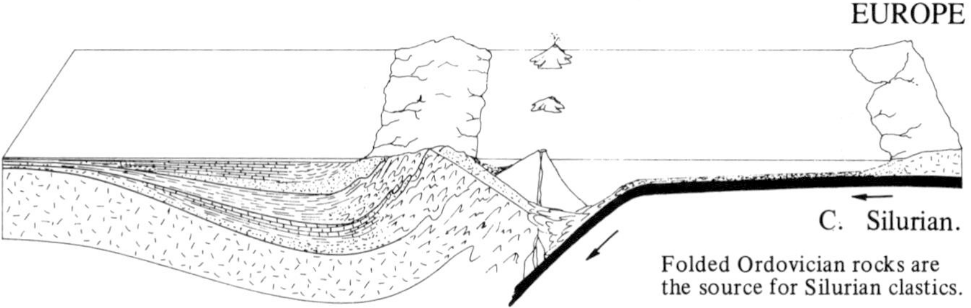

FIGURE 7.8
Development of the eastern geosyncline (geocline). A. At the end of the Cambrian, the geosyncline consisted of a continental shelf, slope, and rise. B. A volcanic island arc formed in the early Ordovician, and the Cambrian rocks were deformed. C. Renewed activity deformed the Ordovician rocks.

northeast, volcanic rocks occur with clastics, many of which are nonmarine. The seas withdrew from the geosyncline in early Devonian time.

West

The geography of the west remained much the same, that is, deposition of carbonates on the shelf and shale with some clastics to the west on the continental slope. Thick sections of clastics and volcanics of this age are found at many places, but these rocks were accreted to the North American continent in later collisions.

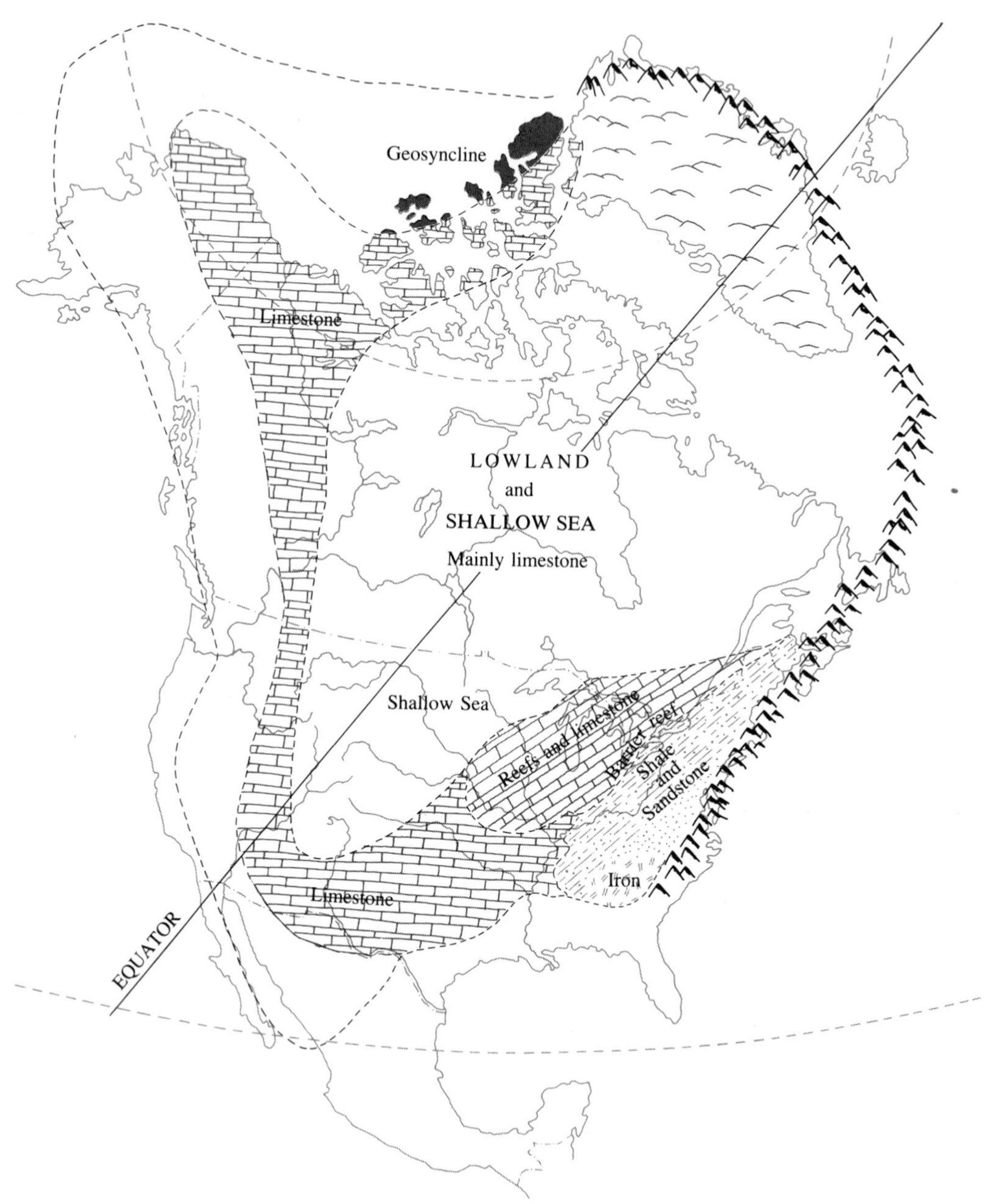

FIGURE 7.9
Silurian paleogeography. There were subduction zones on both sides of the continent.

Far North

The Arctic islands contain thick Ordovician rocks, as described earlier. These rocks grade from volcanic and clastic rocks in the geosyncline to the north through shale to thin limestone to the south near the shield. The Silurian and Lower Devonian rocks are similar. In at least one place, deformation took place in the Silurian, and Late Silurian and Early Devonian rocks overlie deformed younger Silurian rocks. At other places, Early Devonian rocks are folded.

In northern Greenland, dark limestone and coarse clastics of Silurian age are found.

LIFE OF THE EARLY PALEOZOIC

The first abundant, easily recognized fossils occur abruptly at the base of the Cambrian System. Because well-developed, advanced forms of life appear suddenly at this horizon, the implication is clear that life must have existed for a long time previously. In the last chapter, the evidence for this was discussed. All of the phyla (the major subdivisions of life) (see Appendix D) appear in the Cambrian. (See Fig. 7.10.) This suggests that a great radiation and diversification had occurred in the 500 million or so years since green algae had appeared and that this radiation continued through the Ordovician. Not all biologists accept the idea that new life forms developed to fill the many empty ecologic niches. Accidents of preservation probably bias our knowledge of life. This is suggested by the discovery of a number of excellently preserved, flattened carbonaceous films, even with fine hairs intact, of soft-bodied animals of Middle Cambrian age at a single locality on the crest of the Canadian Rockies near Field, British Columbia. (See Fig. 7.11.) A similar lower Cambrian fauna has been found in Greenland. These occurrences are the only ones of this age in the whole world, and in most cases it is the only place where soft parts of the bodies are preserved. It took unusual conditions to preserve these animals.

Life began in the sea, and the first abundant fossils are marine. *More than half of all Cambrian fossils are trilobites, and they and brachiopods together are about 90 percent of Cambrian fossils.* Trilobites were arthropods. (See Fig. 7.12.) Trilobites underwent a great adaptive radiation during the Cambrian, and they are good index fossils for that system. They were bottom dwellers and swimmers and probably ate algae and soft-bodied creatures that are not preserved. They were probably scavengers, too. Trilobites became less abundant after Early Ordovician time as other invertebrates developed, and the trilobites became extinct near the end of the Paleozoic (Permian Period).

A reconstruction of a Cambrian sea bottom is shown in Figure 7.13. Similar reconstructions are shown for most of the other periods. The crowding together of the many organisms shown in all of these figures probably did not ever occur in nature but they have been drawn this way to show the varieties of life in each period. The reconstruction in Figure 7.13 is based on the collections made at the site shown in Figure 7.11. The fossils are the evidence; the reconstruction is an interpretation.

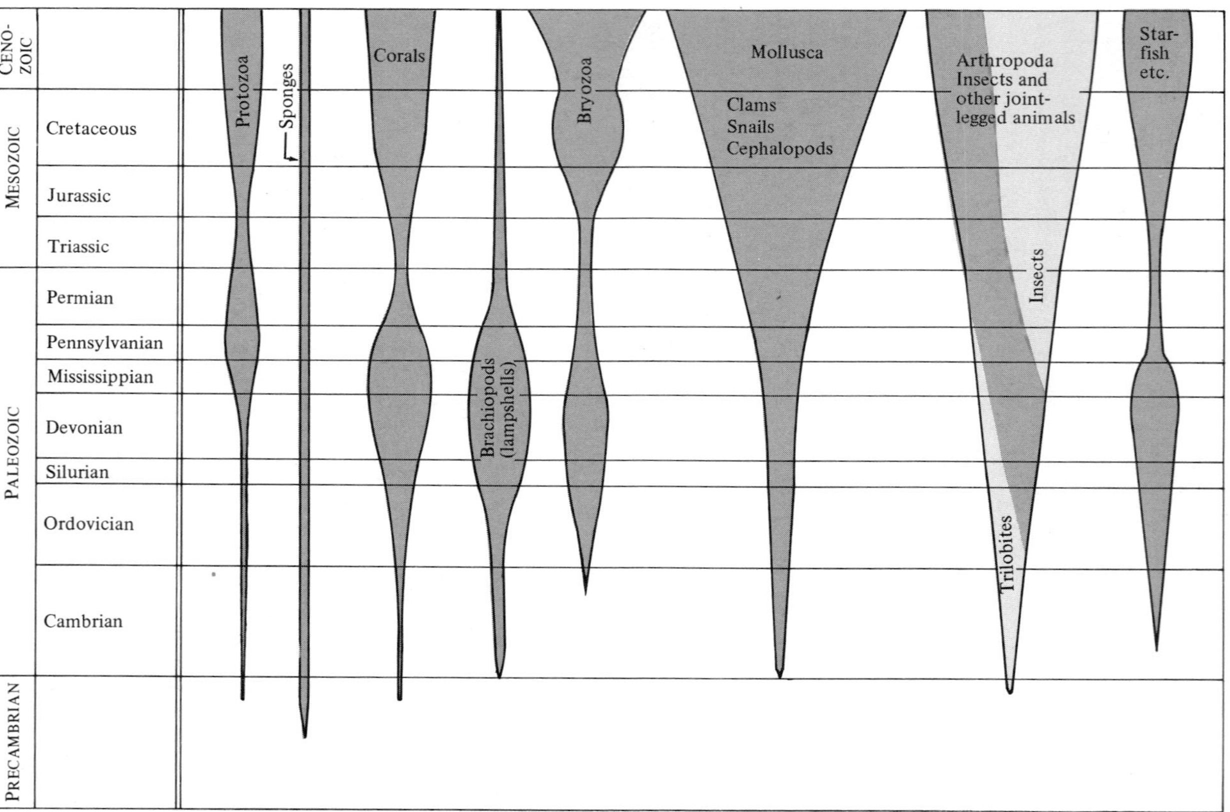

FIGURE 7.10
Geologic occurrence of invertebrate fossils. The width of the areas indicates relative abundance within each group (modern insects are about 90 percent of known species).

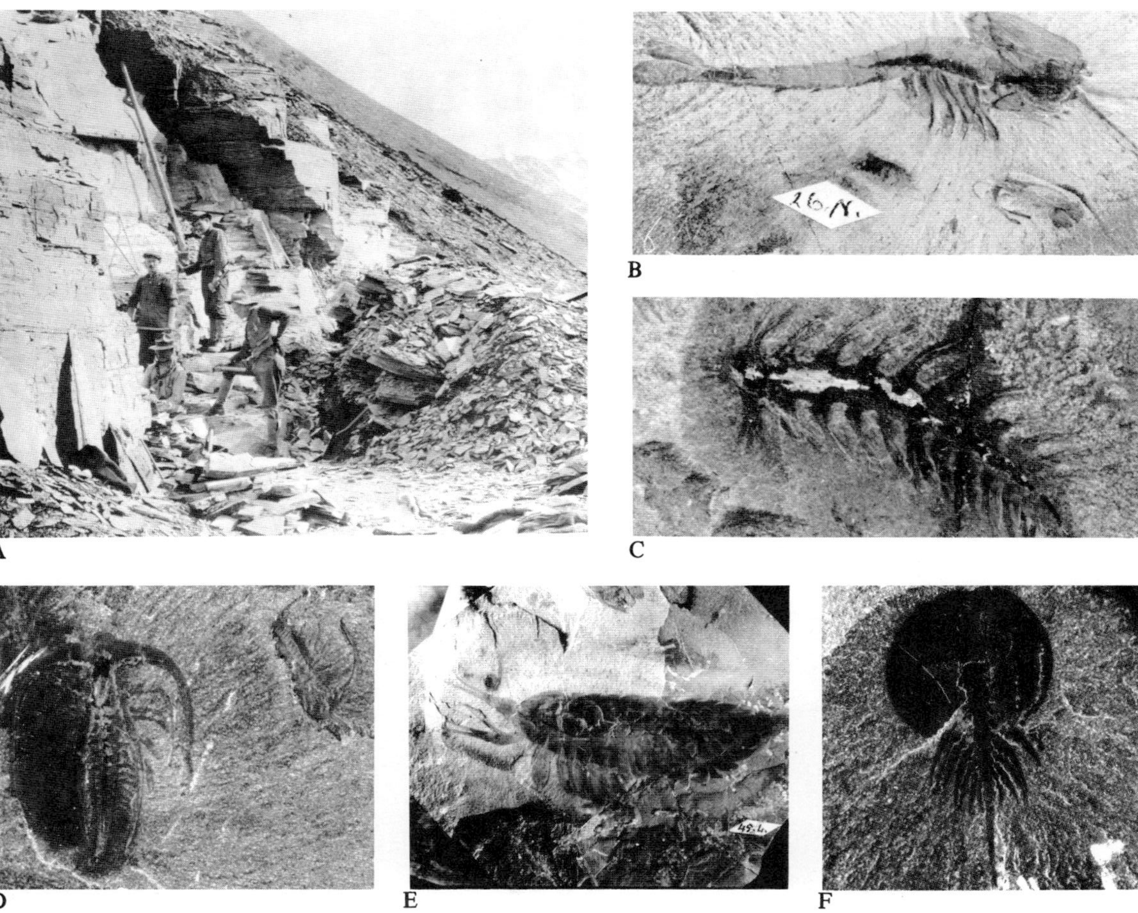

FIGURE 7.11
Burgess Shale on Mount Wapta, British Columbia. This Middle Cambrian formation has yielded extremely well-preserved fossils of 130 species. A. Collecting fossils. The first fossils were found by chance in 1909. B. *Waptia fieldensis,* a shrimplike arthropod. C. *Canadia setigera,* a worm. D. *Marrella splendens,* a trilobitelike arthropod. E. *Leanchoila superlata.* F. *Burgessia bella.* Photos from Smithsonian Institution.

FIGURE 7.12
Trilobite, *Elrathia kingi,* Middle Cambrian, about 2.5 cm (1 in.) long. Photo courtesy Ward's Natural Science Est., Inc.

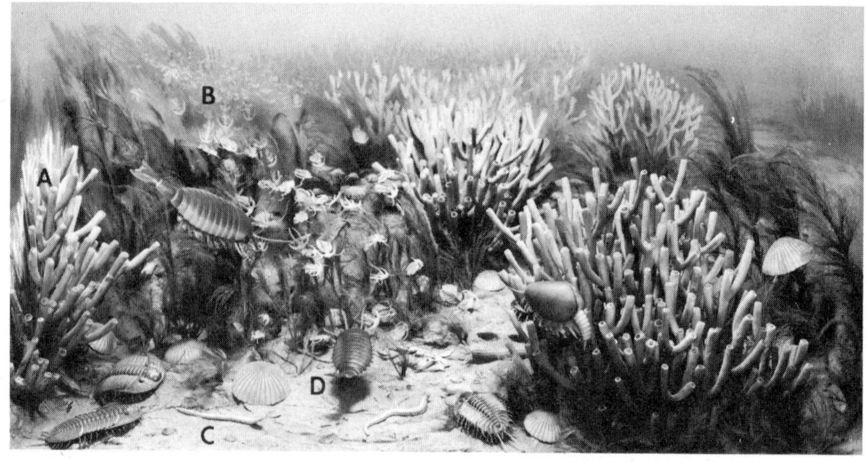

FIGURE 7.13
Reconstruction of Middle Cambrian sea floor in western North America. The fauna shown include the soft-bodied animals in Figure 7.11. A. Spongelike animal. B. *Marrella,* an arthropod. C. A worm. D. A jellyfish. Several different trilobites are also shown. From Field Museum of Natural History. Negative No. GEO.80872.

In Early Ordovician time, the fauna changed somewhat. Brachiopods became more abundant (see Fig. 7.14), as did the other invertebrates such as corals, bryozoans, crinoids, snails, and straight-shelled nautiloids. This change may have occurred in response to the development of predators. Few predators are found in Cambrian faunas, but they are found in Ordovician and later rocks. Straight-shelled nautiloids probably resembled present-day squids, and some could swim rapidly and catch prey with their tentacles. It may be that shells and skeletons were

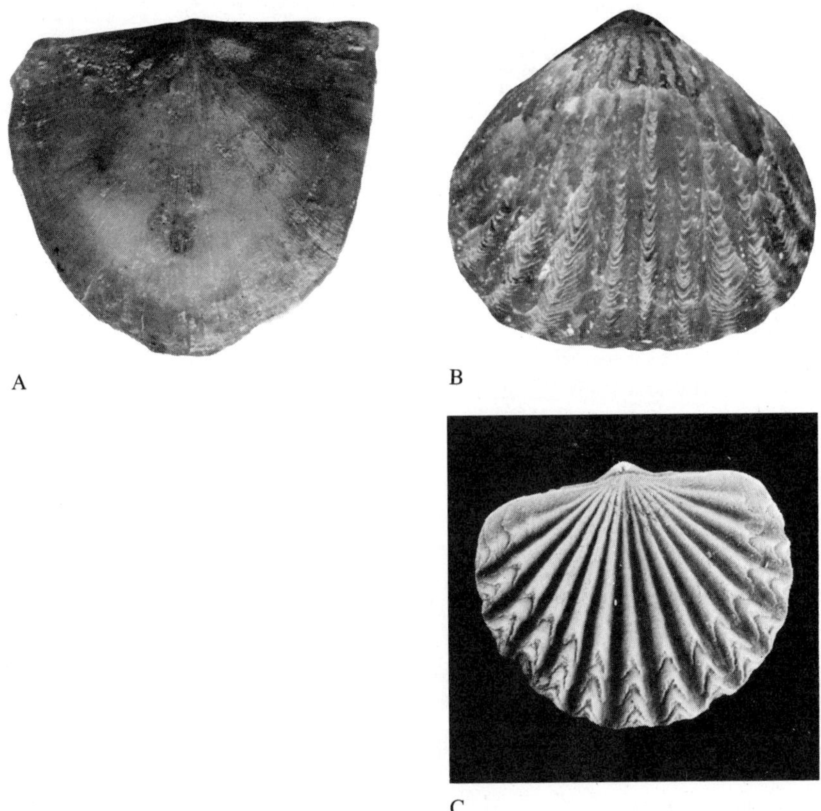

FIGURE 7.14
Brachiopods. Brachiopods are abundant in many Paleozoic rocks. They have two shells, and most were attached to the bottom by a stalk that came out through a hole near the hinge line. They are traditionally figured with the hinge line up. A. *Rafinesquina alternata,* Ordovician, from Ohio, about 55 mm (2 in.). B. *Lepidocyclus capax,* also Ordovician, from Ohio, about 25 mm (1 in.). C. *Eocoelia hemispherica,* Silurian, from Quebec, about 10 mm (0.4 in.). Photos A and B courtesy Ward's Natural Science Est., Inc.; C. from Energy, Mines and Resources (113250-C). Reproduced with the permission of the Minister of Supply and Services Canada, 1990.

FIGURE 7.15
Middle Ordovician sea floor in north-central United States. A. Straight-shelled nautiloids (somewhat like squids). B. Trilobites. C. Snails. D. Colonial corals. E. Solitary corals (with tentacles). F. Brachiopods. G. Bryozoans. H. Seaweed. From Field Museum of Natural History. Negative No. GEO.80820.

developed mainly for support in the Cambrian, but evolved into protection in the Ordovician. The development of spiny shells in late Paleozoic adds credence to this idea.

Figure 7.15 shows a typical Middle Ordovician sea floor. The many

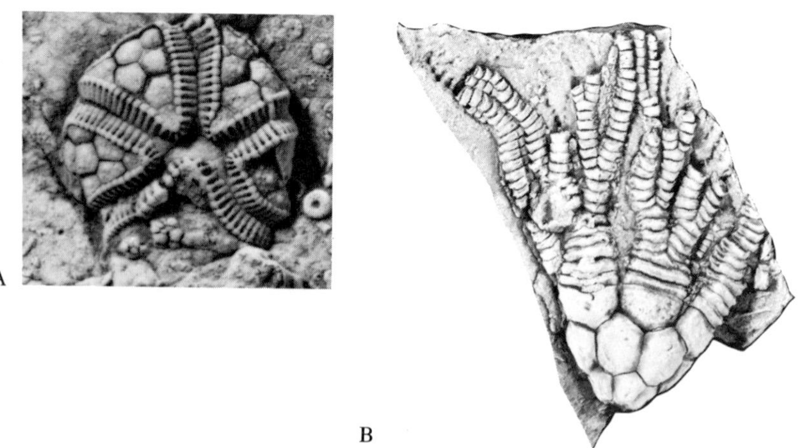

FIGURE 7.16
Ordovician echinoderms. A. *Edrioaster,* a primitive echinoderm that attached to the bottom, about twice natural size. B. *Cupulocrinus jewetti,* a crinoid, natural size. Both from Ontario. Photos courtesy Energy, Mines and Resources (112910-B, 112339). Reproduced with the permission of the Minister of Supply and Services Canada, 1990.

FIGURE 7.17
Bryozoans. These colonial animals build intricate structures on the sea floor in which a separate animal occupies each hole. Some build lacy and others stony or twiglike structures. Fossils are generally fragmentary because the structure is broken after death of the colony. Silurian bryozoans, Rochester Shale, Lockport, New York. A few brachiopod shells can be seen in this slab with several types of bryozoans. Photo from Smithsonian Institution.

differences between this scene and the Cambrian one are obvious. In the Ordovician, corals and bryozoans expanded rapidly, and, to a lesser extent, snails, clams, straight-shelled nautiloids, and crinoids did also. (See Fig. 7.16.) This trend continued into the Silurian, when corals became very abundant, forming reefs as in Figures 7.17, 7.18, and 7.19.

Graptolites are important index fossils in the Ordovician. The floating graptolites evolved rapidly from their first appearance in Late Cambrian to their extinction in the Early Devonian. They are good index fossils because they changed rapidly and, because they floated, are found in many environments. They tend to be found more commonly in black shale, probably because they sank after death onto muddy bottoms where not much other life occurred. For this reason, it is common to speak of graptolite-bearing rocks vs. shell-bearing rocks of Ordovician age. They were colonial animals, with many small individuals living in the stalks that hung down from the float. The fossils look like pencil marks on the enclosing shales (see Fig. 7.20), hence their name. In late Cambrian and early Ordovician, they were multibranched; in middle and late Ordovician, they had two branches; and in Silurian, they had one branch.

Vertebrates first appear in the Cambrian. They, too, began in water. The first evidence of their existence is scales and bone fragments found in an upper Cambrian sandstone. Few other fish fossils are found until Late Silurian, although by Devonian they are dominant. Thus there is little evidence for the stages in the development of the fish. The early fish may have had predators. Their predators may have been straight-shelled nautiloids or, probably more likely, the eurypterids. In any case, the first fish were armored, suggesting that they had predators.

Eurypterids (sea scorpions) became abundant in Silurian rocks. (See Figs. 7.21 and 7.22.) They are not found in normal marine rocks but in what probably were brackish waters or perhaps restricted saline waters. Some were as much as 2.75 m (9 ft) long.

FIGURE 7.18
Middle Silurian sea floor in north-central United States. A. Cystoid. B. *Favosites,* honeycomb coral. C. *Halysites,* chain coral. D. *Syringopora,* tube coral. E. Several types of nautiloids. F. Two types of brachiopods. G. Trilobite. Several other types are also in the view. H. Two solitary corals. From Field Museum of Natural History. Negative No. GEO.80875.

FIGURE 7.19
Chain coral, *Halysites,* Silurian, Louisville, Kentucky. A separate animal lived in each hole in this colonial coral. Photo from Smithsonian Institution.

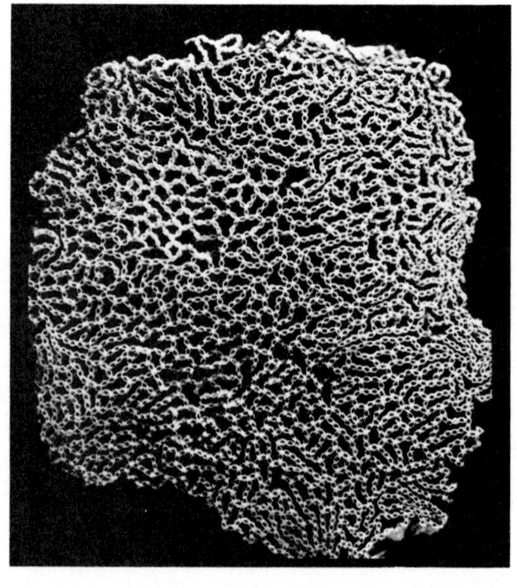

FIGURE 7.20
Graptolites. The fossils of these curious animals look like pencil marks on rocks. They were colonial animals, and a separate animal lived in each of the sawteeth. The stalks hung down from a float, and because they floated, the fossils are found in all environments. They evolved very rapidly: in Late Cambrian and Early Ordovician they had many branches, in Middle and Upper Ordovician they had two branches, and in Silurian they had one branch. They became extinct in Early Devonian. Shown is *Didymograptus,* Middle and Upper Ordovician. Photo courtesy Ward's Natural Science Est., Inc.

The first insects yet discovered are in Silurian rocks. Scorpions are also found in Silurian rocks. (See Fig. 7.23.) *The oldest definitely terrestrial animals known are Early Devonian scorpions.* Scorpions, insects, eurypterids, and trilobites were all arthropods.

Plants on the land must have preceded any animal life. *The earliest reported evidence of land plants so far discovered consists of spores (microfossils) recovered from rocks as old as Middle Ordovician.* The widespread lowlands of Late Silurian time probably were a favorable environment for the development of plants, and the oldest clearly land plants are Silurian in age. Plants were abundant from Devonian on. Beginning in Late Devonian, coal appears in every geologic period. The only earlier coal is in the Precambrian in Michigan, and although it may be of organic origin, it probably was not formed by land plants.

FIGURE 7.21
Late Silurian sea floor in New York. Several types of eurypterids (see Figure 7.22) are shown with snails. Eurypterids probably lived in somewhat unusual environments such as restricted, saline, shallow seas. A. *Carcinosoma,* a scorpion. From Field Museum of Natural History. Negative No. GEO.80819.

FIGURE 7.22
Silurian eurypterid, *Eurypterus remipes.* This specimen came from the same formation as the scorpion shown in Figure 7.23, but from a different part of New York. Photo from Smithsonian Institution.

FIGURE 7.23
Carcinosoma scorpionis, Silurian scorpion. This specimen came from the same formation as the eurypterid shown in Figure 7.22, but from a different part of New York. Photo from Smithsonian Institution.

SUMMARY

In late Precambrian, the continents came together, causing mountain building in Gondwanaland. From Precambrian through Ordovician, the south pole was in northern Africa. During the Silurian, Gondwanaland moved so that the south pole was in southern Africa.

The geologic systems are natural subdivisions only at the places where they were defined.

Geosynclines are important in the history of continents.

Mountain building does not seem to be synchronous over the earth, although the Cambrian and most of Triassic were quiet times, and Ordovician, Devonian, Pennsylvanian, and late Cretaceous were more active.

The interior of North America has experienced times of deposition followed by withdrawal of the seas.

Large extinctions occurred near the ends of the Cambrian, Devonian, Permian, Triassic, and Cretaceous periods.

In the Ordovician, mountains resulted from a subduction zone; in the Silurian, they were caused by a collision with Europe.

In the late Precambrian, continental shelf–slope geosynclines occupied both sides and the north of North America, and at times the interior was a lowland covered by shallow seas.

Late Precambrian to Early Ordovician

Interior: Sandstone grades upward into carbonates; source was land area in northern and central Canada.

East: Clastics grade upward into carbonates; to the north, mainly clastics with some volcanic rocks.

West: Sequence is mainly clastics that transgressed northeastward on the interior.

Far North: Sequence is mainly clastics; very few Cambrian rocks.

Middle Ordovician to Early Devonian

Interior: Sequence starts with widespread thin sandstone; carbonates to west, shale in east; domes and basins form.

East: Mountain building is source of clastics in geosynclines and interior.

West: Sequence is mainly clastics; a volcanic island chain may have formed in Nevada.

Far North: Clastics in the geosyncline grade to limestone in the interior.

Life of the Early Paleozoic

Cambrian life was not primitive, suggesting a long evolution.

Our only knowledge of early soft-bodied animals comes from a single location in the Canadian Rocky Mountains.

Trilobites and brachiopods are the main Cambrian fossils.

In the Ordovician, brachiopods, corals, bryozoans, crinoids, snails, and nautiloids became more abundant.

Graptolites are good index fossils in the Ordovician and Silurian.

The first vertebrates were fish in the late Cambrian.

Silurian scorpions and insects may have been the first animals to breathe air and live on land.

The earliest plants may be Middle Ordovician in age.

QUESTIONS

1. List as many of the assumptions made in interpreting geologic history as you can.
2. Describe the various patterns recognized or assumed in geologic history.
3. Is the Cambrian-to-Ordovician boundary a natural one in North America?
4. What events separate Late Cambrian from Middle Ordovician?
5. What occurred in the eastern part of North America during the Silurian Period?
6. In what continent was the "south" pole in early Paleozoic time?
7. Where was the equator in North America in early Paleozoic time?
8. What is the importance of the Burgess Shale fauna?
9. What are typical Cambrian fossils?
10. In what ways do Cambrian fossils differ from those in the Silurian?
11. Why are graptolites important?
12. What were the first vertebrates, and when did they appear?
13. When did the first land plants appear?

SUGGESTED READINGS

General

Bambach, R. K., C. R. Scotese, and A. M. Ziegler, "Before Pangea: The Geographies of the Paleozoic World," *American Scientist* (January-February 1980), Vol. 68, No. 1, pp. 26–38.

Frazier, W. J., and D. R. Schwimmer, *Regional Stratigraphy of North America.* New York: Plenum Press, 1987, 719 pp.

Morris, S. C., and H. B. Whittington, "The Animals of the Burgess Shale," *Scientific American* (July 1979), Vol. 241, No. 1, pp. 122–133.

Palmer, A. R., "Search for the Cambrian World," *American Scientist* (March-April 1974), Vol. 62, No. 2, pp. 216–224.

Whittington, H. B., *The Burgess Shale.* New Haven: Yale Univ. Press, 1985, 151 pp.

Patterns in Geologic History

Johnson, J. G., "Timing and Coordination of Orogenic, Epeirogenic, and Eustatic Events," *Bulletin, Geological Society of America* (December 1971), Vol. 82, No. 12, pp. 3263–3298.

Kerr, R. A., "Changing Global Sea Levels as a Geologic Index," *Science* (July 25, 1980), Vol. 209, No. 4455, pp. 483–486.

Loutit, T. S., and J. P. Kennett, "New Zealand and Australian Cenozoic Sedimentary Cycles and Global Sea-Level Changes," *Bulletin, American Association of Petroleum Geologists* (September 1981), Vol. 65, No. 9, pp. 1586–1601.

Mussman, W. J., and J. F. Read, "Sedimentology and Development of a Passive- to Convergent-Margin Unconformity: Middle Ordovician Knox Unconformity, Virginia Appalachians," *Bulletin, Geological Society of America* (March 1986), Vol. 97, No. 3, pp. 282–295.

Schleh, E. E., "Review of Sub-Tamaroa Unconformity in Cordilleran Region," *Bulletin, American Association of Petroleum Geologists* (February 1966), Vol. 50, No. 2, pp. 269–282.

Sloss, L. L., "Forty Years of Sequence Stratigraphy," *Bulletin, Geological Society of America* (November 1988), Vol. 100, No. 11, pp. 1661–1665.

Wheeler, H. E., "Post-Sauk and Pre-Absaroka Paleozoic Stratigraphic Patterns in North America," *Bulletin, American Association of Petroleum Geologists* (August 1963), Vol. 47, No. 8, pp. 1497–1526.

8

Late Paleozoic—Devonian, Mississippian, Pennsylvanian, Permian

WORLD GEOGRAPHY IN THE LATE PALEOZOIC

In the late Paleozoic, the continents came together, forming one supercontinent called **Pangaea.** The collisions caused mountain building, and the mountains together with the large landmass profoundly affected the climate and therefore the life of the earth. The evidence that there was only a single continent in the late Paleozoic was very important in the establishment of the reality of continental drift.

In the Devonian, the collision between Europe and North America continued (Fig. 8.1), and in the Pennsylvanian or Permian Period, Africa collided with southeastern North America (Figs. 8.2 and 8.3). These collisions formed the Appalachian Mountains as well as other mountains on all of the continents involved. In western North America, the first of a series of collisions with island arcs or microcontinents began in the Mississippian.

In the northern supercontinent, the mountains formed by the Silurian orogeny shed similar red clastics in Europe, eastern North America, and northern North America.

The Ural geosyncline between Europe and Asia was deformed by the collision of those two continents, forming the Ural Mountains.

A major change that occurred in late Paleozoic is that the location of the south pole changed from North Africa to a position near its present location in Antarctica. This means that the southern continent moved about 90 degrees north during the

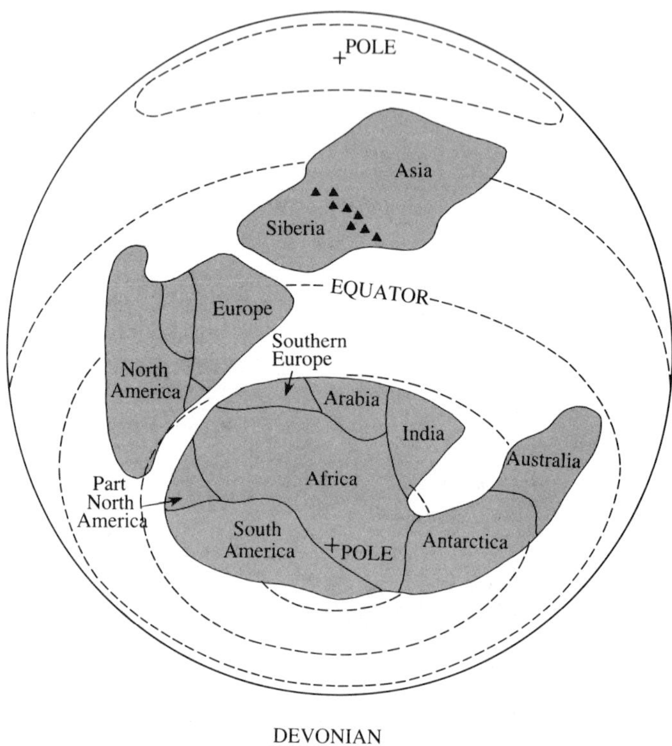

FIGURE 8.1
The continents in the Devonian Period. North America and Europe are in contact.

late Paleozoic, assuming that the geographic pole remained in the same place. The Permian glacial deposits resulted from this movement. North America was rotated in this move, so that the equator ran closer to east-west than to north-south as it had earlier. This change may account for the aridity indicated by the late Paleozoic and Mesozoic rocks in western North America.

MIDDLE DEVONIAN TO LATE DEVONIAN

During the Devonian, North America and Europe collided, causing the **Acadian orogeny.** Clastic fans on both continents record this event. This collision was probably associated with the one that started in the Silurian and would continue at different places through much of the late Paleozoic.

Interior

The seas returned to the interior of North America in Middle Devonian time, and at many places rocks of that age overlie Middle Silurian beds. In the eastern part of the interior, shale was the main rock type deposited. Its main source was highlands

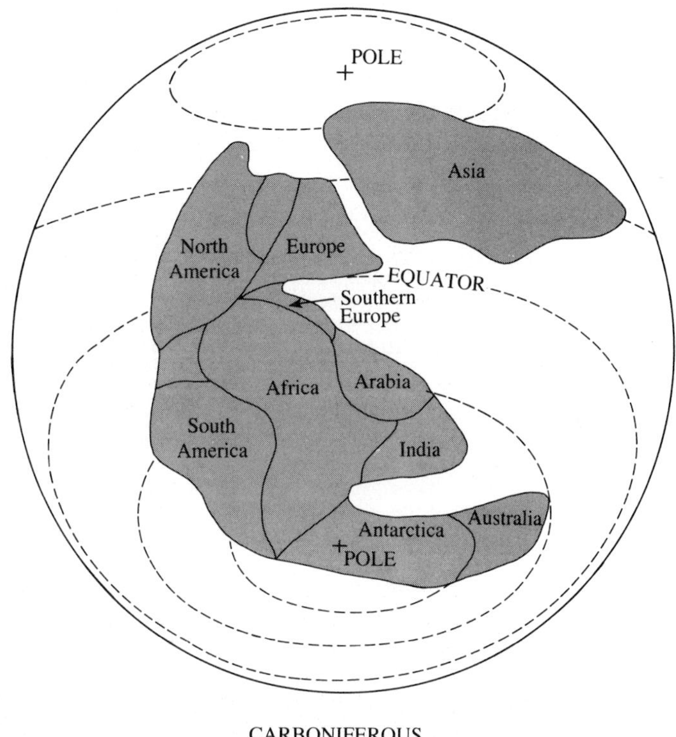

FIGURE 8.2
The continents in the Pennsylvanian Period. Africa and North America collide. Gondwanaland moved so that the southern pole was in Antarctica.

in the eastern geosyncline. Near the domes of the interior and the highlands in the geosyncline, some sand was deposited. The western interior, extending to the Arctic, was the site of limestone and some shale deposition. Reefs developed in this limy sea; in Alberta, these reefs are reservoirs for important oil fields. The domes and basins of the interior became more pronounced. In Montana and Alberta and at other places in Canada, evaporites formed in some of these basins that were cut off from the seas. The seas withdrew in the late Devonian.

East

In Middle and Late Devonian time, orogeny continued in the eastern geosyncline. The collision with Europe that started in the Silurian continued, with mountains forming farther south (Fig. 8.4). The intrusive and metamorphic processes that accompanied this collision transformed the old geosyncline into new continent. The southeastern border of North America was probably a subduction zone because Africa (Gondwanaland) was coming closer.

The area of the old eastern geosyncline became an upland. This upland was rapidly eroded, and the resulting clastic sediments were deposited in the area just

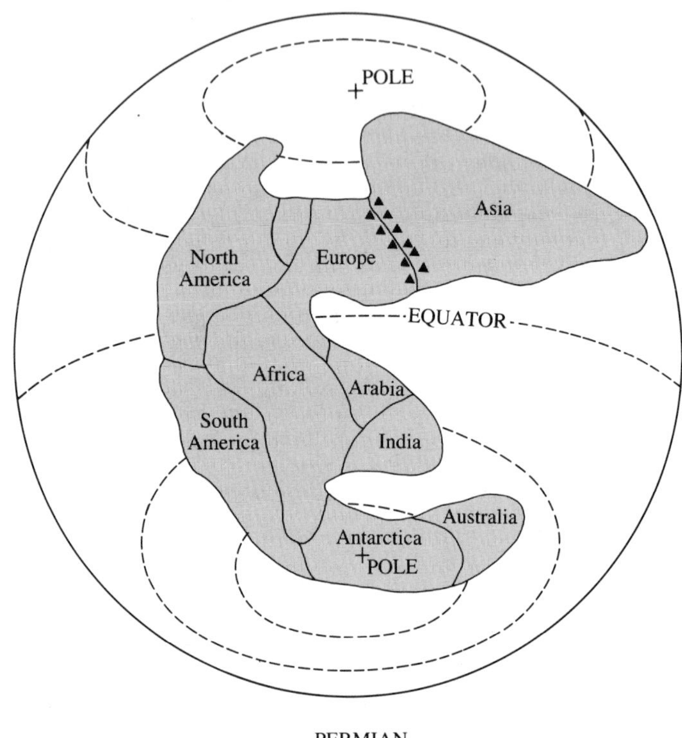

FIGURE 8.3
The continents in the Permian Period. Asia and Europe collide, forming a single continental mass.

west of the old geosyncline. In this depositional area first one place and then another would downsink to trap the debris. Thus, in Middle and Late Devonian, over 3050 m (10,000 ft) of deltaic rocks were deposited in Pennsylvania and adjacent New York (Catskill Mountains) and West Virginia. These deposits were similar to the Ordovician and Silurian deltas but were much thicker and more widespread; this type of deposition continued into the Pennsylvanian. In Mississippian time, the clastic fan spread farther south along the Appalachian Mountains, suggesting that orogeny spread southward. This pattern, too, as we shall see, continued throughout the remainder of the Paleozoic, with the addition of smaller basins and adjacent highlands in the area of the old geosyncline. The seas again withdrew during part of Late Devonian time.

West

In the far west, few Devonian rocks are found, but the history can be reasonably deciphered. A volcanic island arc is believed to have existed to the west of the continental shelf and slope in the vicinity of the California-Nevada border. The evidence for this is that thick sections of clastic and volcanic rocks are found in western Nevada. Farther east, on the continental shelf, carbonates were deposited

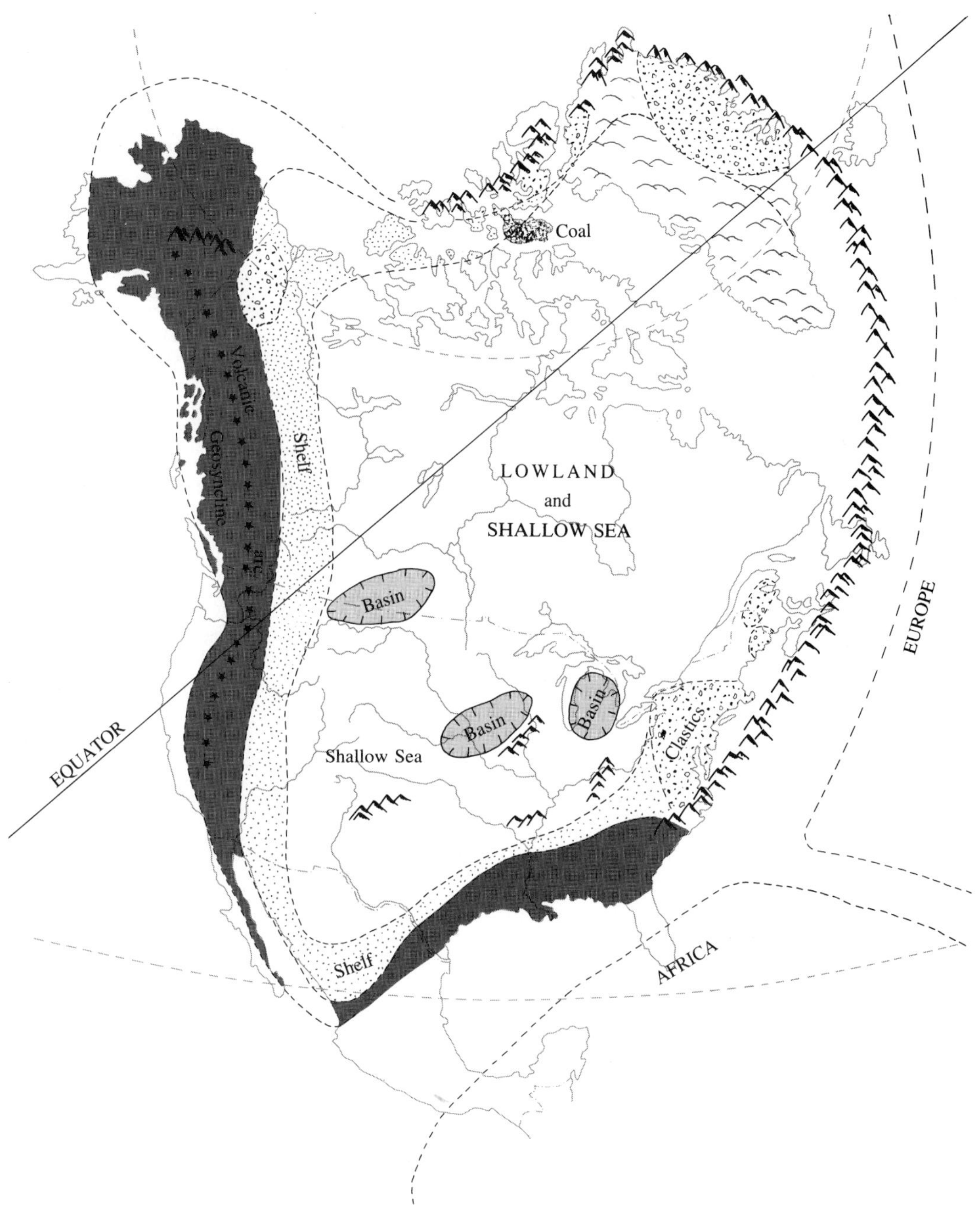

FIGURE 8.4
North America in the Devonian Period. Collision with Europe caused mountains that shed clastics in the east and in Greenland, as well as in Europe.

that extend onto the western interior. Similar sequences are found in the Canadian Rockies in Alberta, where the carbonate rocks are very thick and contain both reefs and evaporites. In latest Devonian time, orogeny began in Nevada.

Far North

The Devonian Period began quietly in the far north. The carbonate deposition of the interior extended to the Arctic. In the geosyncline, a thousand or more meters of carbonates with some shale and siltstone, especially near the top, make up the Middle Devonian rocks. These rocks thin to the south on the interior platform. The Upper Devonian rocks are sharply different. Uplift occurred to the north, forming a highland that shed clastics into the geosyncline. These nonmarine clastic rocks are up to 3050 m (10,000 ft) thick and contain some coal beds. They closely resemble the rocks of the same age in New York and Pennsylvania.

In East Greenland, thick deposits of nonmarine clastic rocks were also deposited in the Devonian. The rocks are up to 6100 m (20,000 ft) thick; they were deposited in basins adjacent to uplands, and so record activity here also. The rocks are red and gray in color and resemble the Old Red Sandstone, a formation of the same age in England. The source of all of these clastic rocks was the highland created by the collision of Europe and North America.

LATEST DEVONIAN TO LATE MISSISSIPPIAN

The Mississippian rocks of eastern North America are similar to those of Devonian age but the collision site moved farther south. In the west, collision occurred **(Antler orogeny)**, probably with an island arc, and this changed the simple geography that had continued since late Precambrian.

Interior

The seas had withdrawn in Late Devonian time, but returned to the interior in latest Devonian time. It may be that at places the seas did not completely withdraw. (See Fig. 8.5.) The Mississippian System was named for the exposures in the area drained by the upper Mississippi River. The Lower Mississippian rocks in that area are black, organic-rich shales. The shale is not the same age everywhere and is rarely more than about ten meters thick; however, it is generally present over a wide area. It is difficult to reconstruct the conditions under which this shale was deposited. The organic content suggests stagnant water, but the fossils and scour channels suggest shallow water. The shale contains enough uranium that it may in the future be an economic resource. The source of the shale was highlands that had formed in the geosyncline to the east and south.

Middle Mississippian limestones overlie the black shale. Some of these limestones are composed largely of fossil fragments and some of spherical calcite grains (oolites). They are quarried for building stone. In Late Mississippian time,

FIGURE 8.5
North America in the Mississippian Period.

limestone and sandstone layers alternate. The source of the sands was highlands to the east. These are the first sandstones in the interior since Ordovician time.

The western interior was the site of limestone deposition. These carbonates extend well into Canada. There the amount of shale increases to the north and in the younger rocks.

The domes and basins of the interior were active and, to a large extent, controlled the thicknesses of all of the units described. The basin in Montana-Dakota was the site of evaporite deposition in Middle Mississippian time.

The seas withdrew in Late Mississippian time. Erosion followed over most of the continent. The erosion was more pronounced in the west than in the east. The east was probably near sea level, setting the stage for the Pennsylvanian coal swamps.

East

The area of the old geosyncline was again uplifted, producing another clastic fan like that of Devonian time. The center was farther south in Pennsylvania, and the total extent was less than in the Devonian. At the same time, in the area of the old geosyncline in New England and northeastern Canada, a number of basins developed. These basins persisted from Mississippian to Permian time and developed great thicknesses of clastic rocks eroded from the surrounding highlands, together with some carbonates, evaporites, and red beds.

In Texas and Oklahoma and extending to Alabama, thick deposits of clastic rocks developed in Middle and Late Mississippian time, suggesting orogeny in that area. Subduction apparently continued there, and Africa came nearer. The collision occurred in late Mississippian or Pennsylvanian time.

West

In Mississippian time, the pattern of the far west changed. Apparently an offshore island arc or a microcontinent collided with the continental shelf, causing the Antler orogeny. In the collision, oceanic sediments were thrust over the carbonates of the shelf on the *Roberts Mountain thrust* fault. In this event, much of what is now western Nevada was accreted to North America. From this time onward, similar collisions occurred along the west coast and Alaska and, by this process, much of western North America was emplaced. (See Fig. 8.6.) The rocks are estimated to have moved almost 160 km (100 mi). The resulting mountain range extended all the way across Nevada and into Idaho, where younger igneous rocks obscure its extent. The mountains were never very high and underwent erosion as they were uplifted, shedding clastics both east and west. They were probably a series of islands connected by shallow water. To the east were deposited about a thousand meters of conglomerate and sandstone that grade into the thinner limestones of the interior. To the west, geosynclinal sediments accumulated. This pattern continued into the Pennsylvanian.

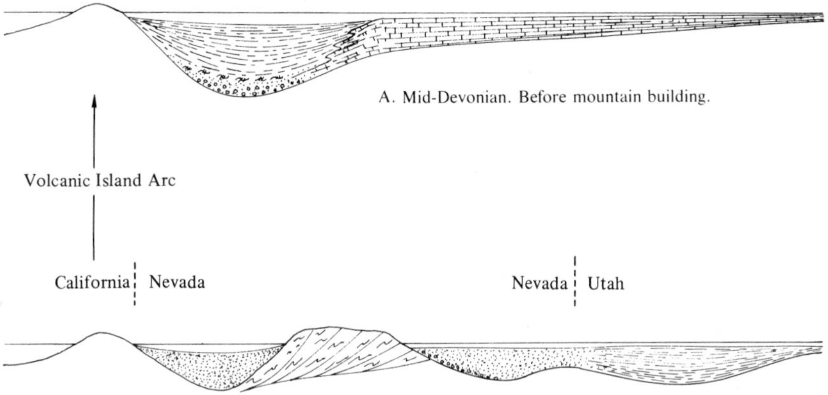

A. Mid-Devonian. Before mountain building.

B. Early Mississippian. After mountain building.

FIGURE 8.6
Devonian and Mississippian events in Nevada. Thrust faulting in Late Devonian and Early Mississippian time created two areas where clastic sediments were deposited. After R. J. Roberts.

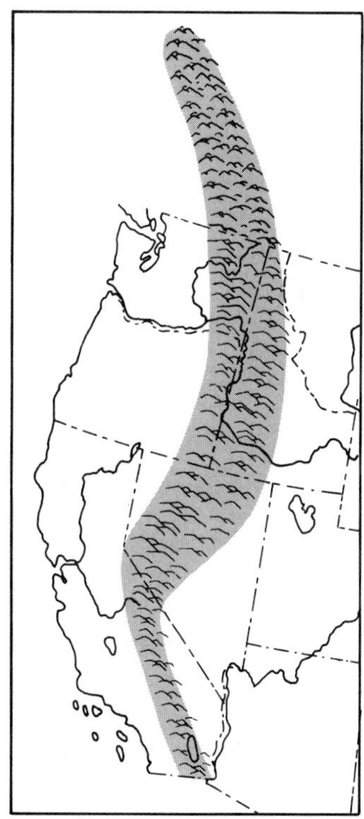

C. Extent of Devonian to Pennsylvanian mountain building.

Far North

In Alaska, uplift caused deposition of thick clastic rocks that grade into the thin limestones of the interior. In arctic Canada, mountain building apparently took place because Middle Pennsylvanian rocks overlie Upper Devonian sediments on a major angular unconformity. The lack of Mississippian rocks makes it impossible to decipher the details. In East Greenland, the deposition of nonmarine clastics that had begun in Devonian time continued.

LATEST MISSISSIPPIAN TO PERMIAN

The seas returned to parts of the interior in very latest Mississippian time, and they remained there through the Permian Period. During this interval, orogeny occurred at many places in the world. In North America, the main orogeny is called the **Alleghenian orogeny.** This was a time of great changes, when all of the continents came together and formed one supercontinent.

Interior

Late Mississippian and early Pennsylvanian shale and sandstone are found east of the Mississippi River valley. These rocks were apparently deposited in the deltas of rivers flowing from highlands to the east and south. By Middle Pennsylvanian time, the seas again covered the interior, and great coal swamps had formed from the Mississippi valley eastward to the Appalachian Mountains. Thicker deposits formed in the basins of the interior. The climate was apparently warm, and the plants that grew in the swamp were buried before they decayed, becoming the very important coal deposits of eastern North America. (See Fig. 8.7.) Both the coal deposits and the marine Pennsylvanian rocks of the interior are remarkable for the great number of repeated sequences of rocks called **cyclothems,** which are believed to be the result of periodic changes in sea level. (See Fig. 8.8.) The cause of these changes in sea level is not known, but they could be due to climatic changes such as glaciation that occurred elsewhere or to vertical movements of the continents or the ocean bottoms. A typical cyclothem begins with coarse, nonmarine sediments that become finer grained as the sea approaches. Coal is deposited during a near-shore swampy time and is followed by marine sediments that grade upward from shale to limestone as the water depth increases. The sea then retreats; then, generally after a short period of erosion, the sea readvances, starting a new cycle.

Permian rocks in the east are few and consist of clastic rocks in the area of the Pennsylvania–Ohio–West Virginia border. The upper Pennsylvanian rocks grade upward into the Permian. These nonmarine Permian rocks, less than 305 m (1000 ft) thick, are the youngest deposits associated with the eastern geosyncline. This will be discussed further in the next section. There are no Triassic rocks in the eastern interior.

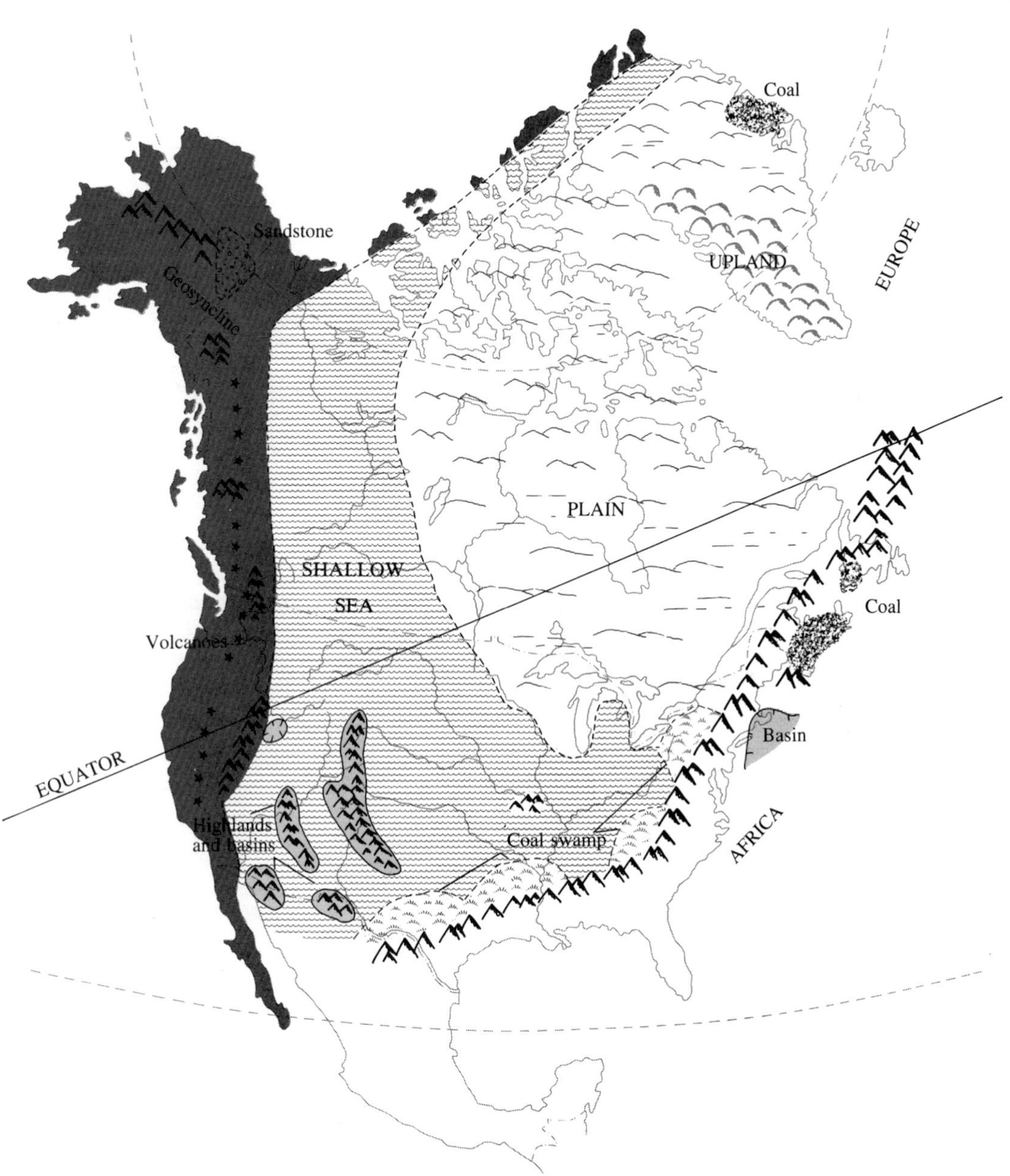

FIGURE 8.7
North America in the Pennsylvanian Period.

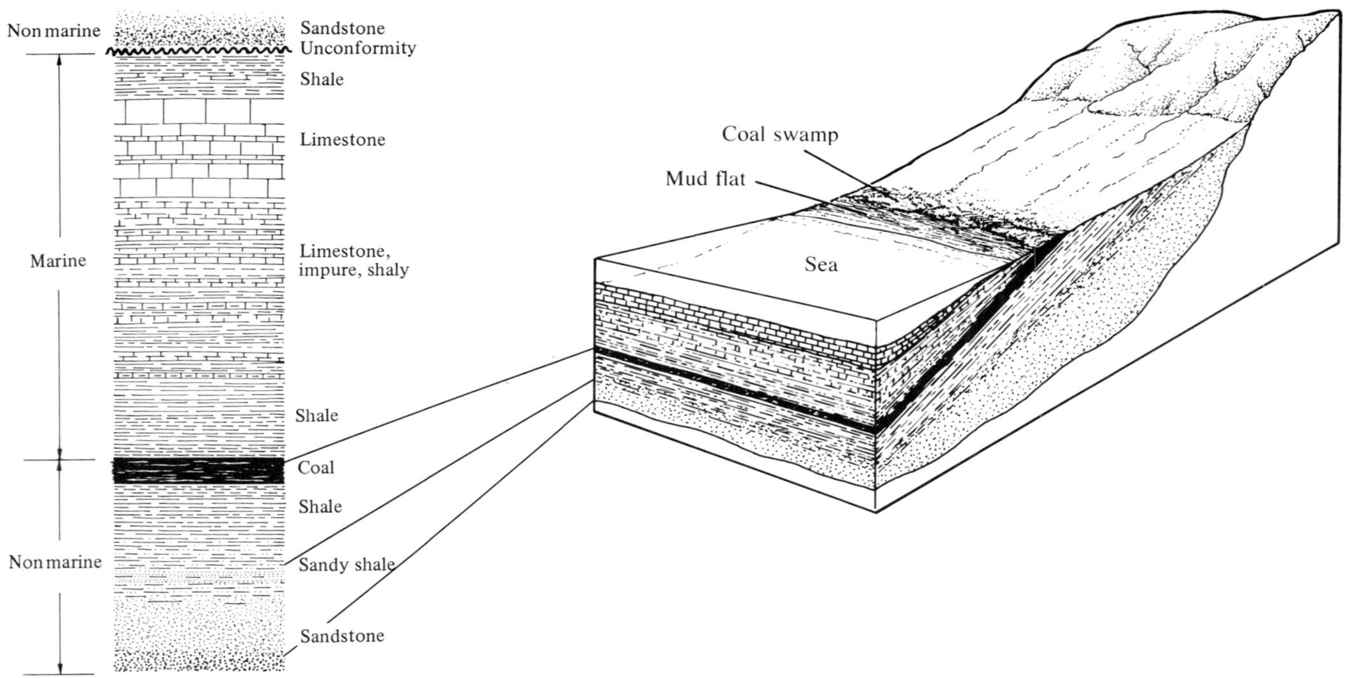

FIGURE 8.8
A typical cyclothem begins with nonmarine sandstone that becomes finer grained as the sea advances. Coal is deposited in a near-shore swamp. As the water depth increases, the marine rocks grade from shale to limestone. The sea then retreats to start the cycle again. Glaciation elsewhere may have caused the changes in sea level.

The paleogeography of the western interior of North America changed drastically in the Pennsylvanian. One of these changes was the development of a number of highlands and deep basins in the general area of the central Rocky Mountains and extending to Oklahoma and Texas. (See Fig. 8.9.) The Pennsylvanian rocks of the western interior are mainly quartz sandstone with some carbonates. In the deep basins adjacent to the highlands, thick clastic sections were deposited. The older broad basin in Dakota-Montana was the site of carbonate and evaporite deposition with some shale. Because of the changes in paleogeography, it will be more convenient to include the later history of the western interior under the heading "West."

East

In the east, the geosyncline was a highland at the start of the cycle, and rivers flowing from the highland brought sediments to the interior as was just described. Basins formed in Massachusetts–Rhode Island and in eastern Canada. Thick clastic sections containing coal accumulated in these basins.

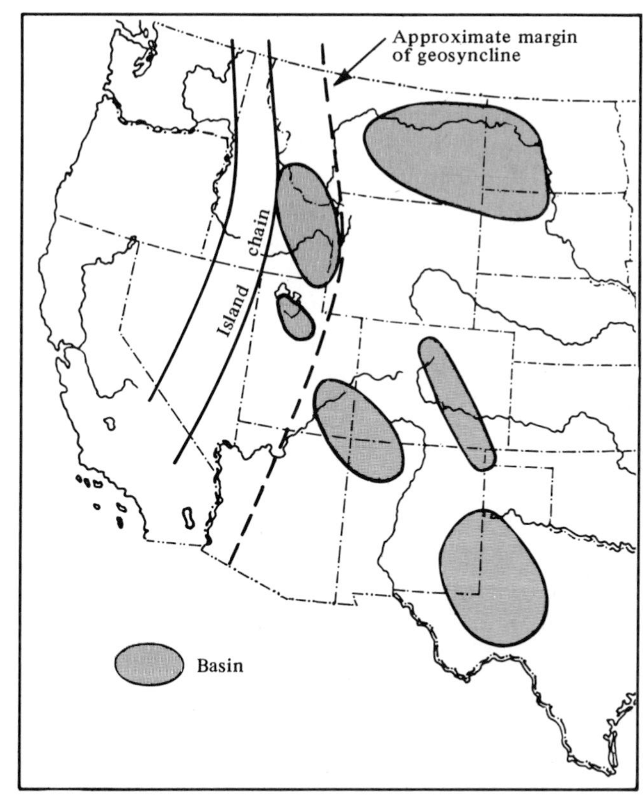

FIGURE 8.9
Some late Paleozoic events in the west. In Late Devonian and Mississippian time, orogenic events produced a chain of islands in the western geosyncline that was the source of clastic sediments. In the geosyncline the Early and Middle Pennsylvanian rocks are clastic west of the islands and limestone to the east. In late Pennsylvanian time, renewed activity in the island chain resulted in clastic sediments east of the islands. Starting in Mississippian, and especially in Pennsylvanian and Permian, a number of deep basins, some with adjacent highlands, developed both in the area of the geosyncline and in the interior sea.

In the southern interior, limestones were deposited, but farther south, near the edge of the continent, thick clastic deposits indicate renewed orogeny as Africa neared North America. In the Texas-Oklahoma area, the geosyncline received thick clastic deposits, suggesting continual orogeny to the south. In this area the geosyncline itself was deformed into a series of linear highlands and basins. These basins received sediments eroded from the adjacent highlands. Another orogeny affected this area, probably in latest Pennsylvanian time, when nonmarine clastic rocks were deposited and, finally, thrust plates moved northward over these rocks. In western Texas, the thrusting was somewhat earlier, and the end of the Pennsylvanian was a time of erosion. (See Fig. 8.10.)

In the late Paleozoic, the Appalachian geosyncline was the site of orogeny that formed the Appalachian Mountains (Alleghenian orogeny). The orogeny was probably caused by the collision of Africa (Gondwanaland) and North America (Fig. 8.11). This collision caused mountain building in southeastern North America and also farther north, where parts of Europe were deformed. Our present Appalachian Mountains contain parts of both North Africa and Europe and perhaps of other much smaller continental plates. (See Fig. 8.12.) The rifting that produced the present Atlantic Ocean began in the Triassic.

West

The west, too, was active in the late Paleozoic. The western geosyncline was dominated by an active subduction zone and its associated volcanic arc. In the western geosyncline, to the east of the island chain that had appeared earlier, the Early and Middle Pennsylvanian rocks are limestones, unlike the Mississippian clastics. West of the islands, conglomerate and sandstone were deposited. In Pennsylvanian time, the island chain was the site of more orogeny and uplift. (See Fig. 8.13.) The Late Pennsylvanian deposits, in part eroded from this uplift, are sandstones and limestones and overlie earlier Pennsylvanian rocks with an angular unconformity. These accumulated in a very thick (12,200 m [40,000 ft]) basin in eastern Utah and adjacent Colorado. (See Fig. 8.9.)

The area east of the geosyncline has, up to this point, been included with the interior; after early Permian time, however, the seas never again covered all the interior. Only the western interior was covered by seas, and because these seas were extensions of those covering the western geosyncline, it is better to include the closely related western interior with the western geosyncline.

The Pennsylvanian rocks of the western interior are quartz sandstones and carbonates. In southwestern Colorado, a highland developed, and thick accumulations of clastic rocks were shed into the adjacent basins. The basin on the Colorado-Utah border (see Fig. 8.9) received Early Pennsylvanian clastics, followed by thick evaporites and Middle and Late Pennsylvanian clastics and carbonates; the total thickness is 3050 m (10,000 ft).

The Permian was an active time in the western geosyncline. In the far west, thick deposits of clastic and volcanic rocks are found. Orogeny occurred in western

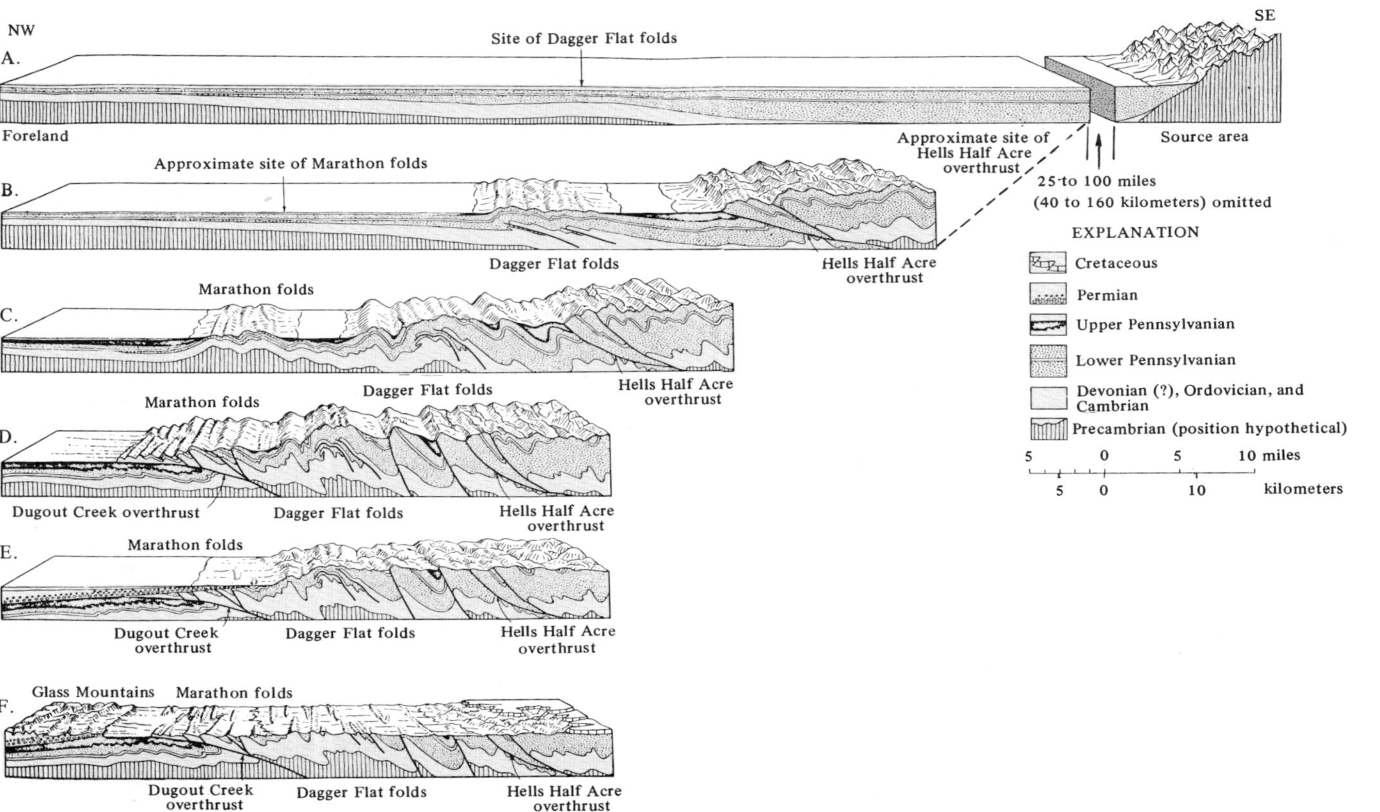

FIGURE 8.10
Hypothetical block diagrams showing the development of the Marathon Mountains in west Texas. Undeformed Permian rocks overlie deformed Upper Pennsylvanian rocks and so date the final orogenic phase. Note the inferred shortening of the rocks. From King, 1938.

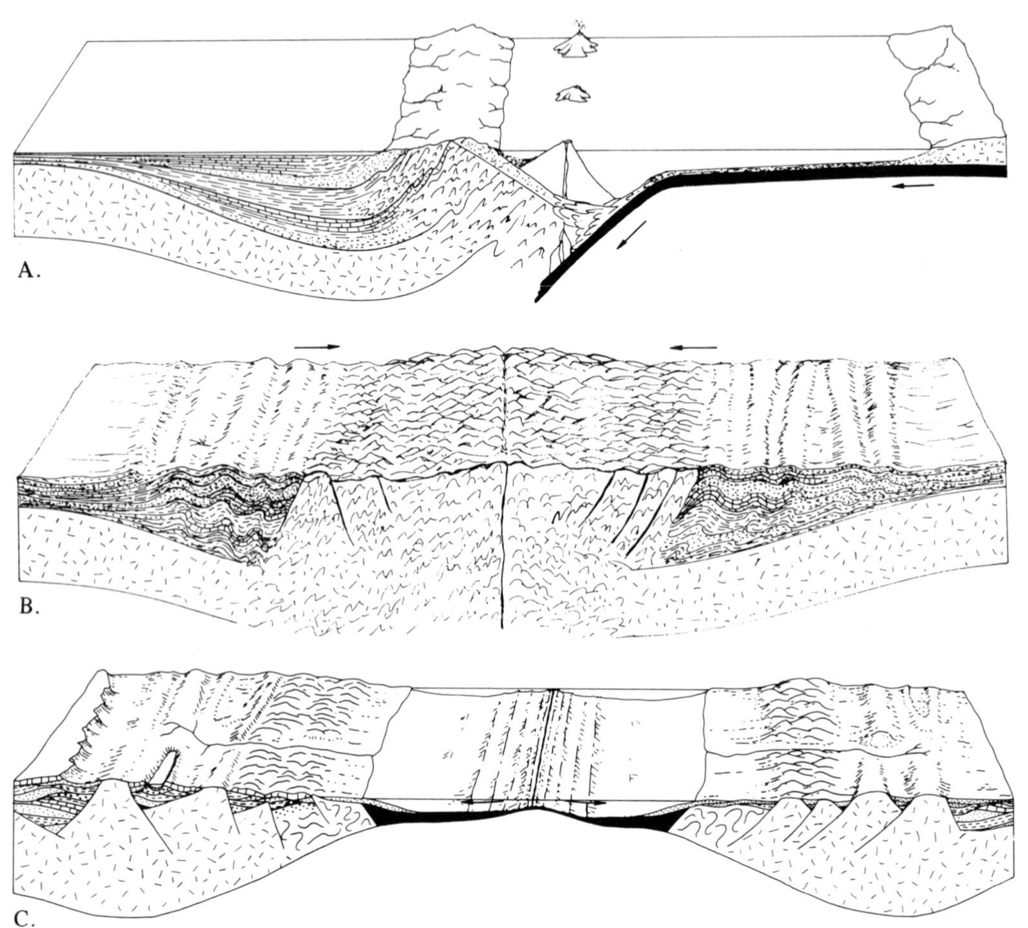

FIGURE 8.11
Formation of the Appalachian Mountains. This figure completes the history shown in Figure 7.8. A. Silurian Period. B. Collision of continents in late Paleozoic time produces the Appalachian Mountains. C. Reopening of the Atlantic Ocean in the Mesozoic.

Nevada; and in Oregon, the Triassic overlies the Permian on an angular unconformity. In central Nevada, the orogenic island chain was uplifted and deformed and was again the site of thrust faulting. (See Fig. 8.13.) East of this area, the main deposits were quartz sandstone and carbonate. In the last half of Permian time, in parts of Idaho, Montana, Wyoming, and Utah, a restricted basin formed in which peculiar phosphate-rich rocks were deposited. They are an important economic source of phosphate. (See Fig. 8.14.) Farther to the east, all of these marine rocks interfinger with clastic red beds whose source was the mountains in Colorado that had formed during the Pennsylvanian.

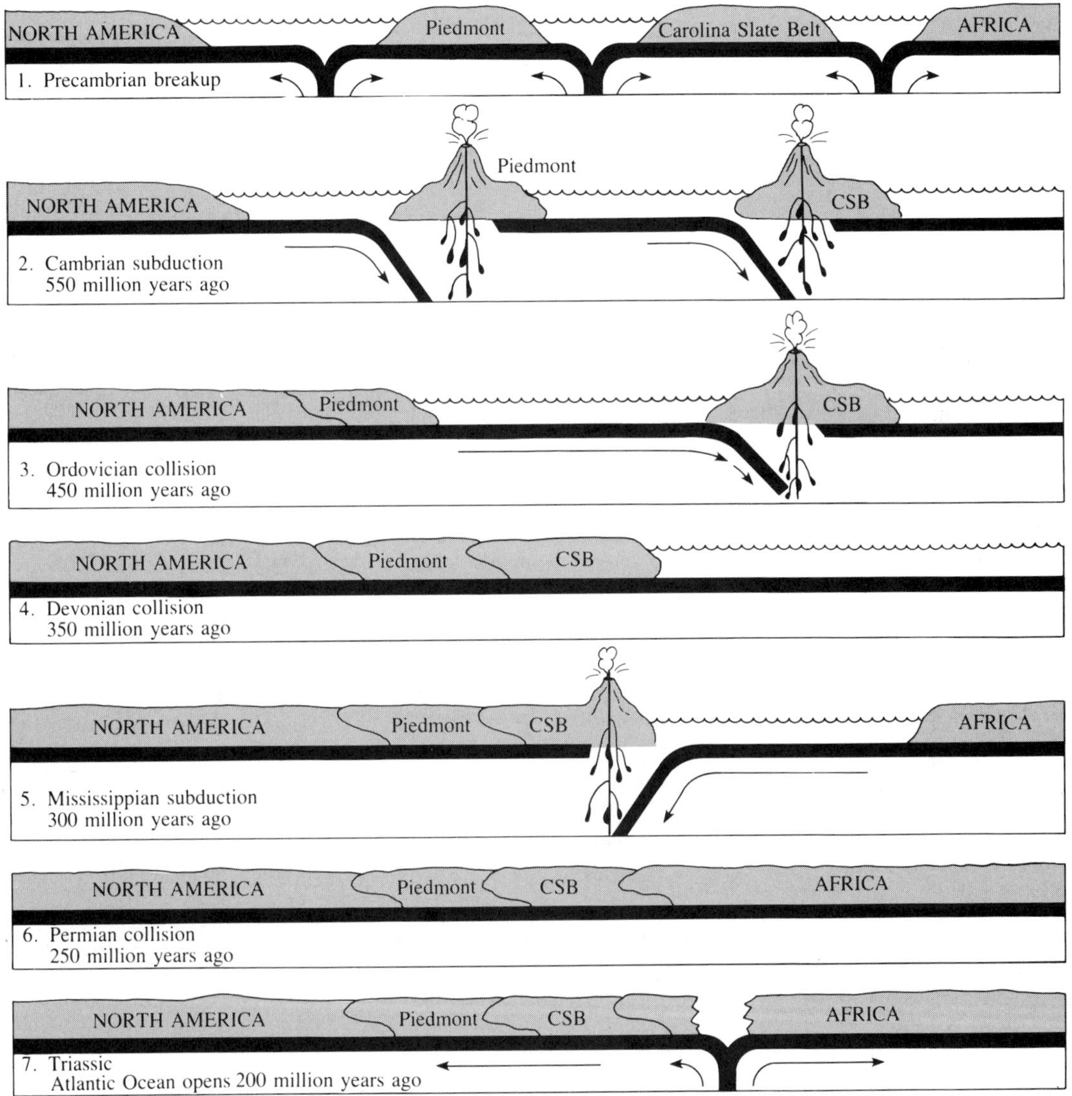

FIGURE 8.12
The development of the southern Appalachian Mountains. After F. A. Cook et al.

Parts of the Permian rocks of the western interior just mentioned have been studied very closely. At the start of the Permian Period, a shallow sea covered the area from Nebraska to western Texas. The main deposits were carbonates and shale along with clastics from the highlands of Colorado and Texas. This sea retreated to the south, so that by late Permian time, only western Texas was the site of marine deposition, and the highlands were largely eroded away by the end of the period.

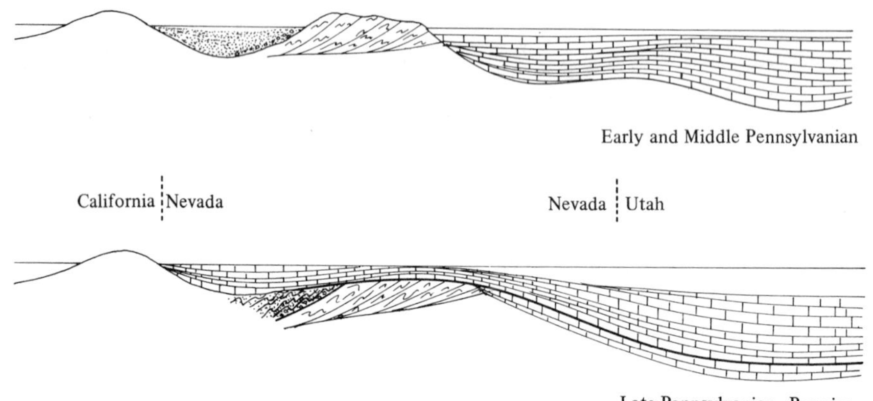

FIGURE 8.13
Pennsylvanian and Permian events in Nevada. These cross-sections continue the history shown in Figure 8.6.

As the sea retreated, marine deposition was replaced by nonmarine clastic red beds. These red beds interfinger with the marine deposits. They suggest that arid conditions prevailed there in Permian time, and similar Mesozoic rocks suggest that this climate persisted into later periods. Because of the retreating sea, almost all of the Permian System is represented by marine rocks in western Texas. It is the most complete Permian section in North America and would be studied for this reason alone; however, many environments are represented in these rocks, making their study even more valuable. This area had a number of basins separated from each other by shallower seas. Reefs developed, evaporites formed in restricted basins, and all of these rocks interfingered with nonmarine red beds. In very Late Permian time, nonmarine red beds were deposited, ending the deposition.

Near the end of the Permian, the seas retreated elsewhere as well. This retreat was not complete in the western geosyncline, and the Triassic rocks reflect a continuation of the Permian pattern.

Far North

The Devonian and Mississippian orogeny formed geosynclinal mountains to the north, composed of metamorphosed volcanic and clastic rocks. To the south of these is a basin in which most of the later rocks were deposited. (See Fig. 8.15.) To the south of the basin is a fold-mountain belt. South of the fold belt is the northern interior, on which basins and arches formed just as in the interior farther south in the United States.

In the basin just described, Pennsylvanian rocks were deposited. They are mainly limestone and chert and are found at widely scattered points. Clastic rocks are also found, and they may be more or less continuous with similar rocks in the

FIGURE 8.14
North America in the Permian Period.

FIGURE 8.15
Oblique aerial view of Melville Island in the Canadian Arctic. The rocks exposed in the center of the prominent anticline are conformable Ordovician, Silurian, and Devonian sedimentary rocks that were folded in the late Devonian. The rocks are overlain unconformably by Pennsylvanian sedimentary rocks, and all of these rocks were folded near the end of the Paleozoic. Photo from Energy, Mines and Resources. Reproduced with the permission of the Minister of Supply and Services Canada, 1990.

western geosyncline. Few Permian rocks have been found. Clastics, carbonates, and 1525 m (5000 ft) of volcanics have been reported, but their relationships are not known. In Triassic time, shale, siltstone, and sandstone were deposited. Their source was to the south.

In Greenland, the Triassic begins with marine rocks and ends with continental deposits, all of which accumulated in faulted basins.

LIFE OF THE LATE PALEOZOIC

The marine invertebrates continued to evolve rapidly, so the life of the late Paleozoic differed from that of the early Paleozoic. Brachiopods (see Fig. 8.16), corals, and bryozoans (see Figs. 8.17 and 8.18) were abundant, and snails and clams were common. Graptolites died out, and trilobites declined, eventually becoming extinct at the end of the Permian. Figure 8.19 shows a late Devonian sea bottom. Note the spines on the trilobite. Fish ruled the Devonian sea, as we will see, and spines developed on trilobites and brachiopods, probably to protect them from the fish. The first spined brachiopods appeared in the Early Silurian. Also shown in this Devonian scene is the coiled cephalopod, which developed by the coiling of a straight-shelled nautiloid. *The coiled cephalopods developed very rapidly and are the most important fossils for dating the late Paleozoic and Mesozoic rocks.* Both the external ornamentation and the shape of the margins of the partitions between the

FIGURE 8.16
Paleozoic brachiopods. A. *Athyris spiriferoides,* Devonian, New York, about 2.5 cm (1 in.). B. *Mucrospirifer mucronatus,* Devonian, about 3.1 cm (1.2 in.). Photos courtesy Ward's Natural Science Est., Inc.

FIGURE 8.17
Honeycomb coral, *Favosites,* Devonian, Arkona, Ontario. Photo from Smithsonian Institution.

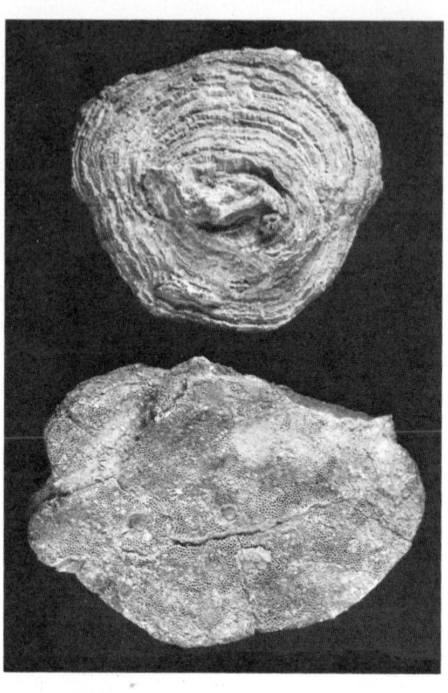

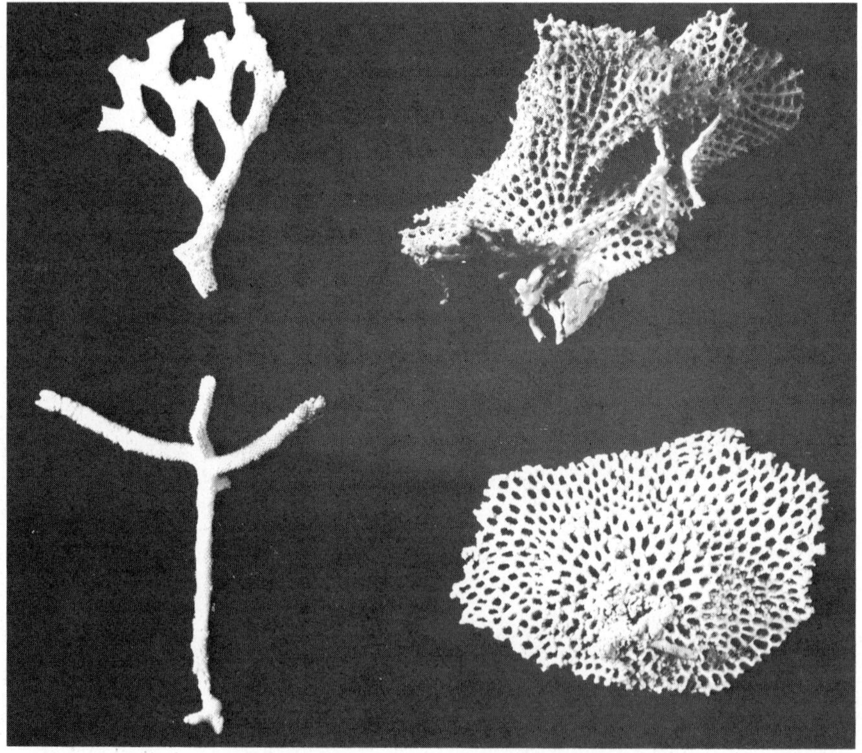

FIGURE 8.18
Permian bryozoans from Glass Mountains, Texas. Photo from Smithsonian Institution.

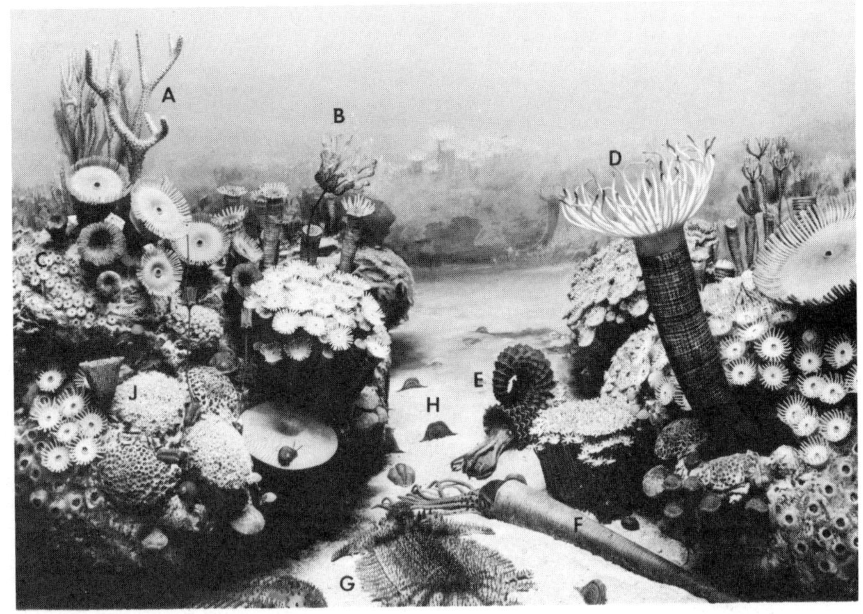

FIGURE 8.19
Late Devonian sea bottom in western New York. A. Bryozoans. B. Crinoid. C. Colonial corals. Several types are shown, and corals are the main element in the fauna. D. Solitary coral. Several types are shown. E. Coiled cephalopod. F. Straight nautiloid. G. Spiny trilobite. Spines probably developed as a defense against fish that had developed by this time. Two smaller trilobites are also in the view. H. Brachiopods. Other types are attached to the coral. J. Sponge to the left of the letter. Also shown are snails. From Field Museum of Natural History. Negative No. GEO.80821.

FIGURE 8.20
Mississippian sea bottom in central United States. Crinoids and blastoids form a local "sea lily" garden. A starfish is also shown. At other places a more balanced fauna is present. From Field Museum of Natural History. Negative No. GEO.80871.

FIGURE 8.21
Blastoid *Pentremites,* Mississippian of Illinois, 1.25 cm (0.5 in.). Members of the phylum Echinodermata are recognized by their fivefold symmetry. Blastoids are attached to the sea floor by a stalk. Photo courtesy Ward's Natural Science Est., Inc.

chambers changed rapidly. The cephalopods will be described further in the next chapter.

The Mississippian scene in Figure 8.20 shows another aspect of late Paleozoic life. Crinoids and blastoids (see Fig. 8.21) expanded greatly in the Mississippian

FIGURE 8.22
Pennsylvanian sea bottom in north-central Texas. A. Sponges, both solitary and colonial. B. Crinoid. C. Solitary coral. D. Snail. Several other types are also shown. E. Brachiopods and clams. F. Cephalopod. Several other types are also shown. From Smithsonian Institution.

FIGURE 8.23
Permian sea bottom, Glass Mountains, Texas. A. Several varieties of sponges. B. Two types of cephalopods. C. Spiny brachiopods. Many other brachiopods and clams can be seen. Corals with tentacles can also be seen. From Field Museum of Natural History. Negative No. GEO.80873.

after some other orders of echinoderms died out in the mid-Paleozoic. At most places, however, the fauna was more balanced.

Pennsylvanian and Permian sea bottoms are shown in Figures 8.22 and 8.23. The development of the cephalopods and spiny brachiopods shows clearly. Some excellently preserved examples of the Permian spiny brachiopods, on which the reconstruction in Figure 8.23 is based, are shown in Figure 8.24. Another aspect of the Pennsylvanian and Permian is the development of *fusulinids.* (See Fig. 8.25.) *They are a type of large foraminifer and were very abundant at that time. They are good index fossils in Pennsylvanian and Permian rocks.*

Near the end of the Paleozoic, many groups became extinct. No completely satisfactory explanation is available for this great dying. Trilobites and blastoids became extinct. Almost no corals or foraminifers survived, and several types each of bryozoans, brachiopods, and crinoids died out. The survivors, however, populated the Mesozoic seas. The only reasonable explanation is that many groups died when their shallow-sea habitat disappeared. An interesting point is that some organisms that died out at most places lived on somewhere. The Permian rocks of Timor contain echinoderms that became extinct elsewhere in the Mississippian.

FIGURE 8.24
Spiny brachiopods from Permian limestones in Glass Mountains, Texas. These fossils were replaced by silica and were recovered, with the spines and other delicate features intact, by dissolving the enclosing limestone with acid. Photo from Smithsonian Institution.

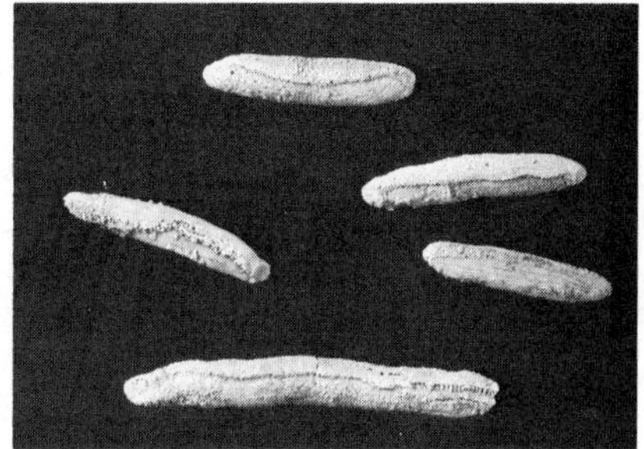

FIGURE 8.25
Fusulinid. Permian *Parafusulina,* Marathon, Texas. About 1 cm (0.4 in.) long. Fusulinids are very large foraminifers that are excellent index fossils in the Pennsylvanian and Permian. They died out near the end of the Permian. Photo from Smithsonian Institution.

Near the end of the Paleozoic, the continents came together, forming a single supercontinent. The evidence for this supercontinent and its subsequent breakup was described earlier. Most of the invertebrates that died out about this time lived in shallow seas, and the joining together of the continents would, of course, destroy much of the area of shallow seas. Thus this major subdivision of geologic eras on the basis of fossils is a natural one. The great change in life was obvious to the early workers who defined the eras. In southwestern United States, deposition continued across this important boundary; however, the seas were very restricted, and most of the rocks are nonmarine.

The Devonian is called the age of fish because fish apparently ruled the seas. As mentioned in the last chapter, the first fish are found in Late Cambrian rocks, and they first had any abundance in Late Silurian. The first fish (ostracoderms) were not at all like those of today. They were less than a foot long, jawless and finless, and were armored with bone. (See Fig. 8.26.) Without jaws, they apparently fed in the bottom mud. Without fins, they were poor swimmers. They probably could rise from the bottom by wriggling their tails like a tadpole, but without fins they could

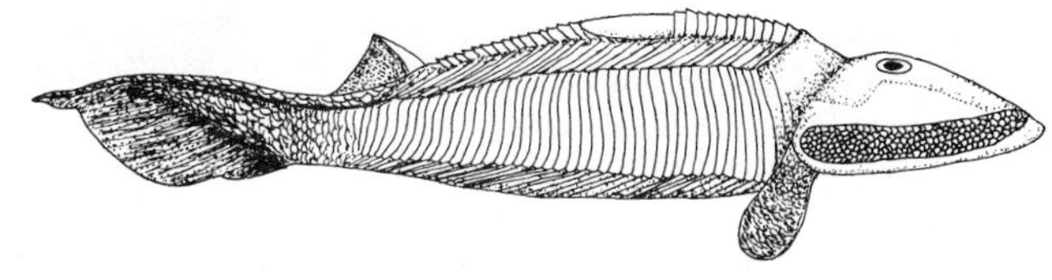

FIGURE 8.26
Early fish. The first fish were the ostracoderms, or jawless fish. The lamprey is the only living close relative. *Hemicyclaspis* of Devonian age is shown here (0.3 m [about 1 ft] long).

not change direction. All of these reasons, plus the armor, suggest that they were essentially bottom dwellers. Almost all were extinct by the end of the Devonian, although some of Pennsylvanian age are known and the living lampreys and hagfish are close relatives.

The jawless fish were replaced by the jawed fish (placoderms) in the Devonian. They, too, were armored, but they had paired fins and so were good swimmers. Some were quite large, and this, together with their jaws and swimming ability, made them fierce predators. (See Fig. 8.27.) The eurypterids that had been the predators declined in the Devonian as the jawed fish developed. The jawed fish spread rapidly because they were so well adapted to predation. Sharks, spiny fish, and bony fish, as well as other types of fish that will be discussed later, all appeared in the Silurian or Devonian. Sharks and bony fish have continued to the present. The placoderms died in the Mississippian, and the spiny fish in the Permian. The relationships are shown in Figure 8.28.

The plant life on the land also developed rapidly in the late Paleozoic. From a few Ordovician occurrences came the abundant flora of the Devonian. (See Fig. 8.29.) The Early Devonian plants were small and primitive and seem to have grown in marshes or swamps. The plants were successful and evolved rapidly. By Middle Devonian time, forests existed. (See Fig. 8.30.) A Pennsylvanian coal swamp is shown in Figure 8.31, and Figure 8.32 shows typical Pennsylvanian plant fossils. The lush Pennsylvanian coal swamps disappeared in the Permian. This was the result of the climatic change to arid or semiarid conditions that is recorded by the red-bed and evaporite deposits and the wind-deposited sandstones. Continental drift may have caused the climatic change. The southern supercontinent had the distinctive *Glossopteris* flora in late Paleozoic; in the northern supercontinent, three floras are found in the late Paleozoic.

The first insects appeared in the Early Devonian, and they became large and relatively abundant during the Pennsylvanian. (See Fig. 8.31.) As noted earlier, arthropods may have been the first animals to breathe air. They have an impervious

FIGURE 8.27
Devonian jawed, armored fish. *Dunkleosteus intermedius* from northern Ohio. This fish is about 1.2 m (4 ft) high. Other members of this group that died out in the Mississippian were up to 9.1 m (30 ft) long with jaws over 1.8 m (6 ft) high. Photo from Smithsonian Institution.

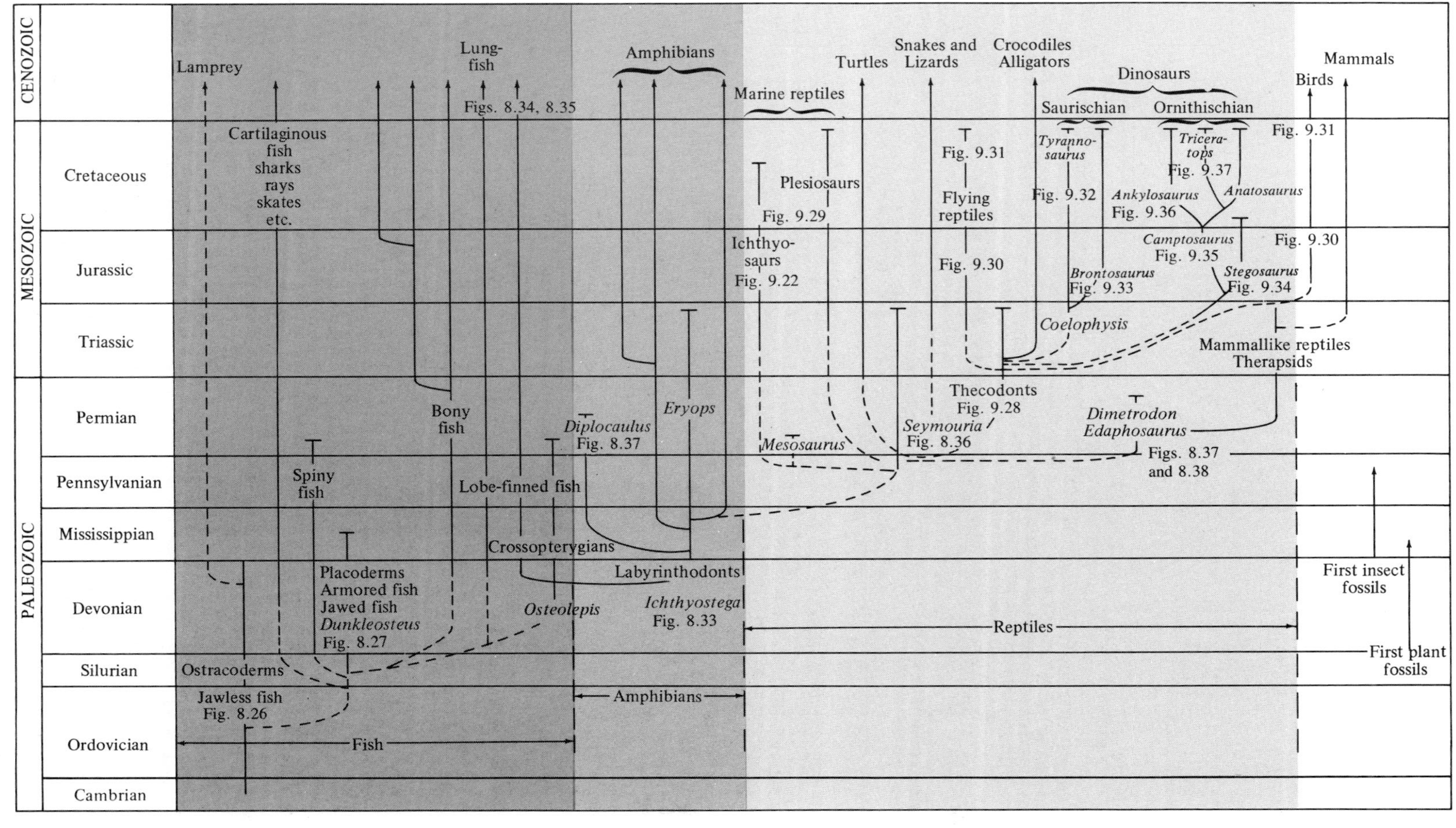

FIGURE 8.28
Evolution of the vertebrates. Although this chart appears formidable, the main theme can be seen along the bottom where progressive developments form a diagonal from primitive fish in the lower left to mammals in the upper right. Only the more important vertebrates are shown. Figures showing these animals are indicated.

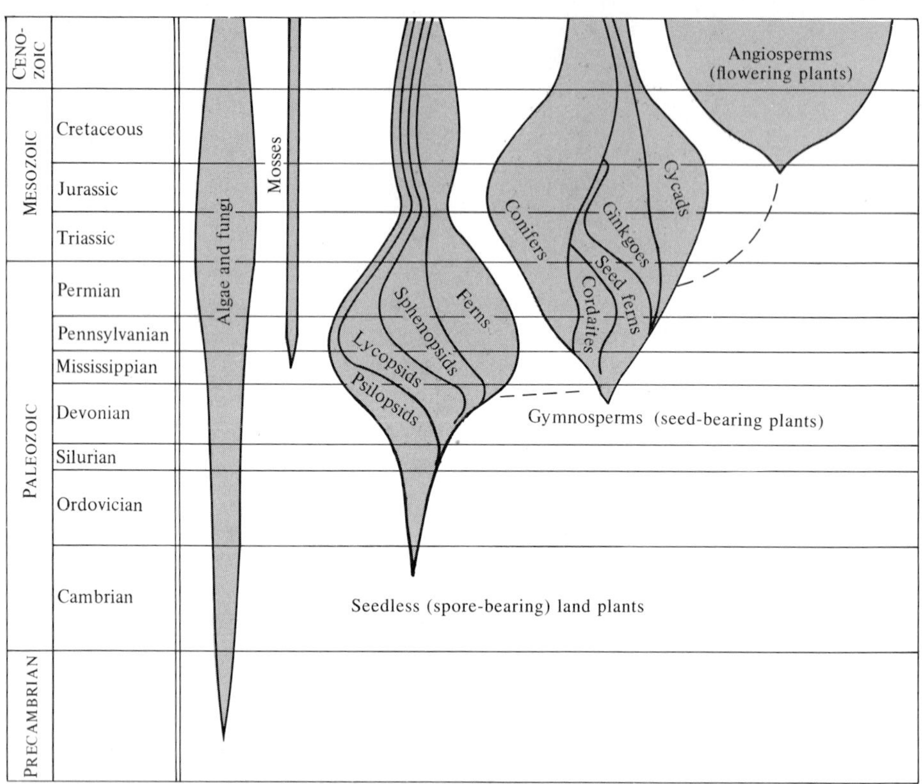

FIGURE 8.29
Development of plants.

FIGURE 8.30
Middle Devonian forest in western New York. The Early Devonian plants were generally small, primitive marsh- or swamp-dwelling plants. A. *Psilophyton* was one of the larger that carried over into the first true forests of Middle Devonian age. B. An early tree that grew up to 15.25 m (50 ft) high. C. Horsetail rushes. D. *Archaeosigillaria*. Painting by C. R. Knight, Field Museum of Natural History. Negative No. CK.73898.

FIGURE 8.31
Reconstruction of a Pennsylvanian coal swamp. A. *Lepidodendron.* B. *Sigillaria.* Note the large cockroach. C. *Calamites.* D. Large dragonfly. From Field Museum of Natural History. Negative No. 75400.

surface and so were among the first to move into fresh water and finally out of water altogether.

The first land vertebrates developed in the Devonian from the lobe-finned fish. This probably occurred in a freshwater environment because the change from ocean to fresh water solved some of the problems of living out of the water. The body fluids of most animals contain about the same amount and type of salts as seawater. Some way to enclose these fluids to prevent dilution was necessary for the animals to move into fresh water. Once that was accomplished, only a way to get oxygen from the air was necessary. It seems likely that the ability to travel out of the water was a great advantage if the pool a fish was in dried out. It could then crawl to the next pool. In the course of moving from pool to pool, it might find something tempting to eat and so begin to live more and more out of the water. In this way, perhaps, the first amphibians developed.

The oldest amphibian fossils so far found are in the upper Devonian rocks of Greenland and eastern Canada. (See Fig. 8.33.) They developed from the lobe-finned fish. (See Figs. 8.34 and 8.28.) The fins of these fish developed into clumsy legs. The lobe-finned fish also developed another adaptation that enabled them to live to the next rainy season if their pool dried up. The present-day lungfish (see Fig. 8.35),

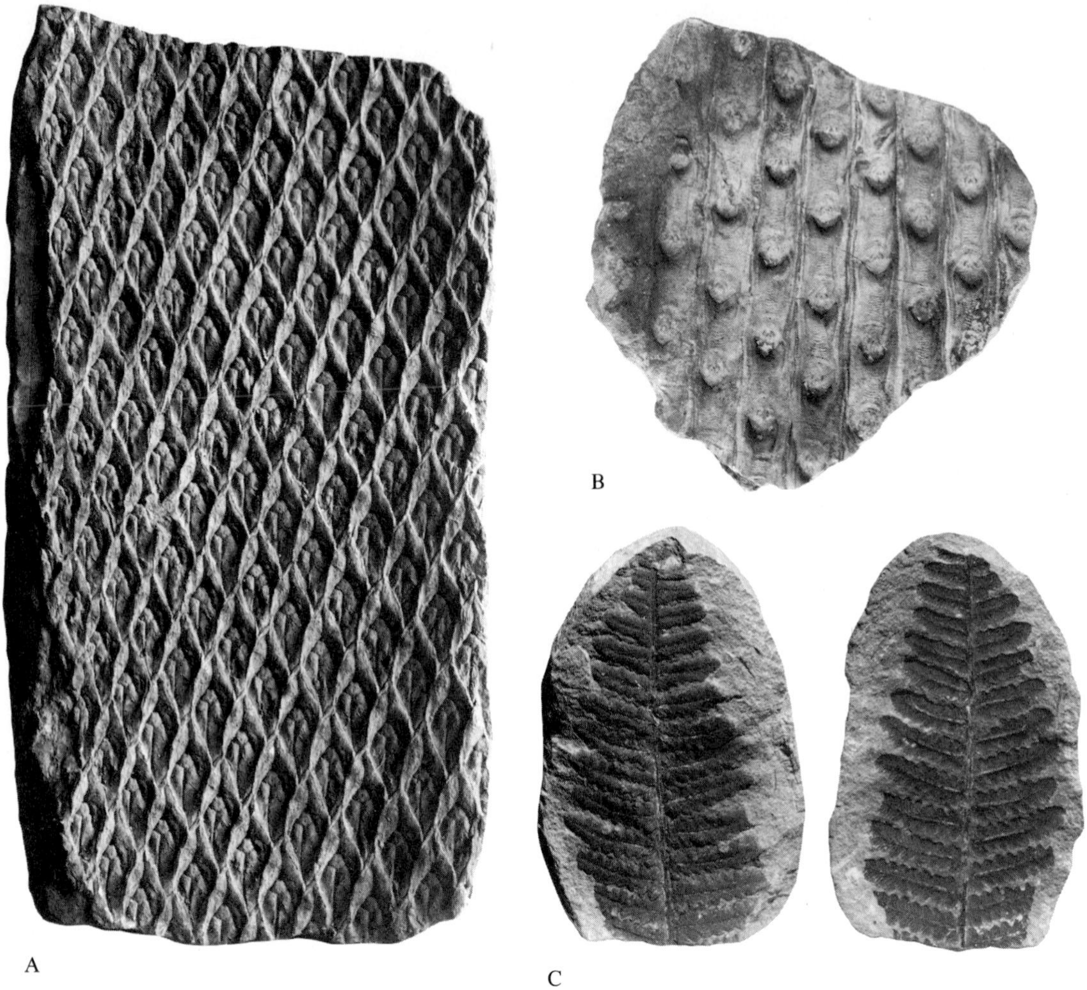

B

A

C

FIGURE 8.32
Pennsylvanian plant fossils. A. *Lepidodendron.* B. *Sigillaria.* Parts A and B are both parts of the trunk showing leaf scars. Photos from Field Museum of Natural History. C. *Asterotheca miltoni,* a fern leaf. Photo courtesy Ward's Natural Science Est., Inc.

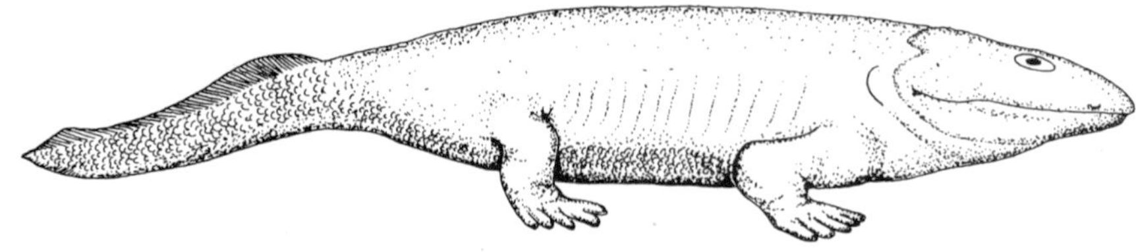

FIGURE 8.33
Ichthyostega, one of the earliest amphibians, from upper Devonian of Greenland. About 60 cm (2 ft) long.

FIGURE 8.34
Latimeria. A "living fossil" found near Madagascar. Lobe-finned fish similar to this developed into the early amphibians. About 1.5 m (5 ft) long. Photo from Field Museum of Natural History. Negative No. Z-87157.

FIGURE 8.35
Australian lungfish. The ability to breathe air shown by lungfish was probably a step in the development of amphibians from fish. Up to 1 m (40 in.) long. Photo from Field Museum of Natural History. Negative No. PCZ.1370.

FIGURE 8.36
Seymouria closely resembles the amphibians and would be classified as an amphibian if eggs were not found in the same rocks. From this stem reptile evolved all other reptiles and higher vertebrates. About 0.75 m (2.5 ft) long. Photo from Field Museum of Natural History. Negative No. GEO.80752.

FIGURE 8.37
Permian "sail lizards." The one showing teeth is *Dimetrodon,* a carnivore. The smaller-headed one is *Edaphosaurus,* a vegetarian. Both were about 2.75 m (9 ft) long. The fin may have been used to control body temperature. In the right foreground is *Diplocaulus,* a 0.6-m- (2-ft-) long, bottom-dwelling amphibian. Painting by C. R. Knight, Field Museum of Natural History. Negative No. CK.73837.

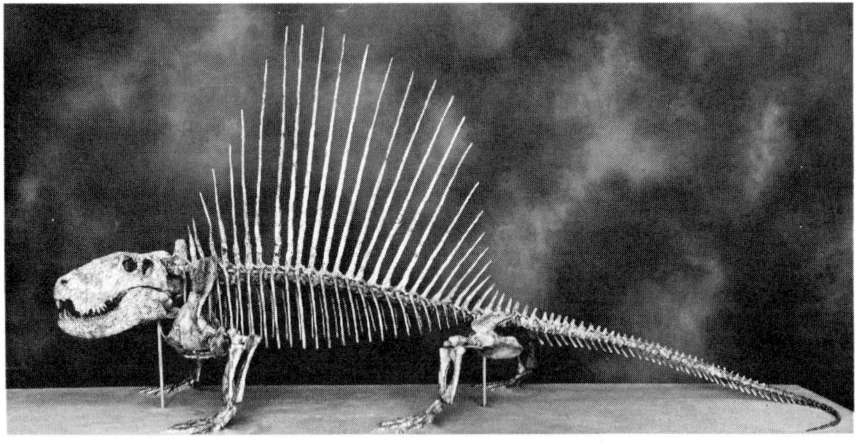

FIGURE 8.38
Skeleton of *Dimetrodon* shown in Figure 8.37. Photo from Field Museum of Natural History. Negative No. GEO.81027.

although not lobe-finned fish, can burrow in the mud and breathe air while in a hibernationlike state.

The development of amphibians was a great step toward inhabitation of the continents, but they were still tied to the water. They had to return to the water to lay their eggs. In spite of this, they expanded very rapidly. *The reptiles were the next step, and they first appeared in the Pennsylvanian.* The reptiles are characterized by the amniote egg that can hatch on land. It is difficult to distinguish between reptiles and amphibians on the basis of skeletons alone because they are so similar. *Seymouria* (see Fig. 8.36), found in Lower Permian rocks in Texas, is thought to be a reptile only because the fossil eggs are found with it. *Seymouria*'s line has now been traced to the Pennsylvanian.

Many branches of reptiles developed in the Late Pennsylvanian and Permian. (See Fig. 8.28.) The Mesozoic is the age of reptiles, and they will be discussed in the next chapter. The Permian "sail lizards" did not continue into the Mesozoic. These unusual animals are mammal-like reptiles (see Figs. 8.37 and 8.38) and had large fins on their backs. The purpose of the fins is not known. They may have been some sort of protection, or they may have been used to regulate the animal's temperature. Reptiles are so-called cold-blooded animals, and temperature regulation is a problem for them.

The land dwellers continued on from the Permian into the Mesozoic and expanded rapidly. Thus whatever caused the great dying among the ocean-dwelling invertebrates had little effect on the land dwellers.

SUMMARY

In the Devonian, Europe and North America collided. In the Pennsylvanian or Permian Period, North America collided with Europe and Africa, forming the Appalachian Mountains.

In the late Paleozoic, all of the continents came together, forming one supercontinent. This affected the climate and therefore the life of the earth.

Middle to Late Devonian

Interior: Clastics were deposited near the domes and the eastern geosyncline, carbonates in the west and north.

East: Mountain building caused by collision with Europe was the source of thick clastic deltas.

West: A volcanic island arc was the source of volcanics and clastics that spread east to interfinger with carbonates on the interior.

Far North: Carbonates, then thick nonmarine clastics, formed a highland to the north.

Latest Devonian to Late Mississippian

Interior: Carbonates were deposited to the west, shale followed by limestones and sandstones to the east.

East: Mountain building was the source of thick clastic fans in Pennsylvania, Alabama, Texas, and Oklahoma.

West: Mountain building in Nevada shed clastics toward the east.

Far North: Clastic rocks record uplift and mountain building.

Latest Mississippian to Permian

Interior: Coal swamps developed to the east. Basins and highlands formed to the west.

East: Mountain building occurred that ended the sedimentation phase of the geosyncline in the Permian.

West: Orogeny occurred in Nevada in the Pennsylvanian and Permian and in the far west in the Permian. A Permian sea extended from Texas to Nebraska.

Far North: Pennsylvanian limestone and chert were deposited in a basin between mountain ranges.

Life of the Late Paleozoic

Brachiopods, corals, and bryozoans were abundant, and snails and clams were common.

Many animals developed spines, apparently to protect them from fish.

Fusulinids are good index fossils of the Pennsylvanian and Permian.

Many invertebrates, such as trilobites and blastoids, and many bryozoans, brachiopods, and crinoids became extinct near the end of the Paleozoic.

In the Devonian, fish became abundant.

Forests developed by the Middle Devonian, and coal swamps were common in the Pennsylvanian.

The first insects appeared in the Silurian, and they became large and abundant in the Pennsylvanian.

The first land vertebrates were amphibians that developed from the lobe-finned fish in the Devonian.

The oldest reptiles are found in Pennsylvanian rocks.

QUESTIONS

1. Where is the thickest section of Devonian rocks in North America? What is the evidence as to their source area?

2. Sketch a hypothetical cross-section through a dome of the type found in the interior during Paleozoic time. Pay particular attention to the changes in thickness of the sedimentary rocks, and contrast these thicknesses with those in a dome formed by a much younger domal uplift.
3. What occurred in the far west during the Devonian?
4. What do the Devonian rocks in the far north resemble?
5. Describe the Mississippian rocks in the east.
6. What occurred in Nevada during the Mississippian?
7. How is the orogeny that produced the Appalachian Mountains dated?
8. What events were occurring in the west during this time?
9. What was the climate in western North America during Permian time? What is the evidence?
10. Where is the most complete section of Permian marine rocks in North America?
11. Describe the movement of the southern continent in the late Paleozoic.
12. Which of the present continents were in contact at the end of the Paleozoic?
13. Why do we know so little about pre-Devonian plants?
14. When did the amphibians first appear? From what did they evolve?
15. When did the reptiles first appear?
16. What are some differences between reptiles and amphibians?
17. When did the first insects appear?
18. In what ways did late Paleozoic life differ from early Paleozoic life?
19. Name some organisms that became extinct near the end of the Paleozoic.
20. Were the jawless fish good swimmers?

SUGGESTED READING

Colbert, E. H., *Evolution of the Vertebrates.* New York: John Wiley & Sons, Inc., 1969, 542 pp.

Craig, L. C., et al., *Paleotectonic Investigations of the Mississippian System in the United States.* U.S. Geological Survey Professional Paper 1010. Washington, D.C.: U.S. Government Printing Office, 1979, pp. 1–559.

Romer, A. S., "Major Steps in Vertebrate Evolution," *Science* (December 29, 1967), Vol. 158, No. 3809, pp. 1629–1637.

Rudwick, M. J. S., *The Great Devonian Controversy.* Chicago: University of Chicago Press, 1985, 494 pp.

U.S. Geological Survey, *The Mississippian and Pennsylvanian (Carboniferous) Systems in the United States.* Professional Paper 1110-A-L. Washington, D.C.: U.S. Government Printing Office, 1979, various paging.

9

The Mesozoic—Triassic, Jurassic, Cretaceous

Mesozoic means "middle life,"—at the time it was named, it was thought to be the middle part of earth history. We now know the great age of the earth and that the Mesozoic is but a part of its late history. The life of the Mesozoic differs markedly from both the Paleozoic and the Cenozoic and was an obvious subdivision of geologic time to the early workers. In North America it was a very active time, as it was throughout the world.

WORLD GEOGRAPHY IN THE MESOZOIC

The Mesozoic was a time of drifting apart of the continents. By the end of the Paleozoic, the continents had all joined to form Pangaea, setting the scene for the great breakup and drifting apart that formed the present earth. Dating of the present ocean floors reveals when the breakup occurred.

The oldest rocks found so far on the sea floors are Jurassic, suggesting that in Triassic and Jurassic time, the present oceans, such as the Atlantic, began to form. A probable sequence of drifting is shown in Figures 9.1, 9.2, and 9.3. During the Mesozoic, Africa and South America separated, as did Antarctica from Africa. India moved farthest during the Mesozoic. Geosynclines existed along the western coasts of North America, South America, and Antarctica. Another important geosyncline that was active during the Mesozoic was between Africa and Eurasia in the vicinity of the present Mediterranean Sea.

FIGURE 9.1
The continents in the Triassic Period.

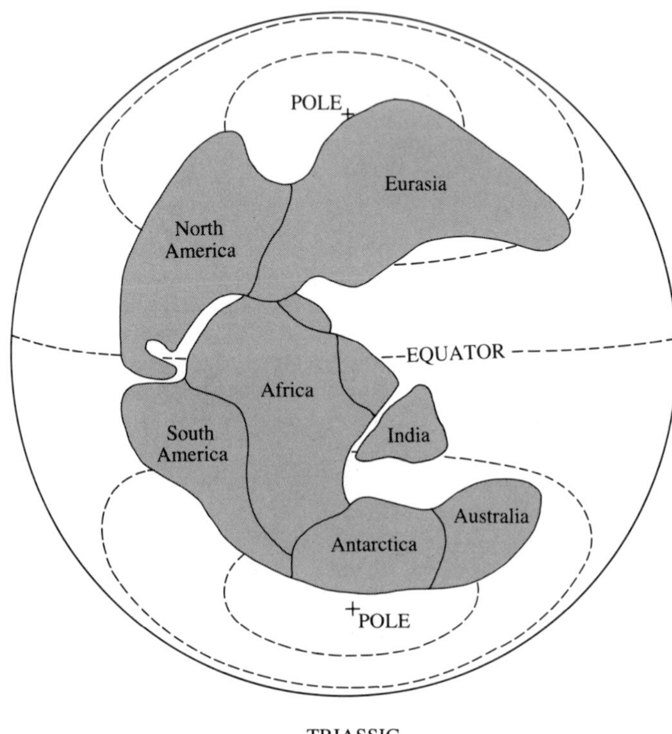

The history of the Atlantic Ocean can be deciphered from study of recovered sediments and paleomagnetic studies of the oceanic crust. North America began to separate from Europe about 200 million years ago, in the Triassic. In the Jurassic Period, a narrow sea existed, in which some evaporites were deposited. The south Atlantic began to form in the Jurassic when South America moved away from Africa. That the Atlantic was still narrow in early Cretaceous time is suggested by some apparently stagnant-water deposits. The Atlantic continued to widen to its present size during the rest of the Cretaceous and the Cenozoic. Sometime during the Cretaceous, Greenland apparently separated from Labrador.

TRIASSIC OF WESTERN NORTH AMERICA

Starting in late Permian and culminating in early Triassic, the **Sonoma orogeny** occurred near the western margin of the continent. This event was similar to the Devonian Antler orogeny—oceanic sedimentary rocks were thrust eastward by collision with a volcanic arc or microcontinent.

Near the beginning of the Triassic, the seas had retreated from the interior and from a large part of the geosyncline as well. (See Fig. 9.4.) In part of the geosyncline in Idaho, Lower Triassic marine rocks overlie the Permian conformably. The

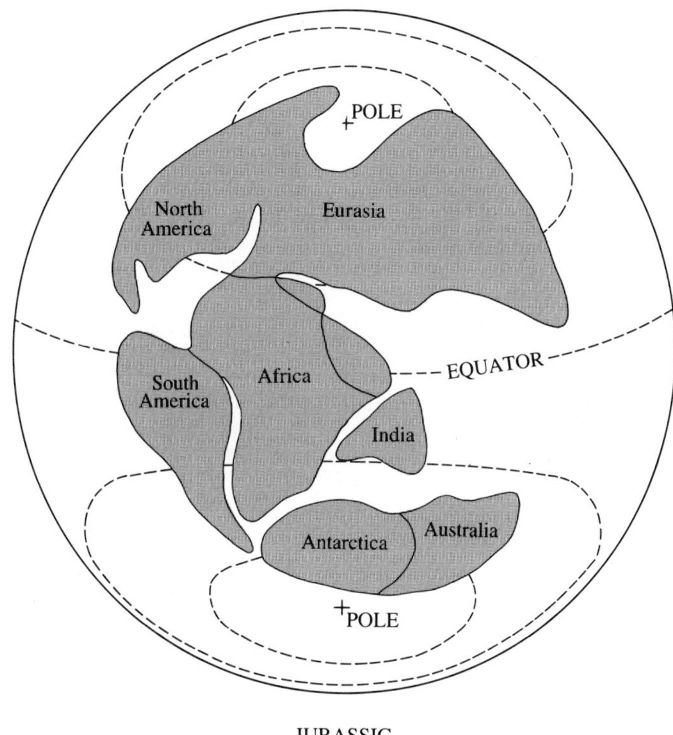

FIGURE 9.2
The continents in the Jurassic Period.

geosyncline had changed somewhat from its configuration in the Permian. The island chain in central Nevada that had been the site of Paleozoic orogeny was not present. In the far west, great thicknesses of Triassic marine clastic and volcanic rocks were deposited. At some places, these rocks lie unconformably on Permian rocks and so record the Sonoma orogeny. Farther east, in the less active part of the geosyncline, marine rocks were deposited. In western Nevada, both clastic and carbonate sections ten thousand meters thick are found. At the edge of the interior, the marine rocks interfinger with nonmarine red clastics.

Lower Triassic marine rocks are found only in the geosyncline but are known from southern Nevada to the Arctic, although in the United States, outcrops are limited. In southern Nevada, they are clastic rocks about 2440 m (8000 ft) thick, and in Idaho they are limestone and shale 1525 m (5000 ft) thick. These rocks interfinger with red sandstone and siltstone that in the northern Rocky Mountains are rarely over 305 m (1000 ft) thick. The Triassic sea in that area advanced and retreated a number of times and apparently had an irregular shoreline with large embayments. Highlands in the interior were the source of some clastics. The nonmarine red beds overlie similar Permian red beds, making it difficult to place the boundary between the two systems because of the paucity of fossils.

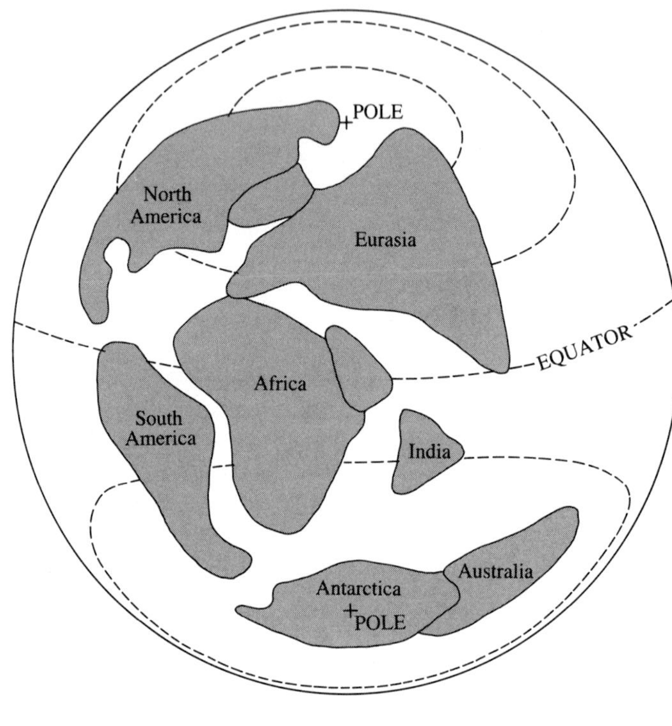

FIGURE 9.3
The continents in the Cretaceous Period.

In middle Triassic time, the seas retreated farther west so that only the active part of the geosyncline was marine except in Canada, where the whole of the geosyncline accumulated marine rocks. Evaporites formed in a restricted basin in Montana-Dakota. With retreat of the sea, some erosion apparently occurred in the areas of earlier deposition as well as elsewhere.

The area of maximum deposition of nonmarine rocks was in Arizona and New Mexico; here, Lower Triassic rocks are unconformably covered by Upper Triassic rocks with conglomerate at the base. This suggests uplift to the north. In this area, some of the Upper Triassic and Lower Jurassic sandstones appear to be wind deposited. These rocks are well displayed at Zion National Park. Because of the lack of fossils in these rocks, the boundary between Triassic and Jurassic is difficult to place. These wind-deposited rocks and the many red beds suggest that arid conditions prevailed over much of the United States.

In Middle and Late Triassic time, a volcanic arc was active in California and batholiths were emplaced in the Sierra Nevada. This activity was the forerunner of later intrusions and metamorphism.

The Jurassic began as a continuation of the Triassic. In the far west, thick volcanic and clastic rocks were deposited, especially in a deep trough in the geosyncline in western Nevada. In Oregon, Jurassic rocks lie with an angular unconformity on Upper Triassic rocks, indicating orogeny in that area. The pattern,

FIGURE 9.4
North America in the Triassic Period. Except in the far west, the sediments are nonmarine.

however, was soon to change, and the trough in western Nevada was the site of thrust faulting during deposition in Early Jurassic time. (See Fig. 9.5.) To the east of here, an uplifted area developed as the seas withdrew.

JURASSIC THROUGH CRETACEOUS

West

When the seas returned in Middle Jurassic time, the shape of the continent had changed. Orogenic events had formed an upland area in the eastern part of the geosyncline. (See Fig. 9.6.) This upland underwent erosion during most of the Jurassic and Cretaceous and shed clastics into seaways on both sides. The somewhat simpler history of the area east of the upland will be described first.

Seas in the area east of the upland covered the region of the present Rocky Mountains and Great Plains. This sea finally withdrew almost completely from the continent near the end of the Cretaceous Period. This was the last time that seas occupied the interior of North America. The Middle Jurassic seas in the northern Rocky Mountain area came from the north, and, at the margin of the Canadian Rockies, Early as well as Middle Jurassic sedimentary rocks were deposited. The shallow sea at first came as far south as southern Utah and covered most of Montana and North Dakota. Sand, silt, and clay, with some red beds and evaporites, were laid down; the greatest thicknesses are found in Wyoming. At the maximum, the seas extended to northern Arizona and New Mexico, and possibly to the Gulf of Mexico. In Late Jurassic time, the sea retreated to the north. As the sea withdrew, lake- and river-deposited rocks accumulated. These rocks are only about a hundred meters thick but are of great interest because they contain dinosaur fossils and some rich uranium deposits.

The seas returned to the area of the Rocky Mountains in Early Cretaceous time. This was the last great inundation of the sea on the North American continent. (See Fig. 9.7.) The seas extended from the Arctic to the Gulf of Mexico, and even along the Atlantic coast, sediments were deposited. In the Rocky Mountains and Great Plains areas, great thicknesses of clastic sediments were deposited. Up to 6100 m (20,000 ft) of clastics are found at some places. The source was the orogenic area that extended from western Utah and eastern Nevada to northern Canada. The geosyncline sank as these sediments were rapidly deposited, and at times the sediments accumulated above sea level. The source area was repeatedly uplifted, causing floods of conglomerate and sandstone to be deposited near the western margin of the basin. These rocks gave way to shale farther east and, near the eastern margin, to thin limestone layers. (See Fig. 9.8.) In Late Cretaceous time, the seas withdrew and nonmarine clastic rocks were deposited. These rocks grade into the Cenozoic deposits that will be described in the next chapter. The southern end of a small, short-lived embayment of the sea in North Dakota lingered into the early Cenozoic. The withdrawal of the sea marked the end of the Cretaceous and the last great inundation of the continent.

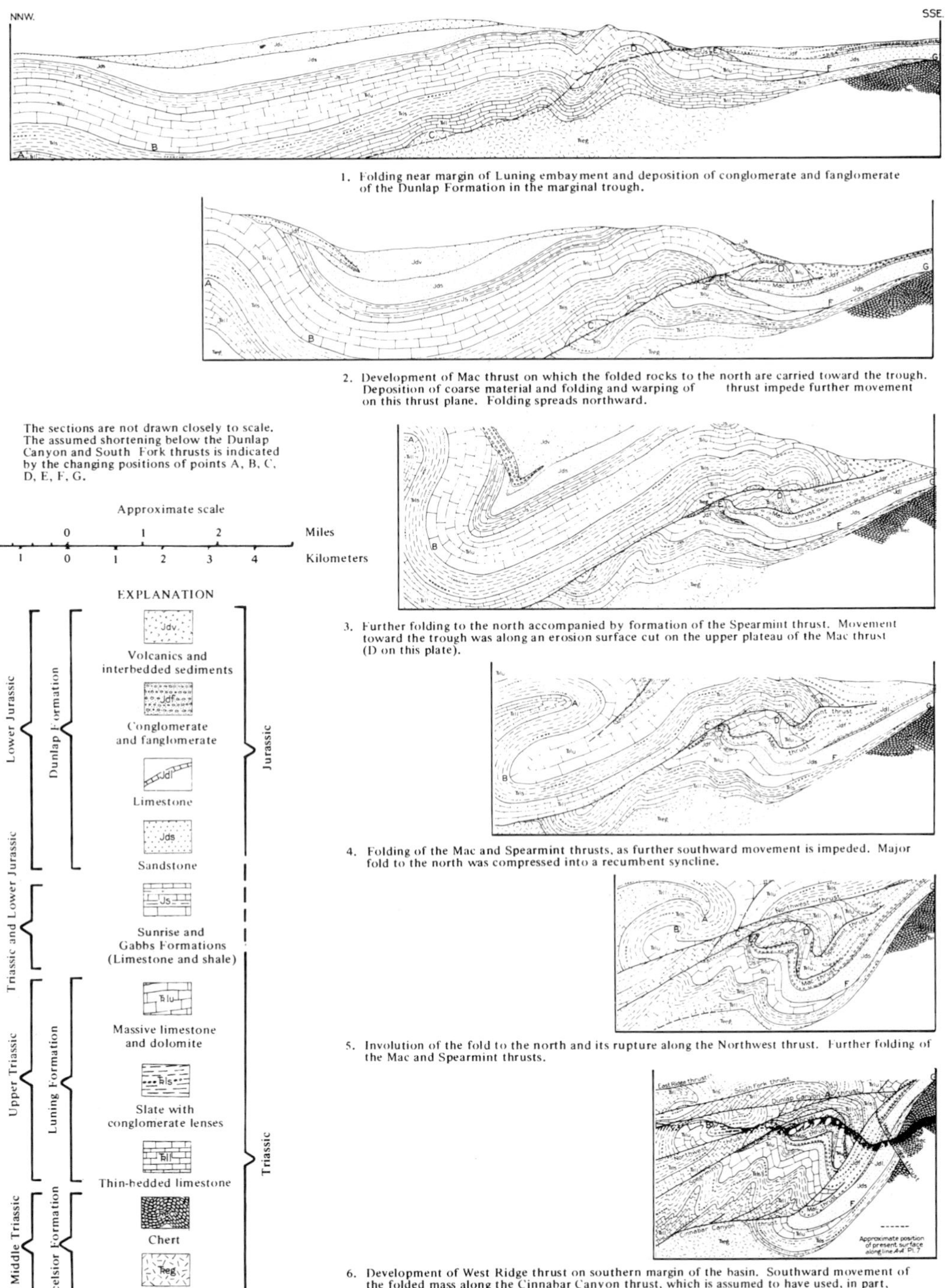

FIGURE 9.5
Diagrammatic sections showing progressive development in Early Jurassic time of complex structure in the northwestern part of the Pilot Mountains in southwestern Nevada. From Ferguson and Muller, 1949.

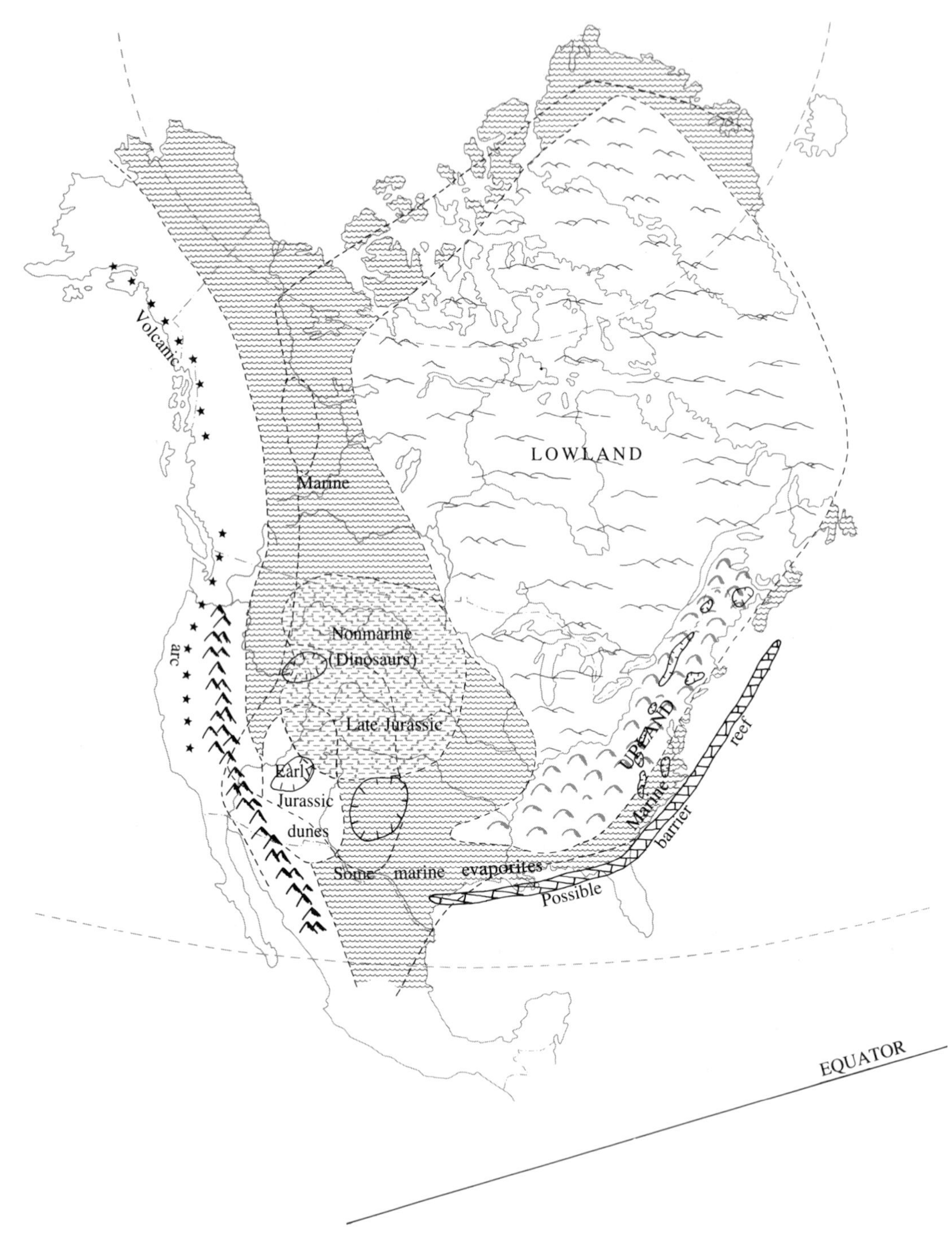

FIGURE 9.6
North America in the Jurassic Period.

FIGURE 9.7
North America in the Cretaceous Period.

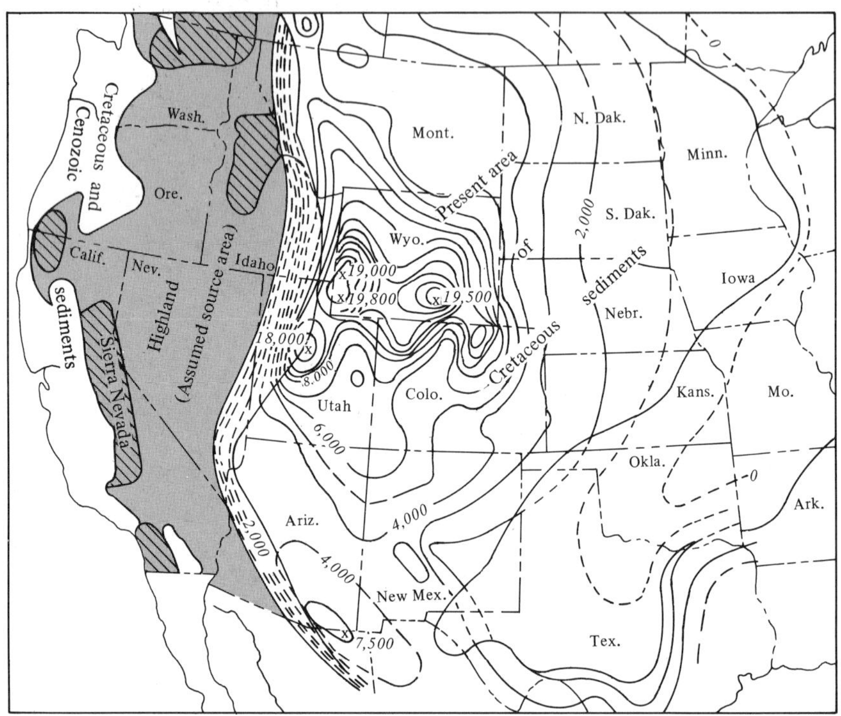

FIGURE 9.8
The distribution and thickness of Cretaceous sedimentary rocks in part of western United States. The lines show the thickness in feet and are dashed where the rocks have been removed by erosion. The rocks are thicker and coarser near the source area. The great volume requires that about 8 km (5 mi) were eroded from the source area, and this does not include the Cretaceous rocks to the west of the highland. The batholiths in the source area are indicated by the diagonal lines. From Bateman and Wahrhaftig, 1966.

Orogeny occurred in the Rocky Mountain area from Alberta to southern California in the Late Cretaceous. This orogeny is called the **Sevier** phase of the Cordilleran orogeny. The Cordilleran orogeny affected the area from the Rocky Mountains to the west coast and is further described in this and the next chapter.

On the western side of the upland, and probably on the upland itself, much activity took place. Jurassic rocks are found at many scattered localities in Oregon, Nevada, northern Washington, and especially California. These rocks are clastics and volcanics for the most part, and some have been metamorphosed. They probably accumulated at a volcanic arc. At some places, such as north-central California, the California-Oregon border, central Oregon, and southern British Columbia, the thicknesses measure ten thousand meters or more.

In northern California, between the coast and volcanic arc (the present Great Valley), at least 4880 m (16,000 ft) of sandstone and shale were deposited in Late

Jurassic time, and these rocks are covered by up to 7625 m (25,000 ft) of similar Cretaceous sediments. Farther west, at the coast, is another sequence, the Franciscan Formation, up to 15,250 m (50,000 ft) thick, that may be of the same age.

In California, some of the events are well dated, but the details remain obscure. The Sierra Nevada is a vast complex of batholiths. The individual plutons have been closely studied, and hundreds of rocks have been dated by radiometric methods. These plutons range in age from Middle Triassic to Late Cretaceous. Late Jurassic rocks are folded and intruded in the Sierra Nevada. In that area, deformation and intrusion are common from late Jurassic to Middle Cretaceous; this is termed the **Nevadan** phase, or first phase, of the **Cordilleran orogeny.** As noted earlier, in central California almost on the flanks of the Sierra Nevada, sedimentation was continuous from Late Jurassic through the Cretaceous. (See Fig. 9.9.) These rocks are largely sandstone, and their source was probably exposed

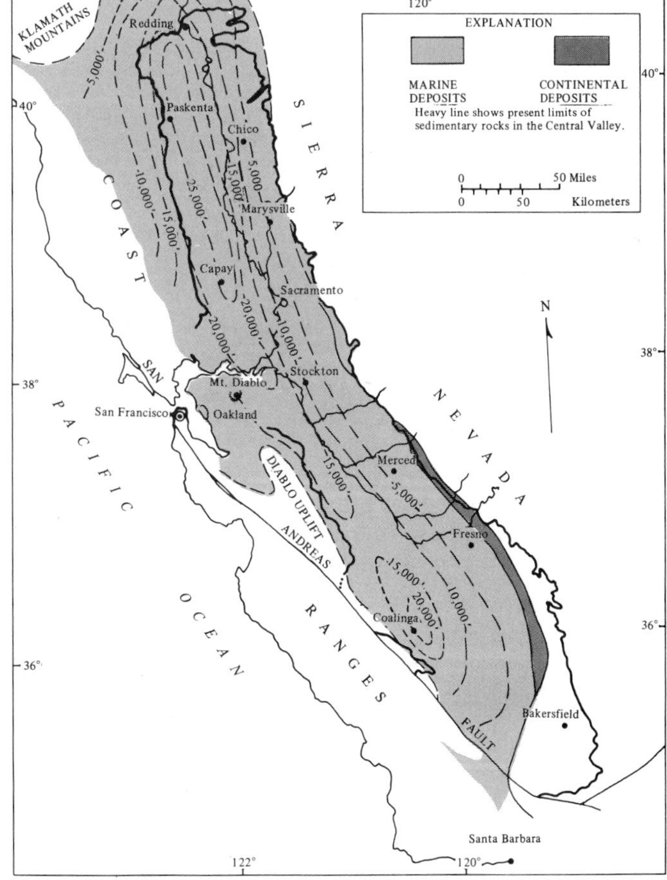

FIGURE 9.9
Distribution and thickness of Cretaceous sedimentary rocks in the Great Valley of California. The lines indicate the thickness that is more than 7620 m (25,000 ft) in places. From Hackel, 1966.

batholiths. Thus it appears that some batholiths were being emplaced as nearby batholiths underwent erosion. (See Fig. 9.10.)

The Mesozoic was a time of batholith emplacement in western North America. Similarly dated batholiths extend from Baja California to Alaska. The Coast Ranges of British Columbia are largely composed of granitic rocks. (See Fig. 9.11.)

Along the Pacific coast, a number of marine embayments existed, especially during the Late Cretaceous. The pattern was probably much like the Cenozoic history of that area. Cenozoic rocks and deformation obscure much of the Cretaceous story. The final phase of the Cordilleran orogeny occurred in the Cenozoic.

During the late Mesozoic, much of western North America was accreted in a series of collision events with island arcs and microcontinents. (See Fig. 9.12.) It is difficult to date these events, which began in the late Paleozoic and may have lasted into the Cenozoic. Each of these events probably caused orogeny, at least locally, affecting rocks that were originally deposited on the North American continent. As

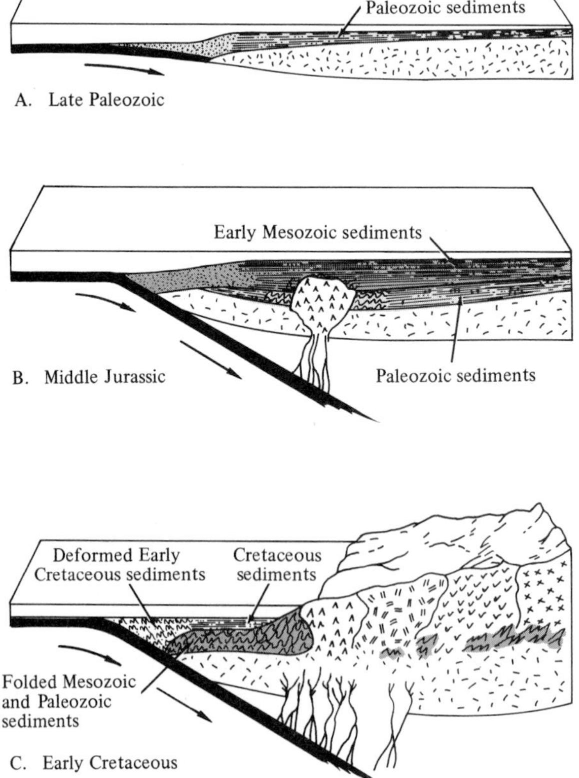

FIGURE 9.10
Hypothetical block diagrams showing the development of the Sierra Nevada batholith and the Mesozoic sedimentary rocks in California.

FIGURE 9.11
Mesozoic batholiths of North America.

this part of the history is studied in more detail, revisions to the above history will have to be made.

Gulf and Atlantic Coasts

The north Atlantic Ocean opened in late Triassic and early Jurassic time. The tension associated with this rifting apparently caused the development of basins along the Atlantic coast. These basins were filled with red clastic sediments interbedded with basalt. These Late Triassic to Early Jurassic rocks accumulated to thicknesses up to 6000 m (20,000 ft) in the area just east of the present Appalachian Mountains. (See Fig. 9.13.)

Deposition in the coastal plain areas of the Gulf of Mexico and the Atlantic coast began near the Gulf in Jurassic time. The Jurassic rocks do not crop out, but they are known from deep drilling for oil. The rocks encountered include evaporites, carbonates, and shale. The salt domes of the Gulf Coast are believed to have come from Jurassic salt deposits.

The Cretaceous deposits are much more extensive, and because of their overlap, the Jurassic rocks are not exposed. In the Gulf area, the Cretaceous sediments are very thick and consist mainly of shale and sandstone. Along the Atlantic coast, thin shale and sandstone beds were the main deposits. The deposition on the Gulf and Atlantic coasts continued into the Cenozoic, and thus the discussion will be continued in the next chapter.

Far North

In the Arctic islands, the Jurassic is represented by a hundred to a few hundred meters of marine shale overlain by nonmarine shale and sandstone. This is overlain by a shale unit of Late Jurassic–Early Cretaceous age that is up to 765 m (2500 ft) thick. In Cretaceous time, a clastic unit up to 1375 m (4500 ft) thick was deposited.

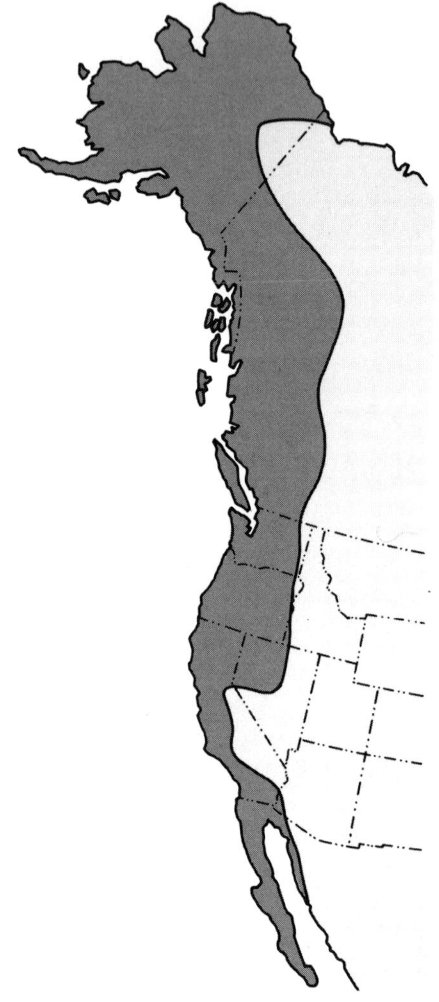

FIGURE 9.12
This part of the west coast of North America was accreted during the Paleozoic and Mesozoic. This area is made of many minicontinents and volcanic arcs.

(See Fig. 9.14.) It is largely nonmarine and contains some coal. Volcanic activity also occurred in the Cretaceous.

In eastern Greenland, Late Jurassic orogeny is recorded by thick clastic rocks of Early Cretaceous age. These are overlain by Cretaceous carbonates and red beds, suggesting a fluctuating shoreline.

LIFE OF THE MESOZOIC

Perhaps the joining together of the continents into one huge continent near the end of the Paleozoic can explain the differences between the life of the late Paleozoic and the early Mesozoic. A single large continent would probably have greater

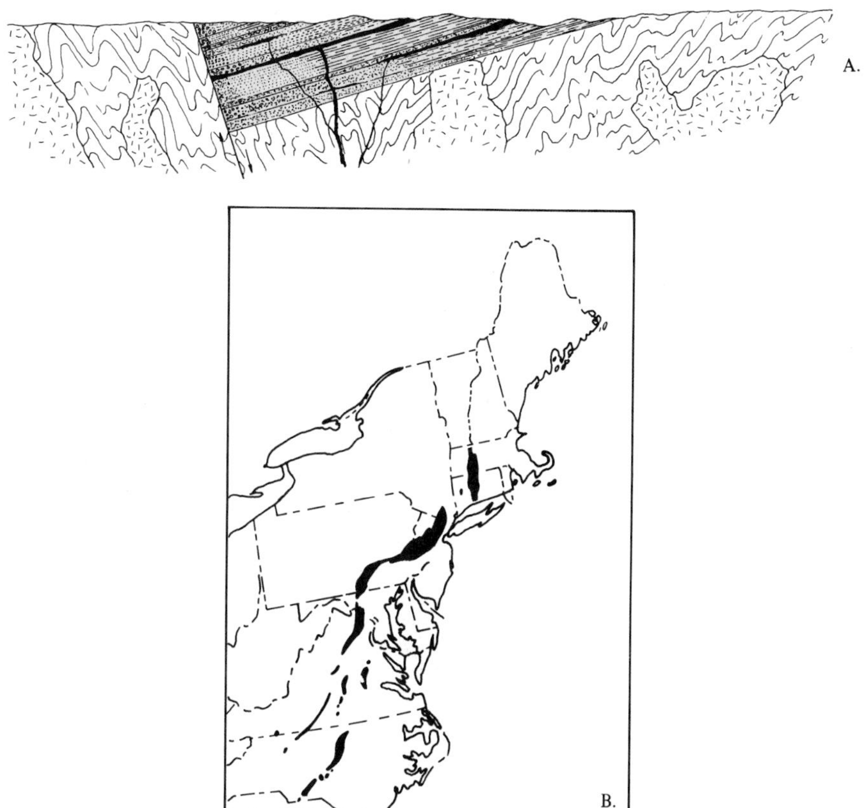

FIGURE 9.13
Late Triassic and early Jurassic sediments formed in faulted basins in the metamorphic rocks east of the present Appalachian Mountains. The rocks are coarse clastic red beds, which grade into finer clastics away from the faults, and basaltic volcanic rocks. Dinosaur footprints are found in these sedimentary rocks at some places. The Basin and Range fault structures described in Chapter 10 are somewhat similar. A. Typical cross-section. B. Map, from U.S. Geological Survey.

climatic extremes, which would affect plants and terrestrial animals. The total shoreline would be less, creating more competition among the marine animals. A single continent should have fewer, more-widespread species than a number of isolated continents. All of these effects can be seen in the life of the Mesozoic. At any rate, after the Permian extinctions, the Mesozoic was a time of rapid diversification comparable to that of late Precambrian to early Paleozoic.

The marine invertebrates of the Mesozoic are very different from those of the late Paleozoic. The extinctions, both in the Permian and at the end of the Paleozoic, were probably caused by the withdrawal of the seas. When the seas returned in the Triassic, the main invertebrates were cephalopods and clams. *Interestingly, no corals,*

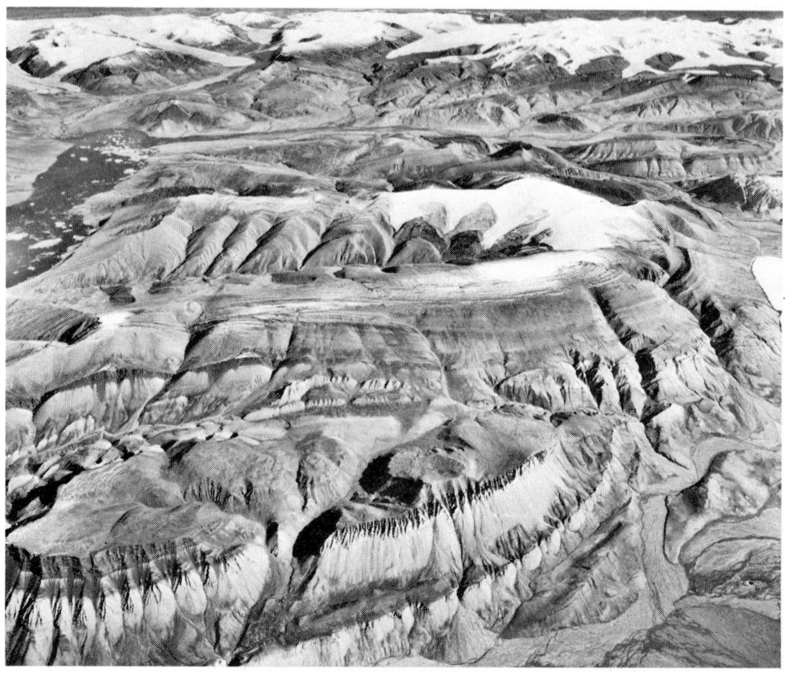

FIGURE 9.14
Oblique aerial view of Axel Heiberg Island in the Canadian Arctic. The folded rocks in the foreground are mainly Cretaceous, and the snow-covered mountains in the background are composed of Triassic sedimentary rocks. The folding occurred in the mid-Cenozoic. Photo from Energy, Mines and Resources. Reproduced with the permission of the Minister of Supply and Services Canada, 1990.

bryozoans, or foraminifers have ever been found in lower Triassic rocks, but they are present in Middle Triassic rocks, so they did live somewhere. The Triassic corals and foraminifers are very different from those of the Paleozoic. (See Fig. 9.15.)

In the late Paleozoic and the Mesozoic, the cephalopods are almost ideal index fossils. They were abundant, are found in many environments, and evolved rapidly. The stratigraphy of this interval is based on these animals. They were similar to the present-day nautiloids. (See Fig. 9.16.) They were swimming animals with coiled, chambered shells. They evolved rapidly, with differences between species being the ornamentation of the shell and the shape of the partitions between the chambers. (See Figs. 9.17 and 9.18.) Because they were swimmers, the fossils are found in all environments. It seems likely that they floated after death as well. Some were very large. (See Fig. 9.19.)

Figure 9.20 shows the stratigraphic range of each of the cephalopod suture types. Notice that they nearly became extinct twice before finally dying out. At the end of the Permian, only one group survived; near the end of the Triassic, the same

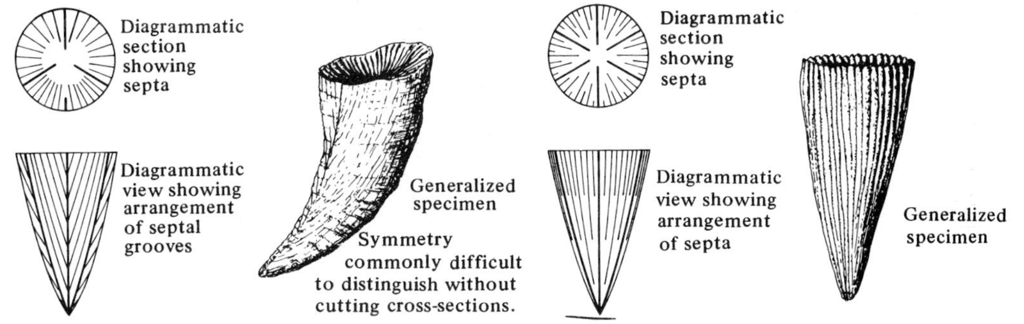

FIGURE 9.15
A. Paleozoic solitary or cup coral, showing fourfold symmetry. B. Mesozoic or Cenozoic solitary coral, showing sixfold symmetry.

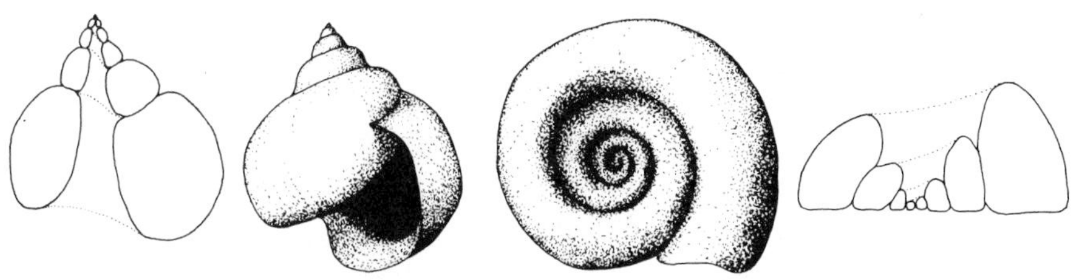

A. Shells and diagrammatic sections through coils of two generalized snails.

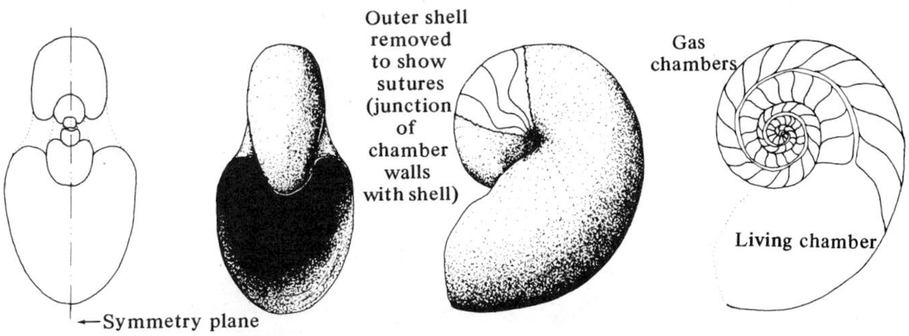

B. Shell and diagrammatic sections of generalized cephalopod.

FIGURE 9.16
Differences between snail and cephalopod. A. Snail has no chambers and no symmetry. B. Cephalopod has chambers, some of complex pattern (see Figure 9.18), and is coiled in a plane, giving a symmetry plane. Some cephalopods are not coiled. The differences shown here are obvious morphology, and there are other, more fundamental, differences.

A B C

FIGURE 9.17

Cephalopods. A. *Baculites compressus.* A 75-mm (3-in.) fragment of a straight-shelled ammonite. B. Two views of *Ophiceras commune,* a Lower Triassic ceratite from Axel Heiberg Island, Arctic Canada. About natural size. C. Two views of *Scaphites depressus,* Upper Cretaceous with ammonite suture from Alberta. About natural size. Photo A courtesy Ward's Natural Science Est., Inc.; B and C from Energy, Mines and Resources (109804 and 109804-A, 113737-K and -H). Reproduced with the permission of the Minister of Supply and Services Canada, 1990.

FIGURE 9.18
Cephalopod sutures. Arrows point toward the aperture. From *Treatise on Invertebrate Paleontology,* Courtesy of the Geological Society of America and the University of Kansas Press.

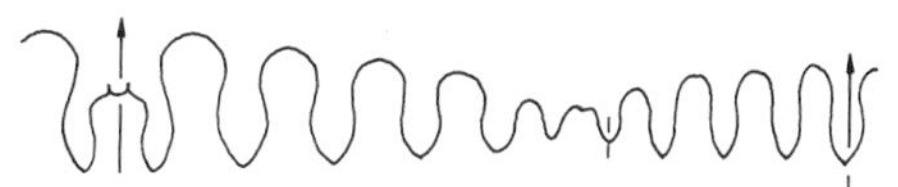

A. Goniatitic

B. Ceratitic

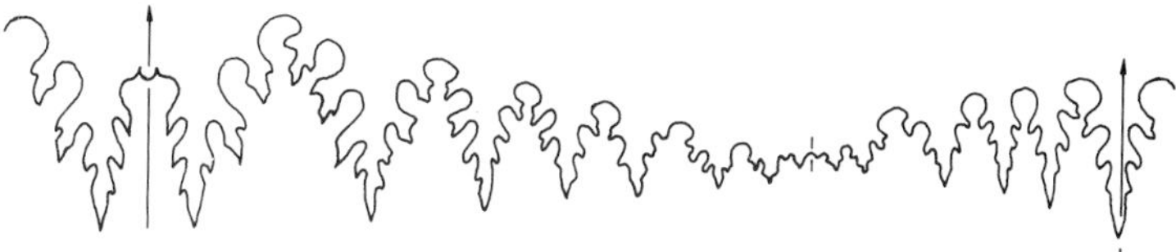

C. Ammonitic

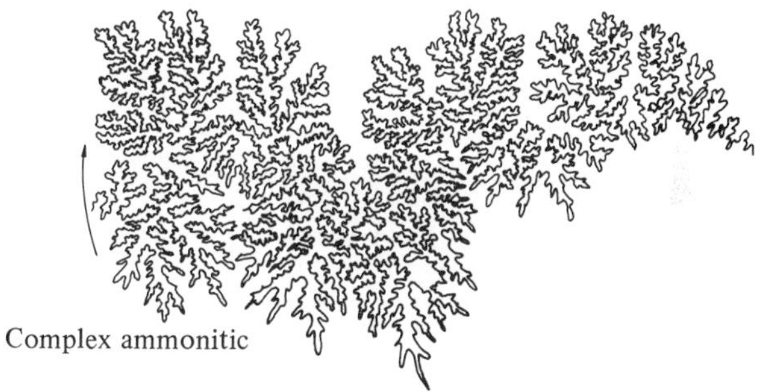

D. Complex ammonitic

FIGURE 9.19
A very large Jurassic ammonite from British Columbia, *Titanites occidentalis,* courtesy Energy, Mines and Resources (203171). Reproduced with the permission of the Minister of Supply and Services Canada, 1990.

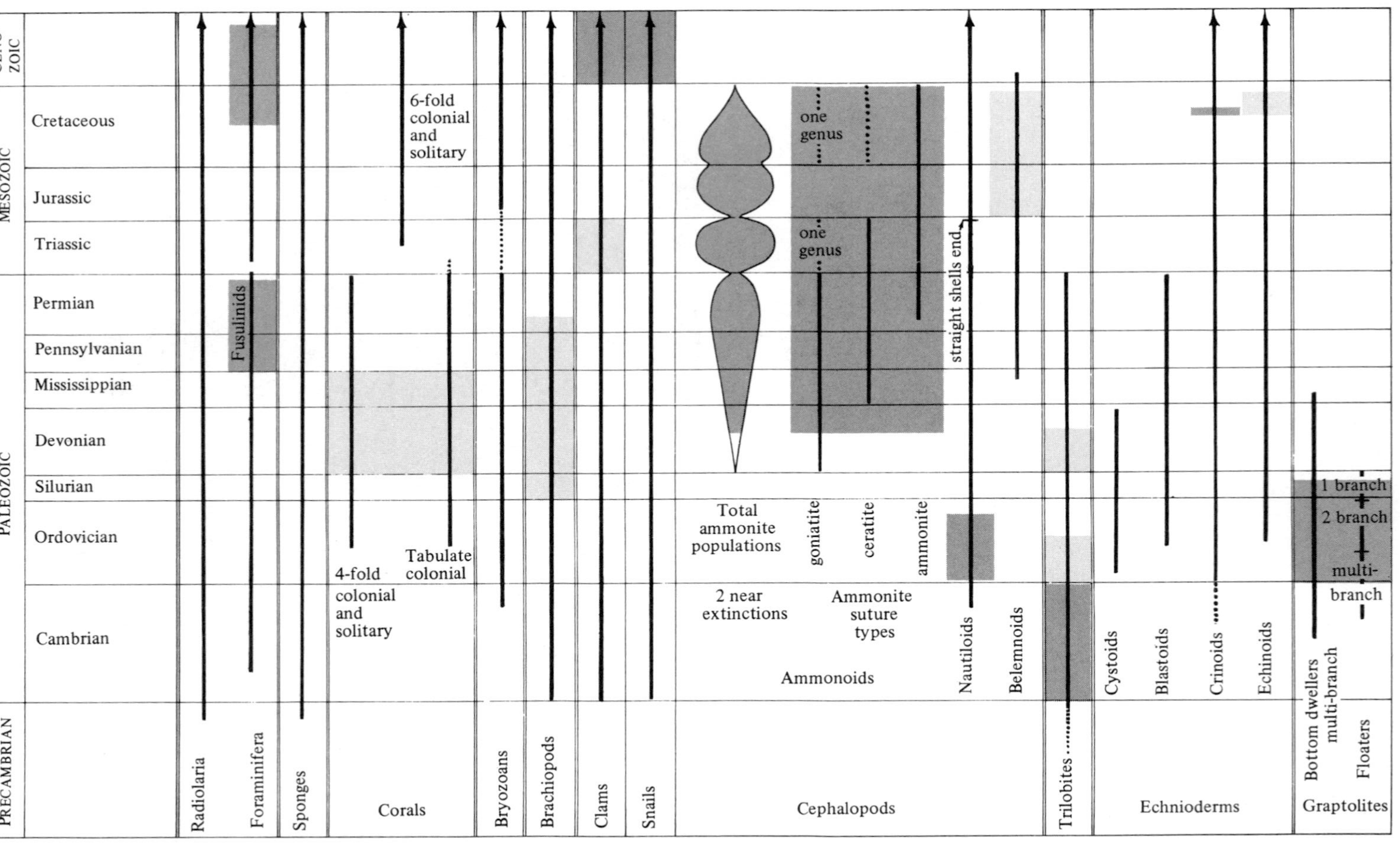

FIGURE 9.20
Chart showing the geologic occurrence of the more important fossil-forming invertebrate animals. The heavy lines show the time span of each animal group. Within some groups, some simple distinctions are indicated that are of value in age determination. Also shown are the parts of the geologic column for which these fossils are useful for worldwide correlation (darker shade) and for correlation of more restricted regions (lighter shade). Correlation data after Teichert, 1958.

thing happened. At the end of the Cretaceous, they did die out. At each of these times, there was a withdrawal of the seas from the interior, but it seems likely that swimmers like these could have lived on in the oceans.

Why they developed such extremely complex partitions is not known, although they probably strengthened the shell much as corrugations strengthen cardboard. Another possibility is that the chambers were used by the animal to adjust buoyancy and for some reason the complex partitions were an advantage. In any case, except for the nautilus, which has simple, smooth partitions, the shelled cephalopods died out at the end of the Cretaceous. Perhaps the more specialized forms could not adjust to the new conditions at the end of the Cretaceous, but this seems unlikely for such a successful group.

In the Jurassic, the seas again expanded, and many new forms appeared. One of these was the belemnoids, which first appeared in the late Paleozoic. They were a type of cephalopod with an internal cigar-shaped shell. (See Fig. 9.21.) They were much like modern squids. (See Fig. 9.22.) Figure 9.23 shows a sea floor in the Late Cretaceous.

In both of the reconstructions, clams, including oysters, are prominent. In the Paleozoic seas, brachiopods were common, but in the Mesozoic and Cenozoic, they are largely replaced by clams. The biologic differences between clams and brachiopods are great, and they can be distinguished easily by the symmetry of the

FIGURE 9.21
Cretaceous belemnoid, *Belemnitella americana,* from New Jersey. Length, 7.6 cm (3 in.). Photo courtesy Ward's Natural Science Est., Inc.

FIGURE 9.22
Jurassic sea floor. Belemnoids (squidlike cephalopods) and oysters are the invertebrates shown. Note the marine reptile ichthyosaur in the shadowy background. From Field Museum of Natural History. Negative No. GEO.80826.

FIGURE 9.23
Late Cretaceous sea bottom near Coon Creek, Tennessee. Coiled ammonite and straight-shelled ammonites *(Baculites)* and a number of snails and clams. From Field Museum of Natural History. Negative No. GEO.80874.

shells. (See Fig. 9.24.) Clams live in many habitats, just as the brachiopods do. Some burrow into the bottom mud, and some are fastened to the bottom. One suggestion for why clams replaced brachiopods is that they burrowed deeper into the bottom mud and so developed a whole new habitat. They took over many habitats and expanded into many environments. They are very useful in determination of the paleoenvironment. The bottom dwellers attached themselves to the bottom by one shell, and some changed in appearance to resemble horn coral. Reefs of these coral-like clams formed at some places. Others evolved into oysters, also losing their symmetry. (See Fig. 9.25.)

The Mesozoic plants, too, were different from those of the Paleozoic, although the change is not apparent until Late Triassic. (See Fig. 9.26.) The lush swamps that produced the Pennsylvanian coal beds disappeared in the Permian. This was the result of the climatic change to arid or semiarid conditions that is recorded by the red bed and evaporite deposits and the wind-deposited sandstones. These desert conditions that persisted into Triassic time resulted in the preservation of very few members of the early Mesozoic flora. The few fossil plants found in early Mesozoic rocks, such as horsetails, are holdovers from the Paleozoic. In Late Triassic, the Mesozoic flora of naked-seed plants such as ginkgoes, conifers, ferns, and especially cycads, all of which differed from Paleozoic forms of these plants, was well established. They appear in the background of Figures 9.30, 9.32 through 9.34, 9.36, and 9.37. *Angiosperms, or flowering plants, appear in the middle of the Cretaceous Period.* The angiosperms, which are important elements of present floras, became

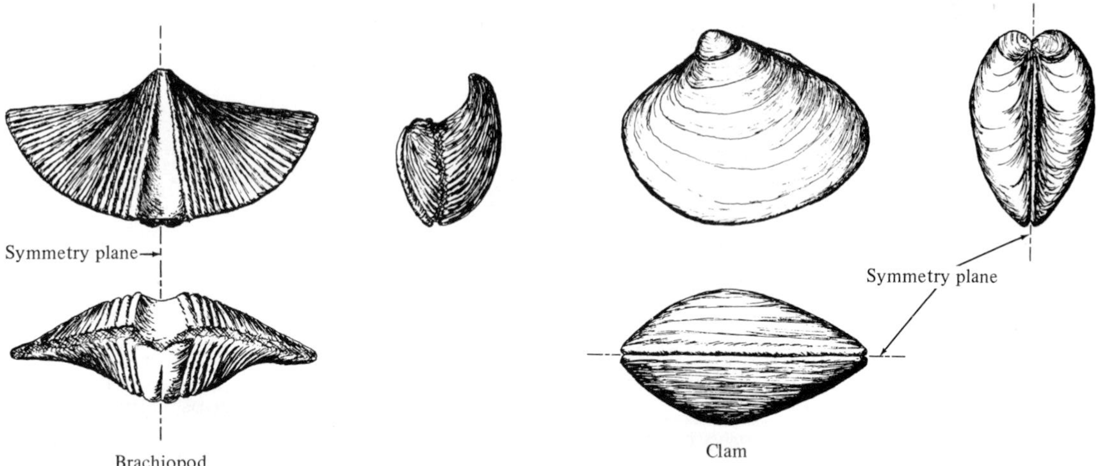

FIGURE 9.24
Differences between clam and brachiopod. A brachiopod has two dissimilar shells. A clam has two shells that are mirror images of each other because the clam has a symmetry plane between the two shells. The symmetry plane in a brachiopod cuts through both shells. Some clams (see Figure 9.25) do not have this symmetry.

abundant and spread to all continents in the Middle Cretaceous. They will be described in the next chapter.

The Mesozoic was the age of the reptiles, and, of course, the dinosaurs come to mind first. The dinosaurs were not the only reptiles. As noted in the last chapter, all of the reptiles have a *Seymouria*-like reptile of early Permian age as a common ancestor. (See Fig. 9.27.) In the Permian and most of the Triassic periods, the dominant group was the mammal-like reptiles. They became greatly reduced in Late Triassic time as the dinosaurs took over. The mammals evolved from the

E

FIGURE 9.25
Mesozoic clams. A. *Buchia piochii,* Jurassic from Prince Patrick Island, Canada. Natural size. B. *Inoceramus labiatus,* Cretaceous from Alberta. Natural size. C. *Trigonia thoracica,* a thick-shelled Cretaceous clam. D. *Exogyra arietina,* Cretaceous, 3.8 cm (1.5 in.). A thick-shelled clam without the symmetry of most clams. E. Cretaceous clams from north-central Texas. Photos A and B from Energy, Mines and Resources (112167-F, 113729-N) Reproduced with the permission of the Minister of Supply and Services Canada, 1990. C from Smithsonian Institution; D courtesy Ward's Natural Science Est., Inc.; E from U.S. Geological Survey.

mammal-like reptiles, and the first mammal fossils are found in Upper Triassic rocks. The mammals were relatively unimportant until the Cenozoic, and so they will be discussed in the next chapter.

Seymouria's group lasted until near the end of the Triassic. The turtles evolved from them and are found from Early Triassic on. Snakes and lizards had the same origin; the first fossil lizards are in Upper Triassic rocks, and the first snakes are in Cretaceous rocks.

The thecodonts (Fig. 9.28) also evolved from *Seymouria*'s group, and they gave rise to the dinosaurs, the crocodiles and alligators, the birds, and the flying reptiles. The thecodonts were a very successful group. They developed legs under the body as the mammal-like reptiles did, and so their locomotion improved. Some developed a two-legged walk. In spite of this, they were generally small animals and were overshadowed by the mammal-like reptiles.

The marine reptiles illustrate how a successful group like the reptiles moved into all environments. They returned to the sea, but they were better adapted than their amphibian ancestors were. (See Fig. 9.29.) They developed a system whereby the unhatched eggs were kept in the mother's body. Some Triassic ichthyosaurs have been found with unborn young in the body cavity.

Birds are similar to reptiles in many ways. They differ in having feathers and being warm-blooded. The oldest bird is Jurassic and is known to be a bird rather

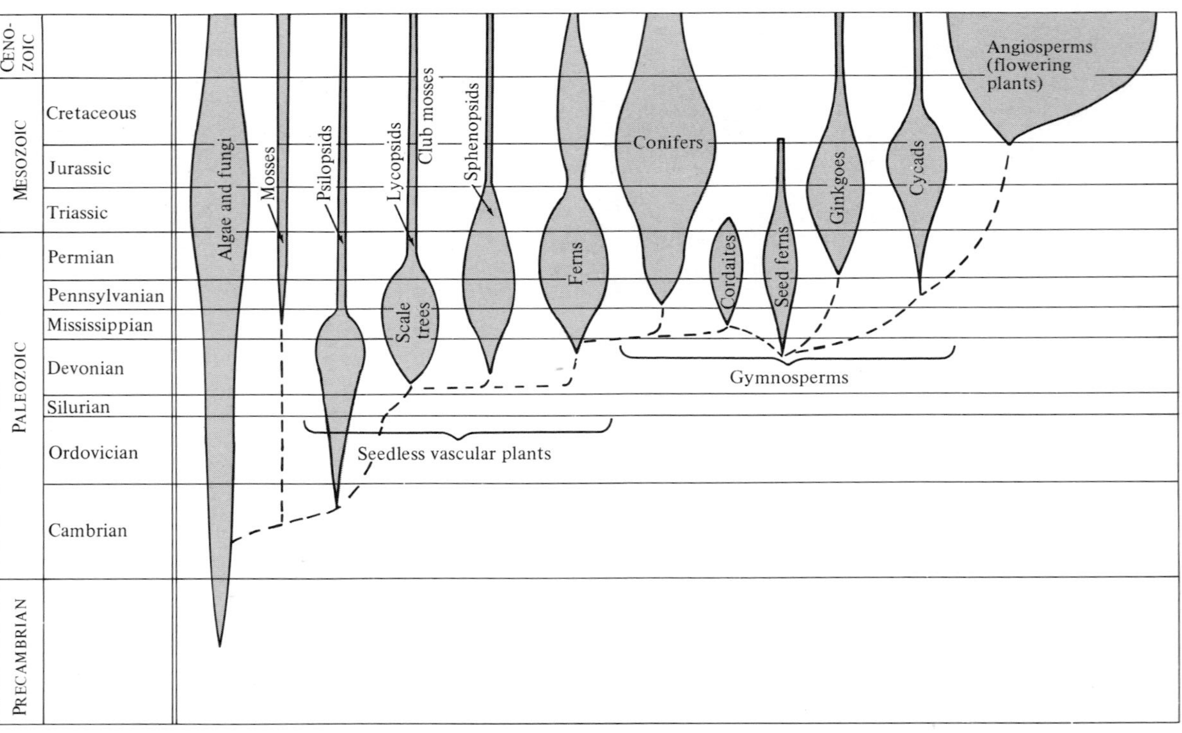

FIGURE 9.26
The development of plants.

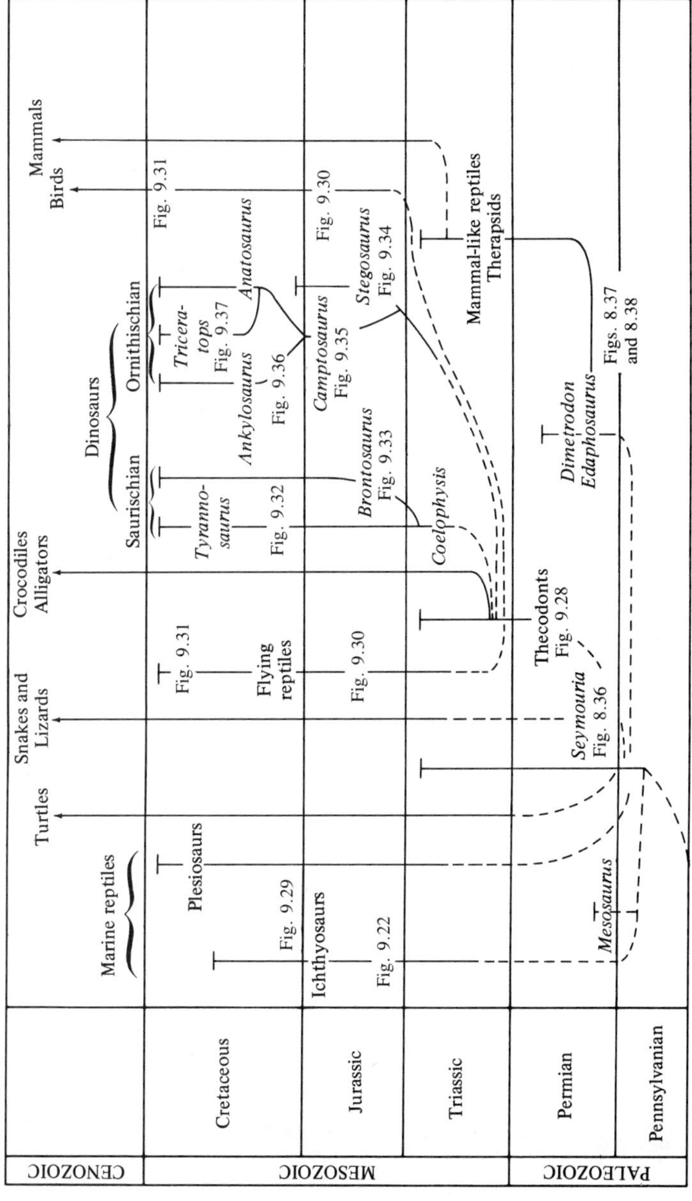

FIGURE 9.27
Evolution of the Mesozoic vertebrates. This figure is part of Figure 8.28.

FIGURE 9.28
Thecodont. This group evolved from *Seymouria,* and the thecodonts gave rise to the dinosaurs, flying reptiles, birds, and crocodiles. Thecodonts had bodies similar to dinosaurs, but were all small, up to 1.2 m (4 ft), and were confined to the Triassic.

than a reptile only because a few almost complete specimens with feathers have been found in fine-grained limestone at a single locality. This bird, *Archaeopteryx,* was about the size of a pheasant and had teeth in its beak. (See Figs. 9.30 and 9.31.) Birds are among the rarest of fossils. The first toothed flying reptiles are found in the Triassic rocks. The largest flying reptile so far found is from Texas and has a wing spread of 15.5 m (50 ft). It has been proposed that from the dinosaurs evolved the flying reptiles and the birds instead of as in the conventional evolutionary scheme shown in Figure 9.27.

The first dinosaurs are found in Upper Triassic rocks. They were small animals, not at all like the huge beasts of Late Jurassic and Cretaceous that were soon to rule the lands. Many of the dinosaurs were small, but it is the big ones that get attention. The term dinosaur means "terrible lizard" and has been applied to two different groups of reptiles. It is an informal but useful term. Dinosaurs were very successful and evolved rapidly. Figure 9.27 shows the relationships among the various dinosaurs.

The difference between the two types of dinosaurs is in their hip bones, one type (saurischians) having typical reptile hips and the other (ornithischians) having

FIGURE 9.29
Jurassic marine reptiles. Plesiosaurs are long necked. Ichthyosaurs are porpoiselike and closely resemble fish. About 2 m (6 ft) long. Painting by C. R. Knight, Field Museum of Natural History. Negative No. CK.72379.

FIGURE 9.30
Jurassic scene. *Archaeopteryx,* the bird, has feathers. The flying reptile, *Rhamphorhynchus,* has no feathers and a wingspread of about 1 m (3 ft). Also shown are tiny dinosaurs and palmlike cycads. Painting by C. R. Knight, Field Museum of Natural History. Negative No. CK.66715.

birdlike hips. The saurischians were of two types, carnivores and herbivores. The carnivores were mainly two-legged, with only small front legs; the largest flesh eater of all times, *Tyrannosaurus* (6.1 m [20 ft] tall, 15.25 m [50 ft] long, weighing 7275 to 9090 kg [8 to 10 tons]), belongs to this group. (See Fig. 9.32.) The saurischian plant eaters were mainly four-legged; some of the largest animals of all belong to this group (*Apatasaurus,* formerly *Brontosaurus*). (See Fig. 9.33.) The ornithischians were apparently all plant eaters, and the bizarre dinosaurs, such as the duckbills, the horned dinosaurs, and the armored dinosaurs, belong to this group. (See Figs. 9.34, 9.35, 9.36, and 9.37.)

A

B

FIGURE 9.31
A. Flying reptile, *Pteranodon.* B. Toothed bird, *Hesperonis.* Both from Upper Cretaceous of Kansas. Photos from Smithsonian Institution.

Most of the carnivores were two-legged, fast-running predators. The plant eaters developed many defenses. Some could run fast, some had armor, some had horns, some were very large, and some lived in water. Much has been written about the small brains of dinosaurs, but compared with other reptiles, their brains were near average. Compared with mammals, their brains were small.

A

B

FIGURE 9.32
Carnivorous dinosaurs. All carnivorous dinosaurs are saurischian. They could move rapidly on two feet and had only small front feet. A. *Ceratosaurus* from Upper Jurassic Morrison Formation near Canyon City, Colorado. About 7.3 m (24 ft) long. Note the small horn on the top of the nose. Photo from Smithsonian Institution. B. *Tyrannosaurus,* the largest of the flesh-eating dinosaurs. About 15.25 m (50 ft) long. Cretaceous. Painting by C. R. Knight, Field Museum of Natural History. Negative No. CK.59442.

It is not clear why some of the dinosaurs became so large. It may be because reptiles are cold-blooded and so it is an advantage to be large because a large animal needs a much smaller percentage of its body weight of food each day to maintain its energy. Most of the big dinosaurs were plant eaters and so may have

A

B

had a problem getting food. Another disadvantage of being large is the difficulty of finding shelter in a storm. Size may, however, have been a defense against predators.

Some dinosaurs seem to have been quite advanced for reptiles. It has been suggested that some of the dinosaurs may have been warm-blooded, but the evidence is not clear. Dinosaur nests complete with eggs have been found at a few places. Study of the embryos suggests that some hatched ready to feed themselves but others were helpless and required parental care. Some of the nests appear to have been used many years.

The reptiles were very similar on all continents during most of the Mesozoic, suggesting that until the Cretaceous, all continents were close enough to allow

C

FIGURE 9.33
Plant-eating saurischian dinosaurs. Mainly very large four-footed dinosaurs that lived in swampy water. A. *Brontosaurus* (now *Apatosaurus*). Late Jurassic. About 21.4 m (70 ft) long, weighing 31,820 kg (35 tons). Eating enough to maintain such a size may have been a problem. Painting by C. R. Knight, Field Museum of Natural History. B. Skeleton of *Brontosaurus.* Photo from Field Museum of Natural History. C. *Diplodocus,* Upper Jurassic, from Dinosaur National Monument, Utah. About 26 m (85 ft) long, 4 m (13 ft) high, weighing 11,820 kg (13 tons). *Diplodocus* is found in Jurassic and Cretaceous rocks. Photo from Smithsonian Institution.

migration. In the Cenozoic, there was less migration; interestingly, 30 orders of mammals evolved during the 65 million years of the Cenozoic, but only 20 orders of reptiles are found in the 200-million-year Mesozoic. This suggests that in the relative isolation of the Cenozoic, more orders formed because each isolated group evolved and radiated.

The orogenic activity that marked the Late Cretaceous and the Early Cenozoic produced a diverse topography that resulted in many climatic variations. These new climatic zones were soon occupied by plants, thus setting the stage for animals to evolve to take over these ecologic niches. The angiosperms, or flowering plants, that appeared near the middle of the Cretaceous include the grains and grasses, and it is the rapid spreading of these to the drier climates that created the new ecologic niches. The plant life of the Cenozoic was essentially modern, consisting mainly of angiosperms and conifers. The many Cenozoic coal beds attest their development.

At the end of the Mesozoic, the dinosaurs, almost all of which had specialized to occupy the many different environments, died out. Some of the dinosaurs were

A

B

FIGURE 9.34
Stegosaurus. These plated dinosaurs lived in the Jurassic and Cretaceous. About 6.1 m (20 ft) long, weighing 2 tons. A. Painting by C. R. Knight, Field Museum of Natural History. B. Skeleton. Photo from Smithsonian Institution. Negative No. CK.59163.

big animals, making food gathering in times of changing climate and plant life a difficult task. At this time, the mammals were small and few in number. They apparently were better able to adapt to the new conditions and so replaced the dinosaurs. The mammals were a more successful group than the reptiles for several reasons. They had bigger brains. They had four-chambered hearts instead of the two-chambered hearts of reptiles and had body hair rather than scales. Both of these features allowed them to maintain their body heat, rendering them less vulnerable to climatic changes. They were better able to preserve their young than the egg-laying reptiles. They also developed jaws and teeth better suited to their food. Their feet underwent a series of changes that better equipped them to their ecology. In spite of all these reasons why mammals were better equipped than

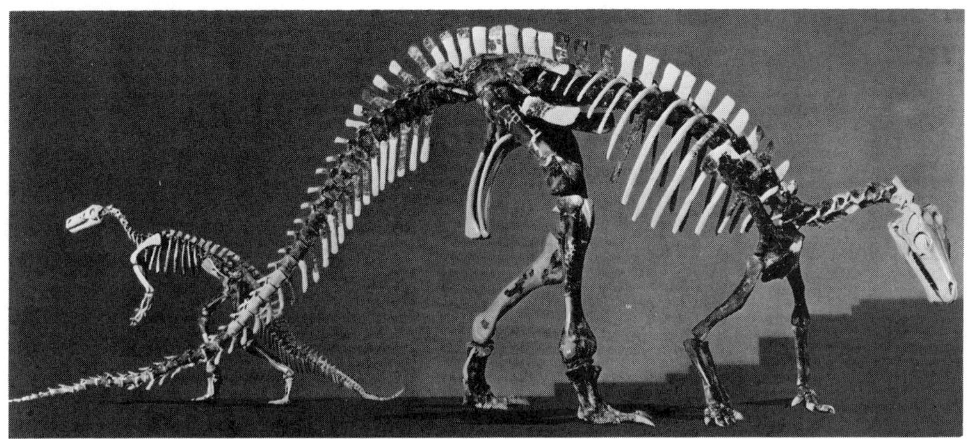

FIGURE 9.35
Duckbilled dinosaur, *Camptosaurus,* Upper Jurassic, Wyoming. *Camptosaurus* was one of the main stems of the ornithischian dinosaurs, and the ceratopsian (horned) dinosaurs evolved from them. *Camptosaurus* lived in the Jurassic and Cretaceous and was 4.6 m (15 ft) long. Photo from Smithsonian Institution.

FIGURE 9.36
Cretaceous dinosaurs. On the right, *Trachodon,* a duckbill. Other duckbills are the web-footed *Corythosaurus* in the water, and crested *Parasaurolophus* in the left background. In the foreground is *Ankylosaurus,* an armored dinosaur. In the center background is the ostrichlike *Struthiomimus,* a saurischian dinosaur. Painting by C. R. Knight, Field Museum of Natural History. Negative No. CK.66427.

FIGURE 9.37
Ceratopsian (horned) dinosaurs. Cretaceous. They evolved from *Camptosaurus*. A. *Protoceratops*. Eggs of this dinosaur were found in Mongolia. Painting by C. R. Knight, Field Museum of Natural History. Negative No. CK.59090. B. *Protoceratops* skeleton. About 2 m (6 ft) long. Photo from Field Museum of Natural History. Negative No. GEO.81446. C. *Triceratops* skeleton. About 7.3 m (24 ft) long. Photo from Smithsonian Institution.

A

B

C

dinosaurs, the dinosaurs were not driven out by the mammals. Rather, for unknown reasons, the dinosaurs died out and the mammals took over.

Many reasons have been suggested for the extinction of the dinosaurs. Most of these reasons could account for the demise of single groups or species but not all of the diverse dinosaurs. Examples are climatic change, loss of food supply because of development of the angiosperms, disease, and predation of their eggs, perhaps by mammals. Because none of these or the many other suggestions can explain the extinction of all of the dinosaurs, some investigators have looked to extraterrestrial causes. Cosmic rays and solar wind, especially at times of magnetic reversals, have been suggested, but most extinctions here and elsewhere in the geologic record do not generally coincide with times of magnetic reversals. Another possibility is that the impact of a large comet or asteroid could have thrown enough dust into the atmosphere to reduce greatly the amount of the sun's energy reaching the earth's surface. Such a loss of sunlight would destroy some organisms in the food chain and so cause extinctions. In support of this theory is the concentration of elements expected in meteorite dust that is found at the Mesozoic-Cenozoic boundary. So far no crater big enough to have been caused by this impact has been found, although a small crater near Manson, Iowa, was formed at the right time. Whatever caused the demise of the dinosaurs, they appear to have been in decline before the end of the Mesozoic.

SUMMARY

During the Mesozoic, the single supercontinent broke up, forming present oceans such as the Atlantic.

A geosyncline formed between Europe and Africa.

Triassic of the West

In the far west, thick marine clastics and volcanics overlie Permian rocks unconformably; to the east are clastics and carbonates.

The marine rocks interfinger with thin red beds in the northern Rocky Mountains.

Nonmarine rocks in Arizona and New Mexico are further evidence of aridity.

Batholiths were emplaced in the Sierra Nevada.

Middle Jurassic through Cretaceous

In the west, the Jurassic begins with an upland area centered in Nevada that shed clastics on both sides.

In the Jurassic, marine deposition occurred in the area of the Rocky Mountains and Great Plains.

The sea retreated north, and nonmarine rocks were deposited.

The seas returned in the Cretaceous, and thick clastics were deposited. This was the last great inundation of North America.

The area west of the upland was active throughout the cycle.

Batholiths were emplaced throughout the far west.

Marine deposition began in the Gulf Coast in the Jurassic and spread to the Atlantic Coast in the Cretaceous.

In the far north are Jurassic marine clastics and Cretaceous nonmarine rocks.

Life of the Mesozoic

Cephalopods and clams are the most abundant marine invertebrates.

Cephalopods are almost ideal index fossils because they evolved rapidly and are widespread.

The Mesozoic flora is cycads, ginkgoes, conifers, and ferns.

Angiosperms became abundant in the Middle Cretaceous.

During Permian and Triassic, mammal-like reptiles were the dominant group. They evolved into the mammals that are first found in Upper Triassic rocks.

Turtles, lizards, and snakes all evolved during the Mesozoic.

Birds and flying reptiles first appeared in the Jurassic.

The first dinosaurs were small and are found in Upper Triassic rocks.

The reptiles are similar on all continents, suggesting that until the Cretaceous, the continents were close enough to allow migration.

The dinosaurs died out at the end of the Cretaceous and were replaced by the mammals.

QUESTIONS

1. Describe the Triassic rocks of the western United States.
2. What was the extent of the Jurassic seas in North America?
3. Why is it difficult to date Cretaceous events between the Rocky Mountains and the West Coast?
4. Where were batholiths emplaced in North America in the Mesozoic?
5. Describe the Cretaceous rocks of the Rocky Mountains.
6. When did deposition start in the Gulf Coast area?
7. Which continents drifted apart during the Mesozoic?
8. Is there a relationship between continental drift and the locations of Mesozoic geosynclines?
9. Describe the fauna of the Triassic.
10. When did the first flying reptiles appear? The first birds?
11. How do Mesozoic corals differ from Paleozoic corals?

12. How are clams distinguished from brachiopods? Does this distinction hold true for all Mesozoic clams?
13. How are snails distinguished from cephalopods?
14. Outline the history of cephalopods.
15. How are Mesozoic plants different from Paleozoic plants? From Cenozoic plants?
16. What was the time of the dinosaurs?
17. Describe the various types of dinosaurs and other reptiles of the Mesozoic. Which ecologic areas or niches did they occupy?
18. What advantages do mammals have over reptiles?

SUGGESTED READINGS

Alvarez, L. W., and others, "Extraterrestrial Cause for the Cretaceous-Tertiary Extinction," *Science* (June 6, 1980), Vol. 208, No. 4448, pp. 1095–1108.

Bakker, R. T., "Dinosaur Renaissance," *Scientific American* (April 1975), Vol. 232, No. 4, pp. 58–78.

Bird, Peter, "Formation of the Rocky Mountains, Western United States: A Continuum Computer Model," *Science* (March 25, 1988), Vol. 239, No. 4847, pp. 1501–1507.

Buffetaut, Eric, "The Evolution of the Crocodilians," *Scientific American* (October 1979), Vol. 241, No. 4, pp. 130–144.

Desmond, A. J., *The Hot-blooded Dinosaurs: A Revolution in Paleontology.* New York: Dial Press, 1975, 238 pp.

Hallam, A., "Continental Drift and the Fossil Record," *Scientific American* (November 1972), Vol. 97, No. 5, pp. 56–66.

Howell, D. G., "Terranes," *Scientific American* (November 1985), Vol. 251, No. 5, pp. 116–125.

Hoyle, Fred, and Chandra Wickramasinghe, *Archaeopteryx, the Primordial Bird: A Case of Forgery.* Swansea, Wales: Christopher Davies (Publishers) LTD, 1986, 135 pp. Presents a different viewpoint from that of the present book.

Imlay, R. W., *Jurassic Paleobiogeography of the Conterminous United States in Its Continental Setting.* U.S. Geological Survey Professional Paper 1062. Washington, D.C.: U.S. Government Printing Office, 1980, 134 pp.

Kielan-Jaworowska, Zofia, "Late Cretaceous Mammals and Dinosaurs from the Gobi Desert," *American Scientist* (March-April 1975), Vol. 63, No. 2, pp. 150–159.

Langston, Wann, Jr., "Pterosaurs," *Scientific American* (February 1981), Vol. 244, No. 2, pp. 19–136.

Mash, Robert, *How to Keep Dinosaurs—The Complete Guide to Bringing Up Your Beast.* New York: Viking/Penguin, 1983, 72 pp.

Mulcahy, D. L., "The Rise of the Angiosperms: A Genecological Factor," *Science* (October 5, 1979), Vol. 206, No. 4414, pp. 20–23.

Valentine, J. W., and E. M. Moores, "Plate Tectonics and the History of Life in the Oceans," *Scientific American* (April 1974), Vol. 230, No. 4, pp. 80–89.

Ward, Peter, Lewis Greenwald, and O. E. Greenwald, "The Buoyancy of the Chambered Nautilus," *Scientific American* (October 1980), Vol. 243, No. 4, pp. 190–203.

10
The Cenozoic

WORLD GEOGRAPHY IN THE CENOZOIC

The present distribution of continents on the earth formed during the Cenozoic. Figure 10.1 shows the drift that occurred. North America separated from Europe, and the Atlantic Ocean formed. Greenland separated from North America, and Australia left Antarctica. The south Atlantic widened. India collided with Eurasia.

Orogeny occurred from Alaska to Antarctica. Africa and Europe drifted closer together, causing orogeny in the geosyncline between them. This mountain-building event produced the Alps in Europe and a mountain chain across Asia. Parts of Turkey and Iran were transferred from Africa to Asia in this collision. The Himalaya Mountains were formed when India collided with Asia.

North America moved somewhat northward, largely by rotation. This may have caused a cooling of the climate, reflected in some of the late Cenozoic life. Alaska and Siberia may have collided, causing some deformation.

The Gulf of California and the Red Sea probably formed late in the Cenozoic. The Bay of Biscay opened by the rotation of the Iberian Peninsula. Iceland formed by volcanic activity on the Mid-Atlantic Ridge in late Cenozoic time.

THE CENOZOIC

Following the withdrawal of the seas near the end of the Cretaceous, the seas never again occupied large areas of the interior, but they did cover most coastal areas during parts of the Cenozoic. The geologic history of the Cenozoic in most of North

FIGURE 10.1
Continental drift during the Cenozoic. Cenozoic mountain building is shown stippled. Dotted positions are the end of the Cretaceous; solid line is the present. The movement of Australia and South America is exaggerated by the map projection used here. Based on Dietz and Holden, 1970.

America is concerned with nonmarine rocks. During the Cenozoic, most of the present topographic features of the earth were formed, making its history of great interest to the traveler. It will be possible to describe only a few of the more important areas.

Atlantic and Gulf Coastal Plains

Marine sedimentation began in the Atlantic and Gulf coastal plains areas in Mesozoic time and continued throughout most of the Cenozoic. The history is one of repeated advances and retreats of the sea. The Gulf area is one of the most closely studied areas in the world because of the great value of the oil recovered from these rocks. The rocks are penetrated by thousands of oil wells, and some of these are very deep. On the south Atlantic coastal plain, three major advances are recognized. They are mid-Cretaceous, Late Cretaceous to early Cenozoic, and Oligocene to middle Miocene. During much of the Cenozoic, Florida was an area of limestone deposition, separating the Atlantic from the Gulf coastal plain. On the Gulf, eight major advances and retreats are recognized, as well as many minor ones. The maximum advance occurred in the Eocene, when the Mississippi Valley was invaded as far as southern Illinois. This was the major advance of the Cenozoic, and the early Cenozoic rocks overlie the Cretaceous unconformably, indicating that the Cretaceous seas had retreated in late Cretaceous time. A typical advance of the sea is recorded by clastic rocks at the bottom of the sequence. The rocks grade from nonmarine sand or clay to similar brackish-water sediments to marine sandstone or shale that thickens greatly toward the Gulf. The advances and retreats of the sea are probably related to uplifts in the Appalachian Mountains, but it has not been possible to correlate these events. Much of the area drained by the Mississippi River was undergoing erosion throughout the Cenozoic and provided debris to the Gulf area. The sedimentary rocks on the Gulf coast are very thick, up to 15,250 m (50,000 ft), with 4575 m (15,000 ft) of Pleistocene at some places. Figure 10.2 shows a typical cross-section and shows the rise in the basement rocks offshore.

Through the Cenozoic, limestone was deposited in Florida. It is believed that this area was covered by shallow, warm water. Under such conditions, limestone is being deposited now in the Bahama Banks just east of Florida. Florida was awash during most of Cenozoic, but at times islands were present.

Appalachian Mountains

The Appalachian Mountains were above sea level and undergoing erosion during the Cenozoic. Their history cannot be learned from sedimentary rocks except in a very general way because the uplifts of the mountains cannot be related to the coastal plain sediments. Thus we turn to study of the erosion surfaces to determine the history of the Appalachians. This was one of the first places that such studies were made. The culmination of these studies was the classical interpretation shown in Figure 10.3. The history shown in the cross-section starts with high mountains in post-Triassic time (A). These mountains were eroded to a smooth peneplain before Cretaceous time (B); in the Cretaceous, coastal plain sediments were deposited on the peneplain (C). This was followed by a broad uparching (D). Erosion of the upwarped range formed a new peneplain (E). This peneplain was uparched (F), and erosion developed a partial peneplain on the softer rocks (G). Another uplift caused erosional development of another partial peneplain on the softer rocks (H). Another uplift caused further dissection of the partial peneplains and produced the present Appalachian Mountains.

More recent studies suggest that a simpler history could produce the same topography. In this case, the history would be the same up to block E. A single uplift followed by erosion could result in the ridge tops preserving remnants of the old peneplain, at elevations of about 1220 m (4000 ft), and flat valleys cut on weak rocks.

Rocky Mountains and High Plains

Orogeny began in the Rocky Mountains near the end of the Cretaceous and extended to the Oligocene. This is the **Laramide,** or last, phase of the Cordilleran orogeny that began in late Jurassic. The age and intensity of this folding vary from place to place. (See Fig. 10.4). At some places, Paleocene rocks unconformably overlie folded Late Cretaceous rocks; at other places, rocks as young as Oligocene are involved in the folding. In general, the more intense orogeny occurred near the west shore of the Cretaceous sea, where thrust faults are found. The compressive phase was over by middle Eocene except in southern Colorado, where late Eocene thrusts are found. Farther east in Montana, Wyoming, and Colorado, most of the present mountain ranges were uplifted, with folding and faulting that was locally intense. Nearby, on the high plains of eastern Montana, undeformed Late

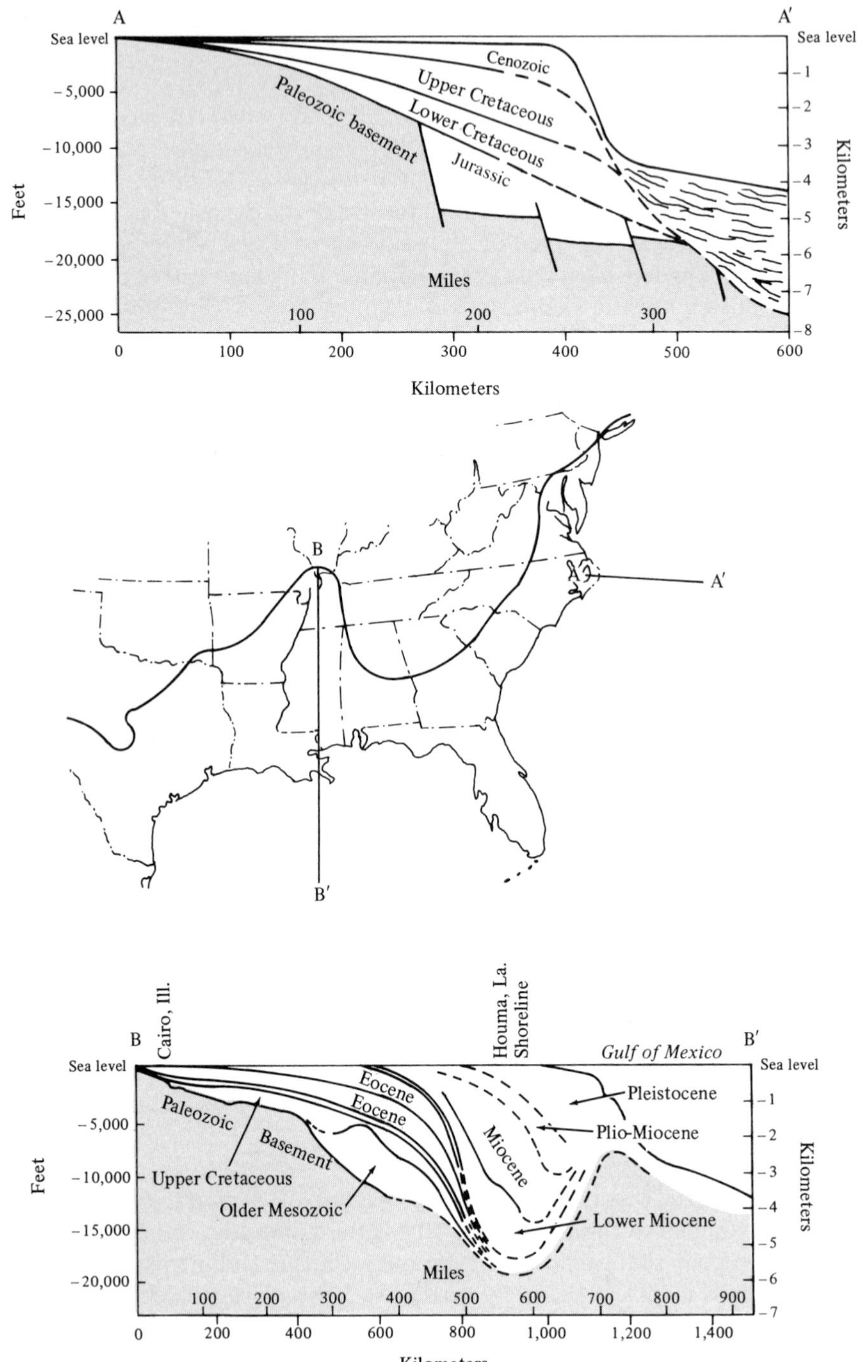

FIGURE 10.2
Cenozoic deposition in the Atlantic and Gulf areas. Note the very thick Cenozoic deposits in the Gulf Coast area. A after Emery and others, 1970; B after Bernard and LeBlanc, 1965.

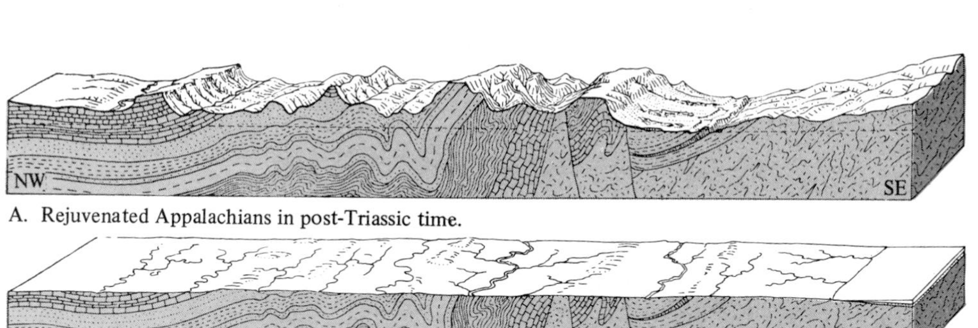

A. Rejuvenated Appalachians in post-Triassic time.

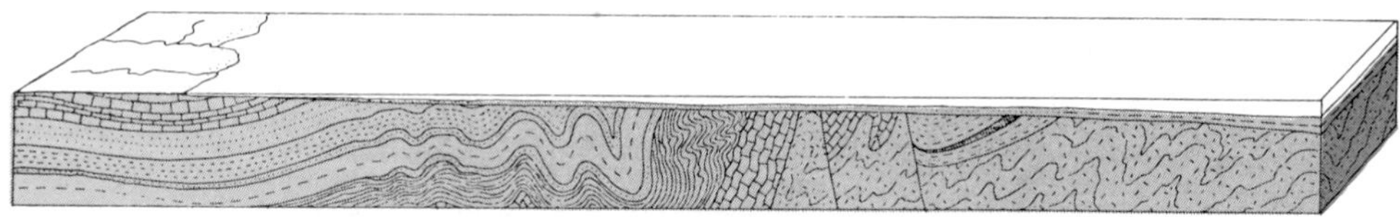

B. Fall Zone peneplain.

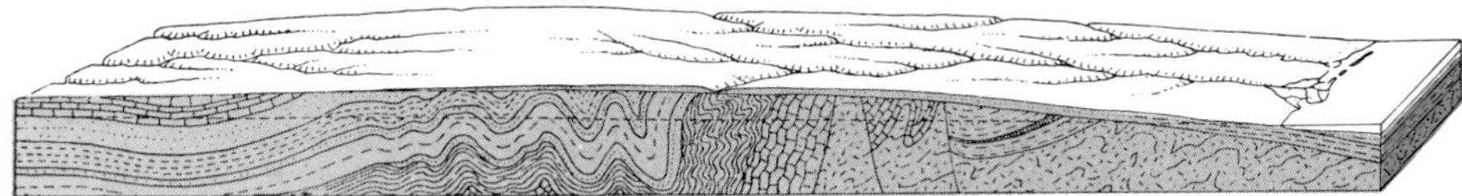

C. Encroachment of Cretaceous Sea and deposition of coastal plain beds.

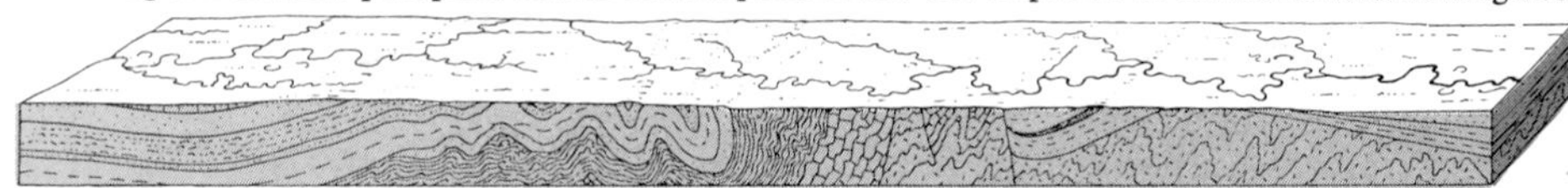

D. Arching of Fall Zone peneplain and its coastal plain cover. Development of southeastward-flowing streams.

E. Schooley peneplain.

F. Arching of Schooley peneplain.

G. Dissection of Schooley peneplain and erosion of Harrisburg peneplain on belts of nonresistant rock.

H. Uplift and dissection of Harrisburg peneplain and erosion of Somerville peneplain on weakest rock belts.

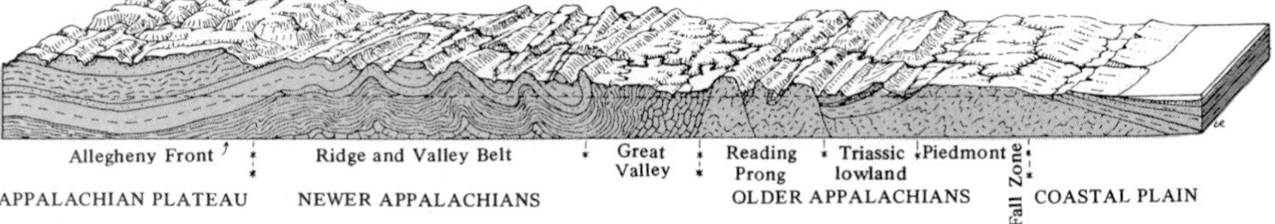

FIGURE 10.3
The development of the Appalachian Mountains. The history shown in the diagrams is the classical interpretation. New concepts suggest that the present form of the Appalachian Mountains could have developed from a single uplift. From Johnson, 1931.

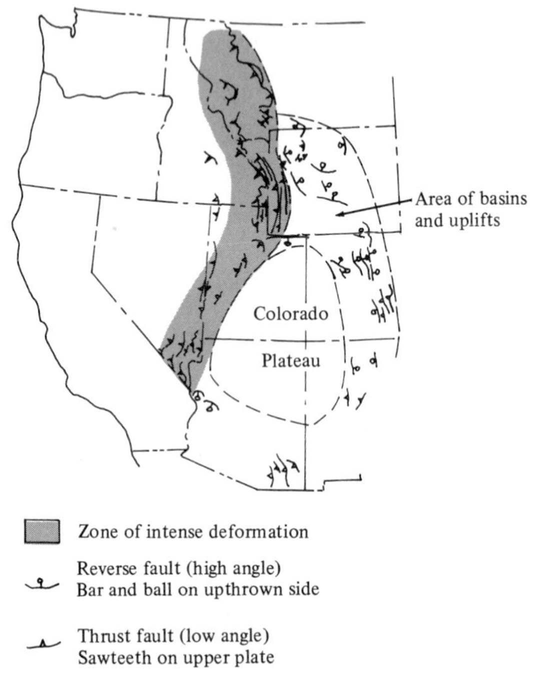

FIGURE 10.4
Late Cretaceous and Early Cenozoic deformation in the Rocky Mountain area. In the western zone of intense deformation, folding and thrust faulting are common. In the eastern area of basins and uplifts, a number of basins sank and were filled by detritus from rising highlands. Many of these basins were folded with local thrust faulting during this interval. Very little deformation took place in the Colorado Plateau, an anomalous situation that has not been satisfactorily explained. The west coast was an area of active basins. Between the west coast and the intensely deformed zone, few Cretaceous rocks are present, so deformation during this interval cannot be recognized. Structural details after Gilluly, 1963.

Cretaceous nonmarine sediments are conformably overlain by similar Paleocene rocks, indicating that the orogeny was localized. This sequence of rocks is of great interest because it records the last of the dinosaurs and the rise of mammals. Farther east in North Dakota and adjacent areas, a remnant of the Cretaceous sea remained until the Paleocene.

With the uplift of the mountain ranges in the Rockies, basins were formed between the ranges. (See Fig. 10.5.) These basins were filled with nonmarine sediments during the early Cenozoic. Some volcanic sediments were deposited in some of the basins, and in the Eocene and Oligocene, great thicknesses of volcanic rocks were emplaced in northwestern Wyoming in the Absaroka Range. Volcanic rocks were deposited throughout most of the Cenozoic in the San Juan Mountains of Colorado. Erosion of the ranges filled the basins, mainly in Paleocene to Oligocene time.

From Oligocene to Pliocene time, the streams from the area of the present Rocky Mountains spread fine clastics over the Great Plains. Then, because of either a change in climate or uplift of the Rocky Mountains, the streams began to erode. It is uncertain whether the uplift occurred at this time or earlier. This erosion

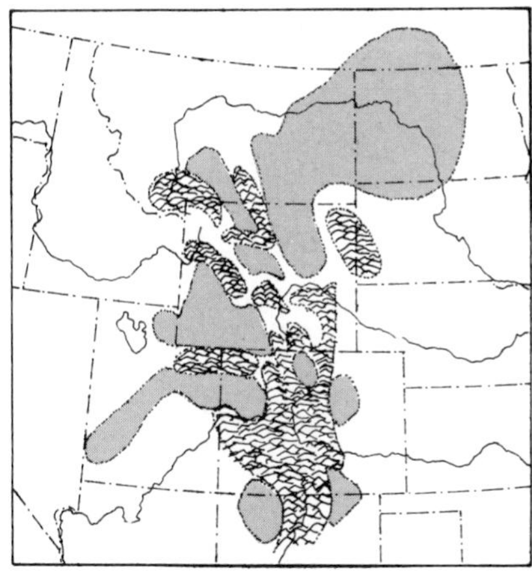

FIGURE 10.5
Early Cenozoic basins and highlands in the Rocky Mountain area.

removed much of the earlier deposits, forming the present topographic mountains. Before this downcutting, the rivers had been flowing on a relatively smooth surface, in part eroded and in part constructed by the rivers. The courses of the rivers on this surface bore no relationship to the rock types at depth, for they had developed their courses over tens of millions of years. Thus the downcutting rivers at most places removed the soft basin-filling sediments, but at some places cut narrow, deep canyons through the mountain blocks composed of Precambrian metamorphic rocks. Examples of this include Big Horn River Canyon in the Big Horn Mountains, Laramie River in the Laramie Mountains, North Platte River, Sweetwater River, and many others. (See Fig. 10.6.)

Colorado Plateau

The events on the Colorado Plateau are related to those in the Rocky Mountains. In early Cenozoic time, the Colorado Plateau was at a low elevation, and rivers carrying sediments from the uplifted areas to the north and east deposited them on the plateau. By Eocene time, the plateau had downwarped, and a vast freshwater lake covered much of Utah. Organic-rich shale was deposited in this lake, and the resulting oil shales may someday be an important source of petroleum. Uplift and local volcanic activity occurred during the rest of the Cenozoic. During this time, the Colorado River eroded the Grand Canyon. The rocks on the Colorado Plateau are, for the most part, horizontal, with some faulting and bending. All around this area, similar rocks are folded and deformed. Why the Colorado Plateau escaped this deformation is not known.

Basin and Range

This area is known variously as the Great Basin, the Basin-Range, or the Basin and Range area because, with the exception of the Colorado River, it is a basin with no

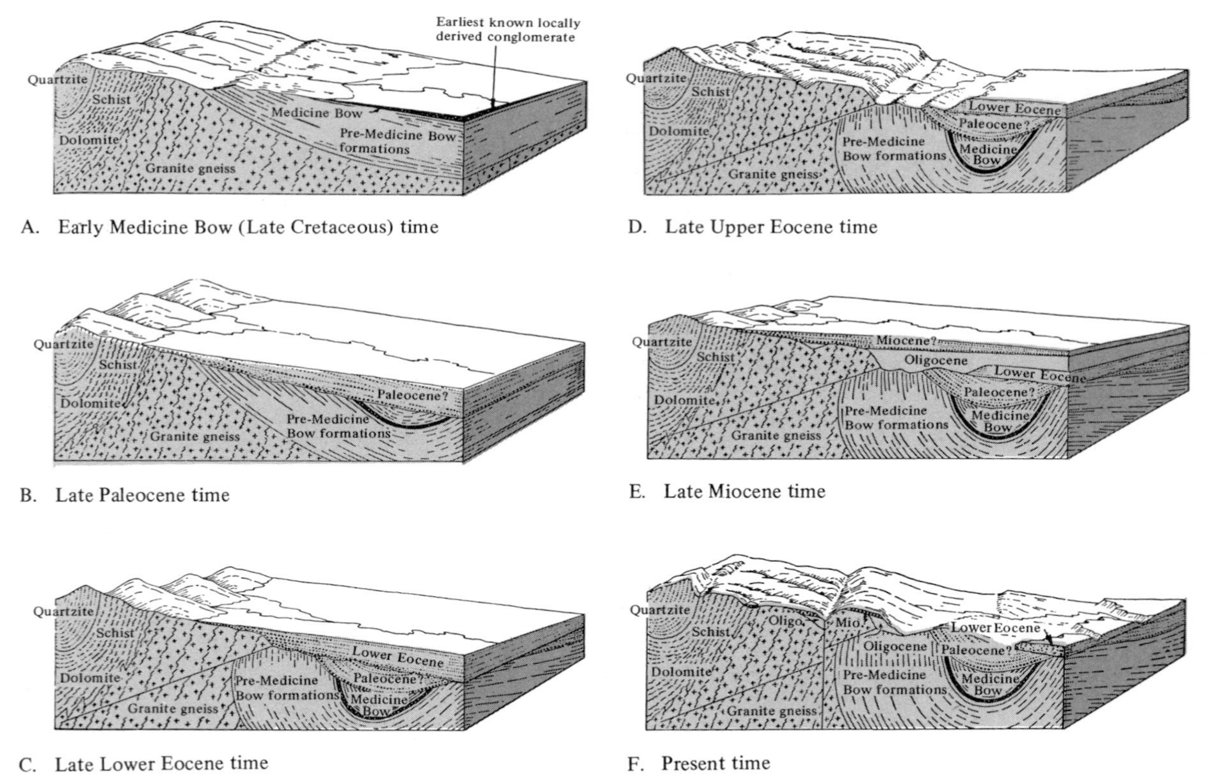

A. Early Medicine Bow (Late Cretaceous) time

B. Late Paleocene time

C. Late Lower Eocene time

D. Late Upper Eocene time

E. Late Miocene time

F. Present time

FIGURE 10.6
Development of the Rocky Mountains. The Medicine Bow Mountains shown here are typical of the Rocky Mountains in Wyoming. In early Cenozoic time, deposition and deformation occurred at the same time (Parts A, B, and C). In late Cenozoic time, the basins were filled to the level of the now-eroded highlands, and the rivers spread sediments in both the area of the present Rocky Mountains and the Great Plains (E). In very late Cenozoic time, uplift or climatic change caused the rivers to erode and downcut (F). The rivers removed much of the soft basin-filling sediments and in many cases cut deep canyons across previously buried ranges, producing the present mountain topography. From Knight, 1953.

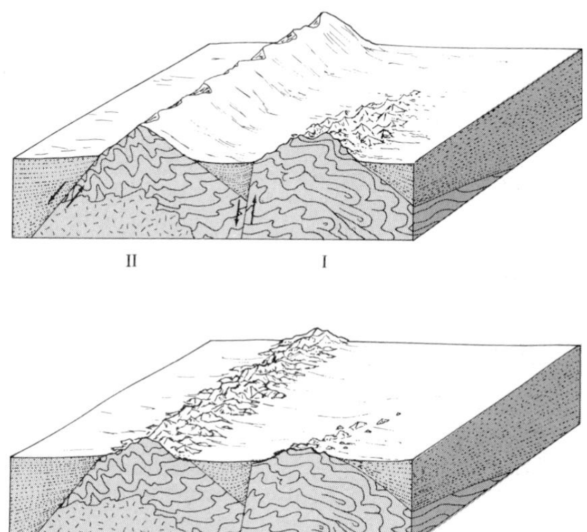

FIGURE 10.7
Typical Basin and Range structure. Both ranges I and II are uplifted fault blocks with intervening basins filled by detritus eroded from the ranges. Range II is younger than range I. The rocks in both ranges were deformed prior to the present uplift. The lower view shows the ranges at a later time.

external drainage to the ocean and because the area is characterized by many fault-block mountain ranges with intervening basins. Like the Colorado Plateau, this is an unusual area. The deformation of the rocks that make up the ranges has been described in the preceding chapters. These deformed rocks have been uplifted in fault blocks to form the Basin and Range area. The faults appear to be normal faults and so differ sharply from the compressional features displayed in the ranges. (See Fig. 10.7.) The faulting and uplifting began in the Miocene and are probably still going on, judging from the recent fault scarps and the number of earthquakes. The best example is probably the westernmost range, the Sierra Nevada in California. (See Fig. 10.8.)

The basins obviously have been sites of deposition since their formation. Prior to the faulting, Cenozoic rocks were deposited in part of the area, and these rocks are exposed in some of the ranges. Volcanic activity occurred at various times in different parts of the Basin and Range area. In some places, thick, widespread welded tuffs can be traced for many miles from range to range.

Northwestern United States

The Northwest was an area dominated by volcanic rocks throughout much of the Cenozoic. Marine sediments and some volcanic rocks accumulated near the Pacific in a number of active embayments. A pre–late Miocene unconformity is widespread. Inland, the Cenozoic rocks are continental and volcanic. They were deposited in many basins, and the number of unconformities shows the crustal unrest. Farther east, in much of eastern Washington and Oregon, these rocks are covered by the spectacular Columbia River Basalt to a depth of at least a thousand meters. (See Fig. 10.9.) Farther east in the Snake River Valley are thick deposits of

FIGURE 10.8
Aerial view of the west slope of the Sierra Nevada, showing Yosemite Valley. Note the relatively smooth surface that was uplifted. The uplifted surface is incised by rivers whose valleys have been modified by glaciation, producing features such as Yosemite Valley. At other places, especially on the skyline, erosion has produced rugged mountain topography. Photo from U.S. Geological Survey.

FIGURE 10.9
The Columbia River Basalt flows in eastern Washington. Photo by F. O. Jones, U.S. Geological Survey.

FIGURE 10.10
Oblique aerial view of Mount Hood in northern Oregon. From northern California through Washington, the high Cascade volcanoes formed on top of a mountain range composed of deformed Cenozoic rocks. Photo from U.S. Geological Survey.

younger volcanic rocks. In Pliocene time, the present Cascade Range was uplifted on a north-south axis that cuts across the earlier structures. The last step was the formation of the present volcanoes, such as Mounts Shasta, Hood, Rainier, St. Helens, and Baker, in late Pliocene and Pleistocene time. (See Fig. 10.10.)

California Coast

A brief description of the Cenozoic history of California will, together with the outline of the Northwest, give an indication of the types of events that occurred along the Pacific coast. This was an extremely active area throughout the Cenozoic, and a detailed history of the Pacific coast would be very long. In Paleocene and Eocene time, a sea occupied the central and southern parts of western California. Eastern California was an upland, and its erosion shed sediments into the sea. Nonmarine clastics were deposited between the sea and the upland. In the area of the present Coast Ranges, several large islands existed. At places, up to 3050 m (10,000 ft) of clastics were deposited in this sea.

During the Oligocene, nonmarine clastics and red beds were deposited inland, with some marine rocks nearer to the coast. The seas returned in the Miocene, and there was much volcanic activity. Orogeny also occurred at this time. In Pliocene time, many basins developed, and some of these were nonmarine lakes. In the Los Angeles Basin, 4575 m (15,000 ft) of sediments were deposited. Deposition continued into the Pleistocene. The San Andreas fault, which is a transform fault, was probably active during much of the Cenozoic. (See Fig. 10.11.) In the Middle

A

B

FIGURE 10.11
San Andreas fault. A. Young sedimentary rocks deformed at the San Andreas fault near Palmdale, California. B. The San Andreas fault cuts across the Carrizo Plains in southern California. Photos from U.S. Geological Survey.

Pleistocene, a major deformation occurred. At this time, the Coast Range mountains were uplifted. Similar uplifts occurred all the way to Washington. As a result of this activity, the Pacific coast has mountains along the coast (the Coast Ranges and Olympic Mountains), an inland valley (the Great Valley in California, Willamette Valley–Puget Sound in Oregon and Washington), and a mountain range (Sierra Nevada and Cascade Range). The only exceptions are southern Washington, where the coastal mountains are low hills; the Oregon-California border, where there is no valley; and southern California.

Arctic

The Cenozoic rocks of the Arctic islands are mainly nonmarine sandstone and shale with some coal. These rocks were apparently deposited in basins, and, although most occurrences are only a few hundred feet thick, at one place they are 2135 m (7000 ft) thick. These rocks were folded some time in the Cenozoic.

Until the Cenozoic, the Arctic islands and Greenland were part of the North American continent. The area apparently sank to form the islands—the waterways among the islands may be, at least in part, a drowned river system.

PLEISTOCENE

The Pleistocene, although generally thought of as the glacial age, is defined by a type section, as are other systems. Glaciers occupied only a part of the earth during the ice age, and they formed and melted at different times at different places.

During the Pleistocene, there were several centers of ice accumulation in Canada. Glaciers moved out from these centers in all directions. Thus, although the ice moved generally southward in the United States, it moved northward in northern Canada. In no sense did the last glacial age result because ice from near the north pole moved south.

At first, geologists in North America recognized four major glacial advances from study of moraines. They were able to do this because at some places fresh, unweathered moraines overlie weathered moraines, which in turn overlie more deeply weathered moraines. Later, more detailed studies have revealed many more advances and retreats of the ice. The details of these oscillations have been deciphered by study of the moraines. Mapping the extent of the moraines led to the recognition of the extent of the glacial advances shown on the map in Figure 10.12.

The most recent continental glaciation of Pleistocene age began about two million years ago. This date is estimated from the degree of weathering of deposits, from the amount of soil produced between glacial advances, from the study of

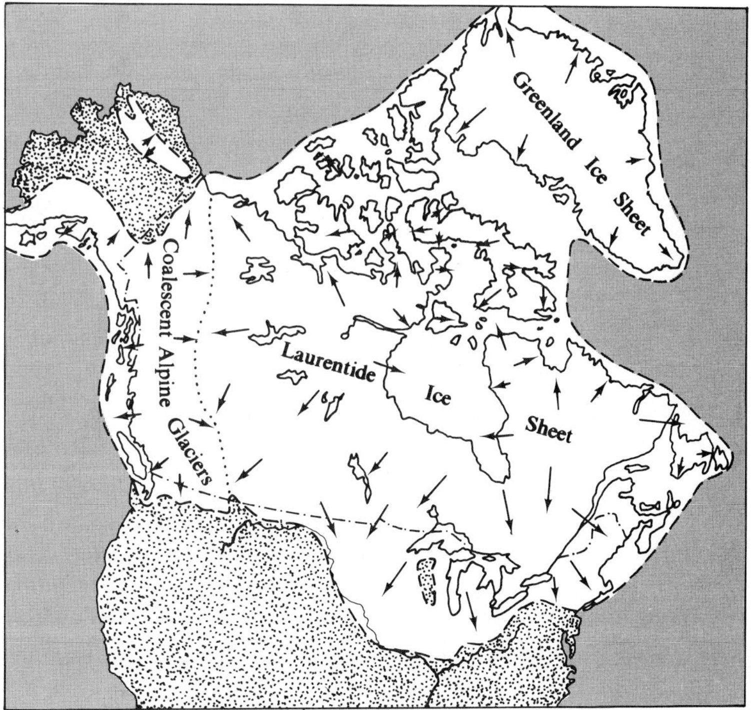

FIGURE 10.12
Maximum extent of Pleistocene ice sheets. Glaciers also formed in the mountains in both eastern and western North America. The Laurentide ice sheet is now known to have consisted of many smaller centers of accumulation, with radial flow outward from each.

fossils, from radiometric dating, and from other information obtained by a variety of methods. The last glacial advance began about 30,000 years ago, reached its maximum about 19,000 years ago, and ended about 10,000 years ago. A minor advance occurred between 1550 and 1850 and is sometimes called the Little Ice Age. These dates are fairly accurately known from carbon-isotope studies. This method is based on the fact that organic carbon has a fixed ratio of carbon isotopes when formed, but this ratio changes after the death of the organism because of radioactive decay of one of the carbon isotopes (carbon 14). The method is not suitable for material more than about 40,000 years old. Radiometric dating methods are discussed elsewhere.

Before radiocarbon was used, other methods, particularly the study of varved clays, were used. This method is much like the study of tree rings—the varves are assumed to be yearly deposits in glacial lakes. The dark, finer-grained part of each varve is deposited during the winter and the light, coarser-grained part in the summer. Thus the thickness of varves records the climatic conditions, and the sequence of relative thicknesses is correlated from lake to lake until the life span of the glacier is covered; then the total number of varves is counted. Apparently because of error in correlation, this method gave too high an age estimate.

The Pleistocene history of North America is a study worthy of a much longer book than this. Only an outline of events can be presented here because almost all of the present landforms were made or greatly modified during the Pleistocene.

The principles involved in Pleistocene geology as outlined here, are

1. erosion and deposition by alpine and continental glaciers,
2. drainage changes caused by glaciers, including damming of rivers by the advance of the glaciers,
3. the effects of the generally rapidly released meltwater from retreating glaciers,
4. the effects of the weight of glacial ice lying on the continents,
5. the effects of the glacial climate in areas away from the ice, and
6. the normal nonglacial geological processes, such as erosion, deposition, and orogeny, that have been discussed in the history of the earlier geologic periods.

The most obvious effects of Pleistocene glaciation are seen in the areas actually glaciated. The continental glaciers covered much of the northern part of the United States. In north-central United States, the long, low morainal hills marking the farthest advance of the last glacier form conspicuous features. The soils of this area and of the Midwest are largely glacial deposits scraped from the now almost bare areas to the north in Canada. In New England and New York, the low mountains were covered by the continental ice. The main effects were a rounding and smoothing of the mountains. Much of the material eroded by the glaciers was deposited nearby, producing the rocky soils and fields of this region. As in other places, the farthest advance of the glacier is marked by terminal moraines, of which Long Island is the most prominent.

In the northern Great Plains in Montana, the Missouri River was diverted by the glacial ice. This river probably flowed northeastward toward Hudson Bay in preglacial time. The presence of the glacier forced it into a southeasterly course to join the Mississippi River. Much of the present course of the Missouri River is within a few tens of miles of the limit of the glacial ice. (See Fig. 10.12.)

Mountain glaciers developed in the high mountains of western United States. These mountain glaciers sculptured the superb scenery of many of our ranges, such as at Yosemite National Park. (See Fig. 10.13.) At some places, so much snow accumulated on the mountains that only the highest peaks were not covered. At places, too, glacial tongues extended some distance out from the mountains. Generally this occurred where several coalescing glaciers extended into the lowland.

The area between the Cascade Mountains and the Northern Rocky Mountains was the scene of a remarkable event in glacial times. A wide path across the

FIGURE 10.13
Yosemite Valley. Note that the present river is very small for the size of the valley. Photo by S. A. Davis, courtesy California Division of Mines and Geology.

Columbia Plateau has been eroded by much rapidly flowing water. This is shown by the stripping off of most of the loose surficial material, eroded stream channels, and large sand bars. This erosion is believed to have been accomplished by glacial meltwater, but obviously some remarkable events must have occurred to channel enough water through the area to do this work. As the glacier retreated in the northern Rocky Mountains in Montana, meltwater filled the northward-sloping valleys, forming a large, irregularly shaped lake. The glacier to the north dammed this lake. When the fragile ice dam failed, the lake emptied very rapidly. This is believed to be the source of the rapidly flowing, large amount of water. This water flowed across Idaho and met the Columbia Plateau near Spokane. The Channeled Scablands, as the area eroded in this unusual way is called, extend south from Spokane to Grand Coulee Dam. Apparently many floods of water reached the Columbia River in this way as the ice dam was reformed by forward-moving glacial ice and destroyed by the overtopping meltwaters again and again. Remember that even in a retreating glacier the ice moves forward. These floodwaters could not follow the Columbia River Valley because the Columbia was also dammed by glacial lobes extending south from the mountains north of the Columbia Plateau. These latter glacial lobes apparently blocked the Columbia River at different places as the various floods reached the Columbia, and they diverted the floods southwesterly across the Columbia Plateau. The diverted floodwaters eroded the scabland and rejoined the Columbia River in southern Washington. (See Fig. 10.14.)

The climatic changes that caused glaciation in high latitudes produced aridity near the equator. The evidence for this is both the type of sediment deposited in the oceans in low latitudes and sand dunes in both northwestern Australia and western Africa that pass under the sea on parts of the shelf that were above sea level when glaciers occupied the northern lands. Nearer the glaciers, the climate was wet.

The Great Lakes of northern United States were formed during the Pleistocene. As long as the front of the continental glacier was south of the divide that separates the drainage of the Mississippi River from that of the St. Lawrence River, the meltwater all flowed out the Mississippi River. When the glacier retreated north of this divide, lakes were impounded between the glacier and the divide. As the water level rose and the glacier retreated, a number of different outlets were uncovered. If nothing else had happened, the lakes would all have been drained when the glacier melted away. However, the melting of the glacier reduced the weight pressing down on the crust, and the crust began to rise, perhaps isostatically. This uplift acted as the glacier had to impound the Great Lakes. The development of these lakes is known in great detail as a result of much study. When the ice first melted from this area, the crust of the earth had been depressed by the weight of the ice so that it was below sea level, and marine conditions prevailed in part of the St. Lawrence Valley until the crustal rise forced the seas out. (See Fig. 10.15.) Other lakes also formed in this region but were drained when the ice retreated and the crust rose. Lake Winnipeg is a small remnant of one of these other glacial lakes.

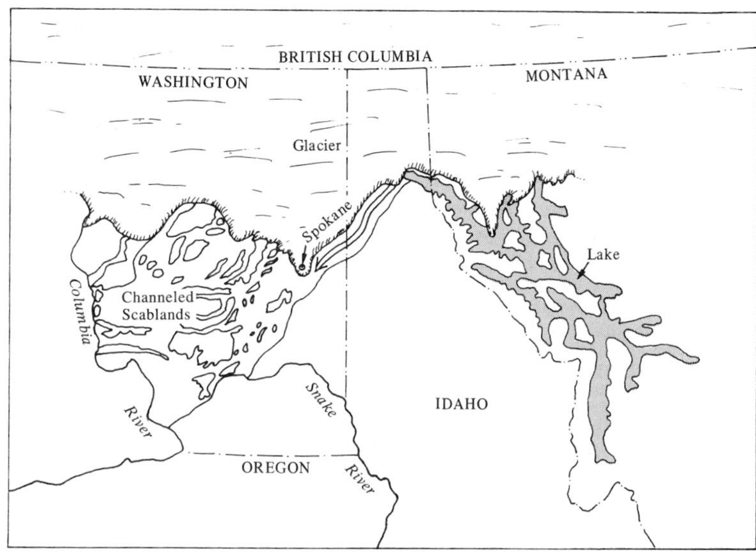

FIGURE 10.14
The development of the Channeled Scablands in eastern Washington. The lake formed from glacial meltwater in the north-sloping valleys in Montana as the glacier retreated. When the glacial ice that dammed the lake was removed by melting, or erosion by water that overtopped it, the lake drained rapidly along a broad valley south of the continental glacier. The water crossed the Columbia Plateau, eroding the Channeled Scablands, and then joined the Columbia River. Based on maps by Bretz, 1956.

Glacial Climate and Glacial (Pluvial) Lakes

The climatic conditions that cause glaciation also had far-reaching effects, even far from the areas of actual glacial accumulation. It is estimated that sea level was lowered about 152 m (500 ft) during the glacial advances and that, if the present-day glaciers melt, sea level will rise between 30 and 46 m (100 and 150 ft).

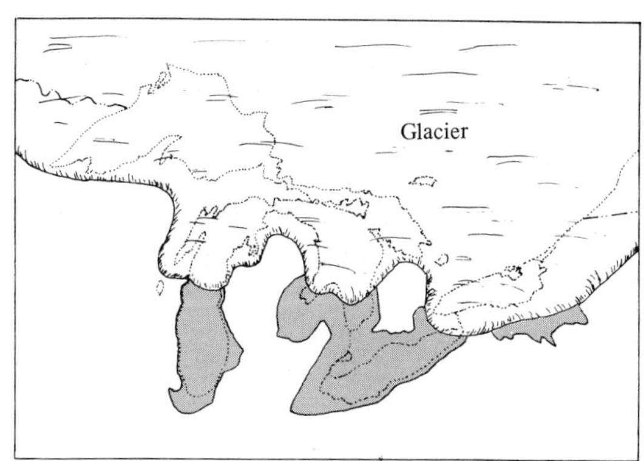

FIGURE 10.15
An early stage in the development of the Great Lakes. The glacier has melted back, exposing large valleys that slope northward. Meltwaters have filled the valleys that are dammed by the glacial ice.

To discover the cause of glacial periods, we must first see under what conditions glaciers form. Average temperatures must be lower than at present, but not greatly lower, and precipitation must be high. These two conditions will cause the accumulation of snow, which will form glaciers. The temperature must be low enough to ensure that precipitation will be in the form of snow, but if it is too low, it will inhibit precipitation. This latter point is illustrated by the lack of glacier formation, because of low precipitation, in many very cold regions today. An average annual temperature decrease of about 6°C will probably cause a new onset of glaciation.

Heavy precipitation during at least parts of the glacial age is suggested by the huge lakes that formed in areas south of the glaciers, where it was too warm for snow to accumulate. On the other hand, a humid climate without much evaporation may also have been involved in the formation of these lakes. These lakes formed in the basin areas of the western United States; the present Great Salt Lake is but a small remnant of Lake Bonneville, which covered much of northwestern Utah. Other lakes covered large areas in Nevada, and some of the present saline lakes, such as Carson Sink, are remnants. (See Fig. 10.16.) These lakes are saline because, without outlets to the sea, dissolved material builds up in them while evaporation keeps their levels fairly constant.

Cause of Glaciation

A number of theories of glacier origin have been proposed, but none is completely satisfactory. A successful theory must explain the change in climate just discussed. It also must account for the multiglacial advances that occur during an ice age, such as the separate advances of the recent ice age, as well as for other times of glaciation at different places on the earth. Many of these glaciations occurred far from the present poles. The main theories are as follows:

1. Changes in the amount of energy received from the sun. Astronomers detect very slight changes in the sun's energy production, very much too small to affect the earth's climate. A variation of this theory is that as the sun moves through space in its trip around the center of the Milky Way galaxy, it may encounter regions of space containing dust.
2. Changes in the amount of the sun's energy reaching the earth's surface, resulting from changes in the earth's atmosphere. Dust from volcanic eruptions could cause cooling of the surface.
3. Continental drift caused by plate tectonics can move continents nearer to the poles and so help to start an ice age.
4. Periodic changes in the earth's motion around the sun affect the amount of the sun's energy received at any point on the earth. This is an old theory recently revived. In the 1930s, Milutin Milankovich, a Serbian mathematician, suggested that the effects of the earth's motions could cause glaciers to form. His calculations were based on ideas that were nearly 100 years old then.

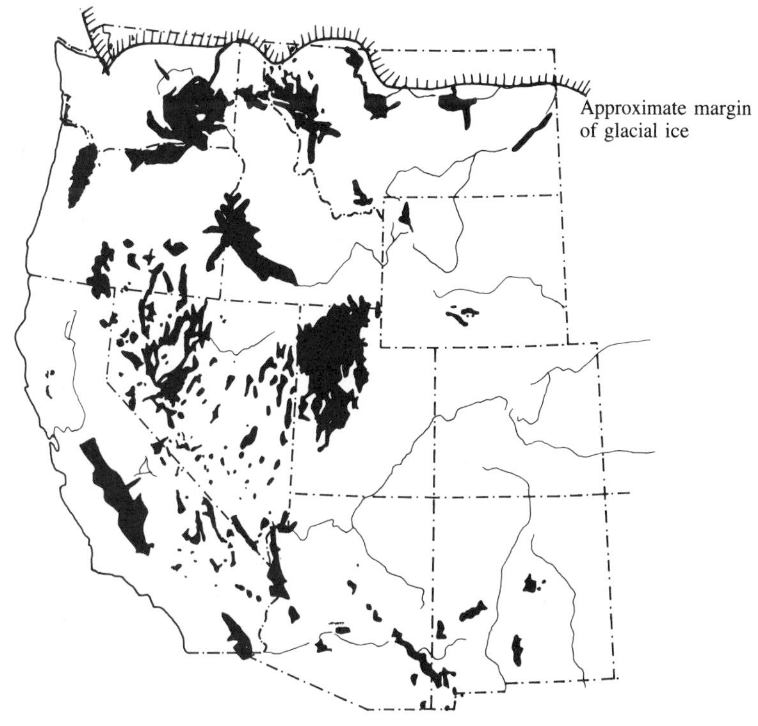

FIGURE 10.16
Map showing the known lakes of glacial times in western United States. From Feth, U.S. Geological Survey, 1961.

FIGURE 10.17
Clams. A. *Clinocardium nuttalli,* Pleistocene from British Columbia. B. *Chlamys hindsi,* Pleistocene from British Columbia. Photos courtesy Energy, Mines and Resources (113556-N and -L). Reproduced with the permission of the Minister of Supply and Services Canada, 1990.

Milankovich recognized three periodicities. The earth's orbit around the sun is not quite circular, and the average distance between sun and earth changes slightly, with about a 105,000-year period. The calculated effect of this period is very small because the total energy input from the sun varies only about 0.1 percent. The seasons are caused by the tilt of the earth's rotational axis relative to the plane of the earth's orbit. The tilt is currently 23.5 degrees, but it varies between about 22 degrees and 24.5 degrees, with a period of 41,000 years. The changing tilt angle changes the contrast between the seasons. The third periodicity is in the direction in which the earth's axis points. This motion is known as the precession of the equinoxes, and the period of this motion is about 26,000 years. The precession determines where on the orbit the seasons occur, that is, whether winter in the northern hemisphere occurs when the earth is closest or farthest or some other distance from the sun. This motion also accentuates the contrast between seasons.

The effects of these cycles can be tested in several ways. Cores of sedimentary rocks that range in age from the present back through much of the last ice age can be obtained in some lakes and in the deep ocean. The pollen in the cores obtained on the continents can be studied, and climatic changes can be inferred from the types of trees and plants. In the marine cores, tiny fossils can also be studied, and the temperature of the seawater can be determined from study of the isotopes of oxygen present. The cores that have been studied revealed climatic cycles of 23,000 years, 41,000 years, and 100,000 years—all in remarkably close agreement with the theoretical predictions. This suggests that these orbital motions do affect climate.

5. Changes in the circulation of the oceans. This is a promising hypothesis that seems adequate for the recent glacial periods. The hypothesis begins with an ice-free Arctic Ocean. With the present pole positions and our present atmospheric circulation, the ice-free Arctic Ocean would cause the heavy precipitation necessary to initiate a glacial advance. By the middle of the glacial advance, the Arctic Ocean would freeze, cutting off much of the supply of moisture. Precipitation would continue until the North Atlantic

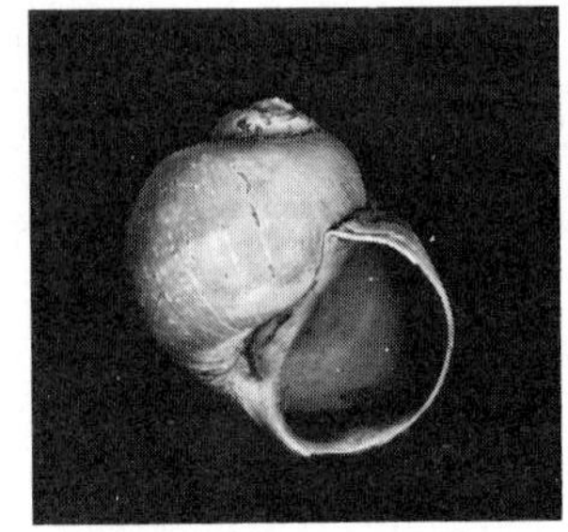

FIGURE 10.18
A Pleistocene snail from Quebec, *Natica clausa.* Photo courtesy Energy, Mines and Resources (113093-B). Reproduced with the permission of the Minister of Supply and Services Canada, 1990.

FIGURE 10.19
Echinoid *Eupatagus floridanus,* Eocene, Ocala limestone, Florida. Photo courtesy Ward's Natural Science Est., Inc.

A

B

C

FIGURE 10.20
Cretaceous and later leaf fossils have a modern aspect. A. *Alnus larseni,* Creede, Colorado. B. *Metasequoia,* Collawash River, Oregon. C. *Mahonia marginata,* Creede, Colorado. All are Miocene. Photos from Smithsonian Institution.

1. Late Paleocene (early and middle Paleocene found only in North America) and early Eocene

2. Middle Eocene

3. Late Eocene

Monkeys do pass barrier

4. Oligocene

FIGURE 10.21
Cenozoic vertebrate migrations. Barriers and migration routes are indicated diagrammatically. Continental drift and changing sea level apparently caused many of the barriers and connections.

became too cold to provide enough moisture. Melting would then begin and continue until the Arctic Ocean was again ice-free. The stage is now set for the cycle to begin again. This hypothesis can account for glacial periods only when the poles are located as they are now. If both poles were over open ocean, then the atmospheric circulation could not develop glacial periods. Thus, to initiate a glacial period, the poles would have to shift. This theory accounts for the multiple glacial advances of the recent glacial period; it can also explain the irregular periods of older glaciation. Recent studies, however, suggest that the present Arctic ice may be much older than the last glacial advance, a finding that casts doubt on this theory. All of the causes of glaciation discussed have both advocates and dissenters because none of them is entirely satisfactory.

5. Early Miocene

6. Middle and late Miocene

7. Pliocene

8. Pleistocene. In Middle and Late Pleistocene climate limits interchange. Most profound interchange occurred in Early Pleistocene.

LIFE OF THE CENOZOIC

In the Cenozoic, all types of life develop into the modern forms, making their study especially interesting. It was an active time, with much mountain making and climatic change. As a result, the changes in life, especially on the land, are great.

In the ocean, the biggest change is the absence of some groups of cephalopods that were so abundant in the Mesozoic. Clams (see Fig. 10.17), oysters, and snails (see Fig. 10.18) were the main elements of Cenozoic seas, along with echinoids (see Fig. 10.19), bryozoans, and foraminifers. The foraminifers underwent a new development and were very widespread. Because they are small enough to be recovered intact in drill cuttings, they have been used extensively to date the rocks encountered when drilling for oil. Limestones composed of large foraminifers were deposited in tropical areas.

The plant life of the Cenozoic was essentially modern. (See Fig. 10.20.) The first fossil angiosperms are found in middle Cretaceous rocks, but such fossils are relatively rare. In Late Cretaceous time, well-differentiated angiosperms became abundant, showing that they had developed somewhere. By the end of the

Cretaceous, they replaced the Mesozoic plants. Their radiation continued throughout the Cenozoic.

The Cenozoic is the age of mammals. Mammals have many advantages over reptiles. They have a more efficient heart and have hair, making them warm-blooded. Their brain is larger, and the senses, such as smell and especially hearing, are better. Specialized teeth and jaws improve feeding, and the digestive system does not require inactive periods after eating, as reptiles require. Teeth are easily preserved and are excellent mammal fossils. Growth of the young within the mother and a long period of nursing make the young more likely to survive. Just as the reptiles expanded to all habitats in the Mesozoic, so did the mammals in the Cenozoic. Bats are flying mammals; seals, whales, and porpoises are marine mammals.

There are two main types of mammals: the *marsupials,* such as the kangaroo and opossum, that carry their young in an external pouch; and the *placental* mammals that carry their young internally. There is one other type of mammal, *monotremes,* living in Australia. Only two of these are known: the duck-billed platypus and the spiny anteater. They have hair and nurse their young but they lay eggs. There is almost no fossil record of these animals, so they may be "living fossils," almost unchanged from the Cretaceous.

The oldest mammal fossils are small shrewlike skulls found in Upper Triassic rocks. Very few fossils are found until the Late Cretaceous. The largest Mesozoic mammals were the size of house cats. The Cenozoic placental mammals evolved from small Late Cretaceous insectivores.

The marsupials were much less successful than the placentals. It is estimated that only about five percent of all Cenozoic mammals were marsupials. The oldest marsupial fossils are middle Cretaceous in age. The marsupials appear to have migrated to Australia and South America in the Late Cretaceous and then became isolated. As a result, the marsupials there did not have to compete with the placentals. They evolved into many types and occupied many habitats. Australia is noted for its marsupial fauna. The same thing happened in South America, but here many of the marsupials became extinct at the end of the Pliocene. This occurred because about this time, the Isthmus of Panama formed, and the placentals from North America migrated to South America. The marsupials could not compete with the placentals. The Cenozoic barriers and migrations are shown in Figure 10.21. During most of the Cenozoic, there was some connection between Eurasia and North America.

In North America, the Paleocene was warm and temperate. The Paleocene mammals were archaic, and most were small, up to sheep size. They lived in forests and near streams. Primates, rodents, insectivores, carnivores, and browsers were present. (See Fig. 10.22.) The browsers had hoofs and four or five toes. European and North American forms were similar, but the South American types were different. The first horses appeared in the Paleocene.

As the Eocene opened, the climate was subtropical and the animals were largely forest dwellers. All of the modern orders of mammals have been found,

FIGURE 10.22
Paleocene scene. *Barylambda,* a primitive hoofed animal with five toes. Western Colorado. About 2.5 m (8 ft) long. Painting by J. C. Hansen, Field Museum of Natural History. Negative No. GEO.84487.

although the species are different from those of the present. Rhinoceroses were present, and in the late Eocene, the deer, pig, and camel appeared. (See Fig. 10.23.) The largest land animal of the time was the rhinoceroslike *Uintatherium.* Toothed whales appeared in the sea, indicating that this habitat was soon occupied after the extinction of the marine reptiles. (See Fig. 10.24.) In late Eocene time, more grasslands appeared as the climate became somewhat cooler and drier. The Oligocene had a mild, temperate climate in North America. The Great Plains was a large floodplain with many rivers and some forests. Figure 10.25 shows the differences in the climate and vegetation. The archaic forms disappeared, and the fauna took on a more modern aspect. The drier climate favored the faster, long-legged, hoofed grazers and browsers. Cats and dogs appeared. The titanotheres, which had begun in early Eocene as small rhinoceroslike animals, reached the size of elephants in the Oligocene and then died out. *Brontotherium* was one of the largest mammals ever to live in North America.

The drying and cooling trend continued into the Miocene in North America, and the grasslands expanded. This favored the rapid evolution of grazers such as horses. (See Fig. 10.26.) The large cats, bears, and weasels also appeared.

In the Pliocene, the climate cooled, remaining relatively dry. As a result of these changes, evolution continued rapidly. Many forms were weeded out, and those remaining became more specialized. The trend continued into the Pleistocene and produced our present life. Figure 10.27 shows a scene on the high plains. Many

Part 1

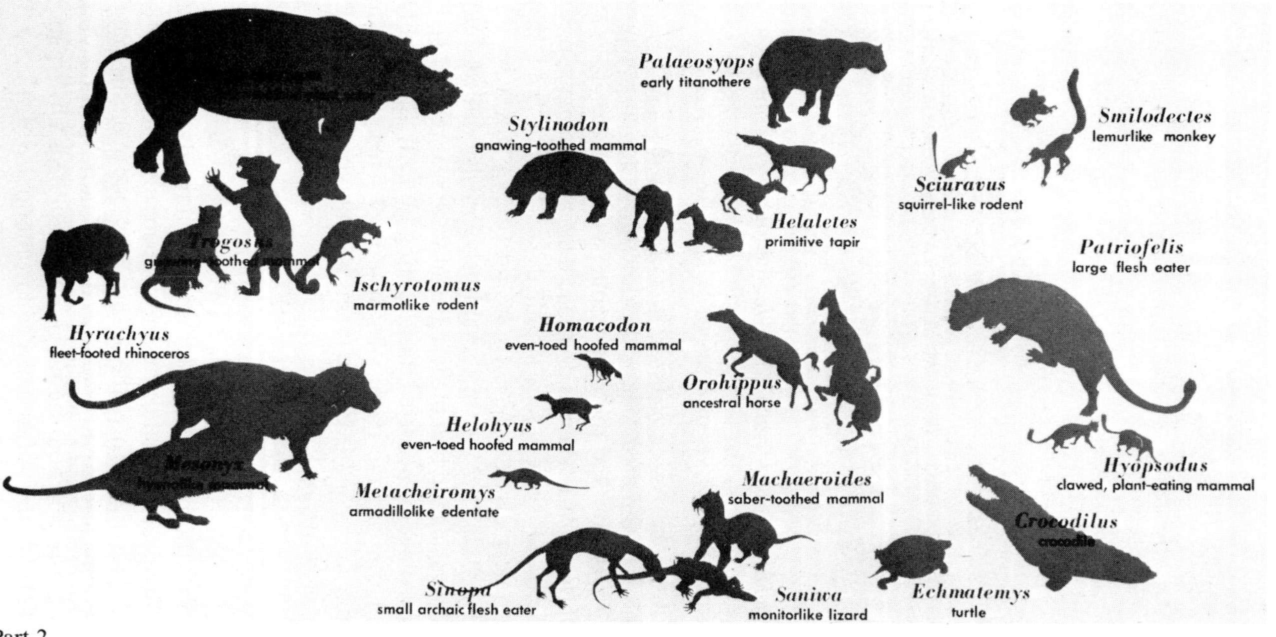

Part 2

FIGURE 10.23
Middle Eocene Bridger Formation of Wyoming. Although the plants and animals are reconstructed as accurately as possible, all of these creatures were probably never so close together. Painting by J. H. Matternes, Smithsonian Institution.

FIGURE 10.24
Eocene whale, *Basilosaurus cetoides,* from Alabama. These toothed mammals quickly took the place of the Mesozoic marine reptiles. The jaws are 1 m (3 ft) long. Painting by C. R. Knight, Field Museum of Natural History. Negative No. CK.63228.

of the animals resemble those in the area today. The differences are conspicuous, such as the shovel-tusked mastodon and the giant camel.

The Pleistocene life of North America was dominated by large animals such as mastodons, mammoths, ground sloths, saber-toothed cats, bears, and giant beavers, as well as horses and camels. (See Figs. 10.28 and 10.29.) This was a time of great climatic change everywhere. Most of the animals just listed died out in North America about 8000 years ago, when the last glacial stage was retreating. No one knows why this great extinction occurred, but hunting by early humans has been suggested. It was not the climatic change because they lived through the glacial times. The present-day fauna of the African plains is somewhat similar to the fauna of the late Cenozoic of North America.

HUMAN FOSSILS

The development of humans, the most successful of the mammals, is relatively poorly known. The humanlike primates and the apes apparently evolved from a common ancestor, but lack of fossils obscures the details. This lack of fossils may in part result from the fact that the apes evolved in the narrow transitional environment between the grasslands and the jungle or forest. This environment probably accounts for the development of two-legged locomotion, which freed the hands for other activities.

The early primates, like other mammals, evolved from small insectivores in the Late Cretaceous. They first adapted to life in the trees. The most important adaptations were the development of an opposable thumb, enabling grasping and swinging from tree limbs, and the shortening of the face, moving the eyes forward for stereoscopic vision. The latter enabled better judgment of distance—a necessity for a climbing animal. The size of the brain also increased. The higher apes came from the trees to the ground, perhaps because the cooling trend caused the forests to be replaced by grasslands. The gorilla and the chimpanzee live largely on the ground.

The oldest apes are early Oligocene, although other primates are found throughout the Cenozoic. Apes appear to have evolved rather rapidly in the Miocene. One, *Ramapithecus,* the oldest tool user, has been dated as Miocene. (See Fig. 10.30.)

Fossil humans are about the rarest fossils. In the past, each new discovery of a bone (complete skeletons are extremely rare) was given a separate name because of disagreement as to where each fossil fits into the evolution of modern humans.

Australopithecus is found with crude stone and bone tools and the remains of increasingly larger animals that it apparently hunted. *Australopithecus* first appears in rocks as old as 3.5 million years. "Lucy," a 40-percent-complete skeleton, was found in 1974, and in 1975, 13 individuals, "the first family," were also found near Hadar, Ethiopia. Early *Australopithecus* was small but evolved into a much larger, more humanlike creature; even the earliest forms walked erect—the first known to do so. The early discoveries led to great confusion because at least two and possibly three species of *Australopithecus* existed at the same time.

Olduvai Gorge, Tanzania, is a unique site in which a stratigraphic record going back 2 million years is exposed; its beds have been dated by the potassium-argon method. The lower and middle beds have revealed evolving *Australopithecus. Homo erectus* has been found in beds dated about a million years ago. (See Fig. 10.31.)

Homo erectus fossils are generally associated with stone tools, and some finds suggest the use of fire. The general characteristics used to distinguish *Homo erectus* are a thick, flat, narrow skull, with an average brain size of 800 to 1000 cm^3 (compared with averages of 1400 cm^3 in modern humans and 500 cm^3 in *Australopithecus* and a modern gorilla), protruding brow ridges, a marked angle for muscle attachment at the rear of the skull, and larger and more primitive teeth and jaw than a modern human's. The legs are close to those of modern humans, unlike the more primitive upright, *Australopithecus.* (See Fig. 10.32.)

Homo erectus has been found in Java, China, and South and East Africa. Tools similar to those associated with those finds have been found in Europe, but so far no fossils, with the exception of a jaw found near Heidelberg. This jaw is dated in the generally accepted mid-range of *Homo erectus,* but it is less primitive than the usual *Homo erectus* fossils, and skull portions that might give more positive assignment as to species are absent; its true position is greatly in doubt. (See Fig. 10.33.)

Although we are now approaching modern humans and the record should begin to clear, it actually does not; there are too many anomalies in the idealized picture. *Homo sapiens* is the species to which modern humans belong. When and

Part 1

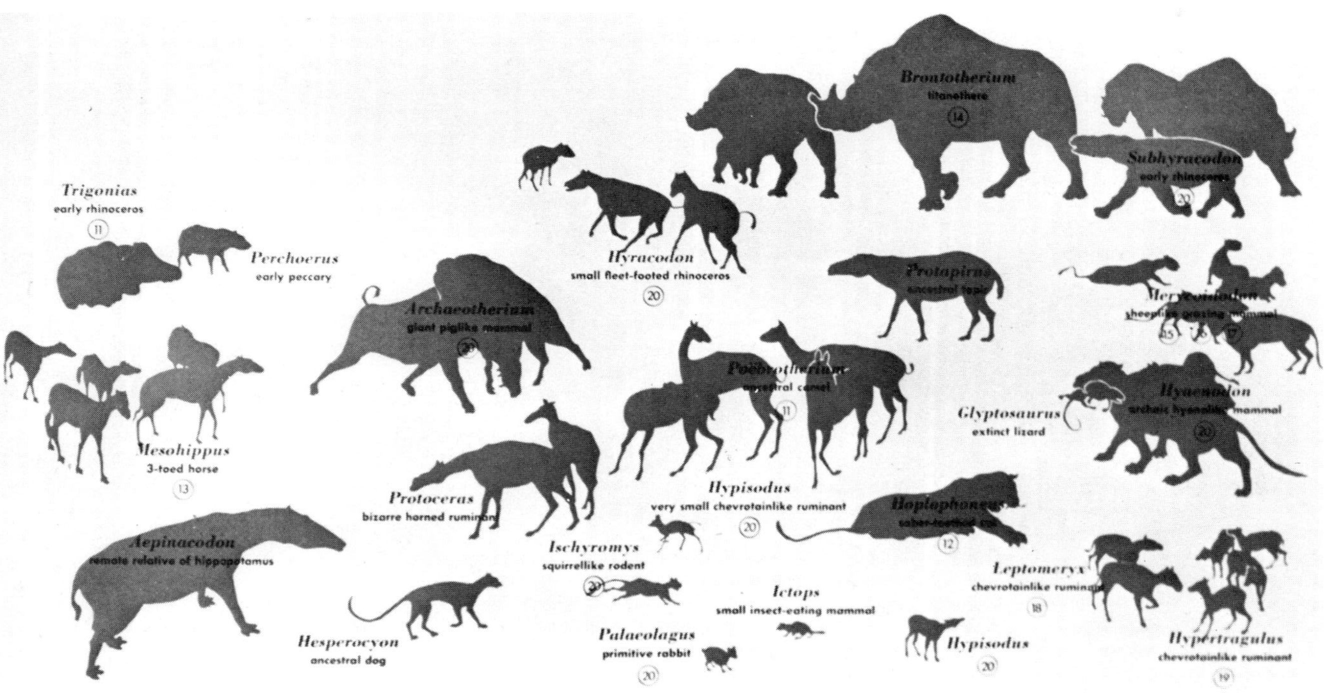

Part 2

FIGURE 10.25
Oligocene of the White River Formation of South Dakota and Nebraska. Painting by J. H. Matternes, Smithsonian Institution.

Part 1

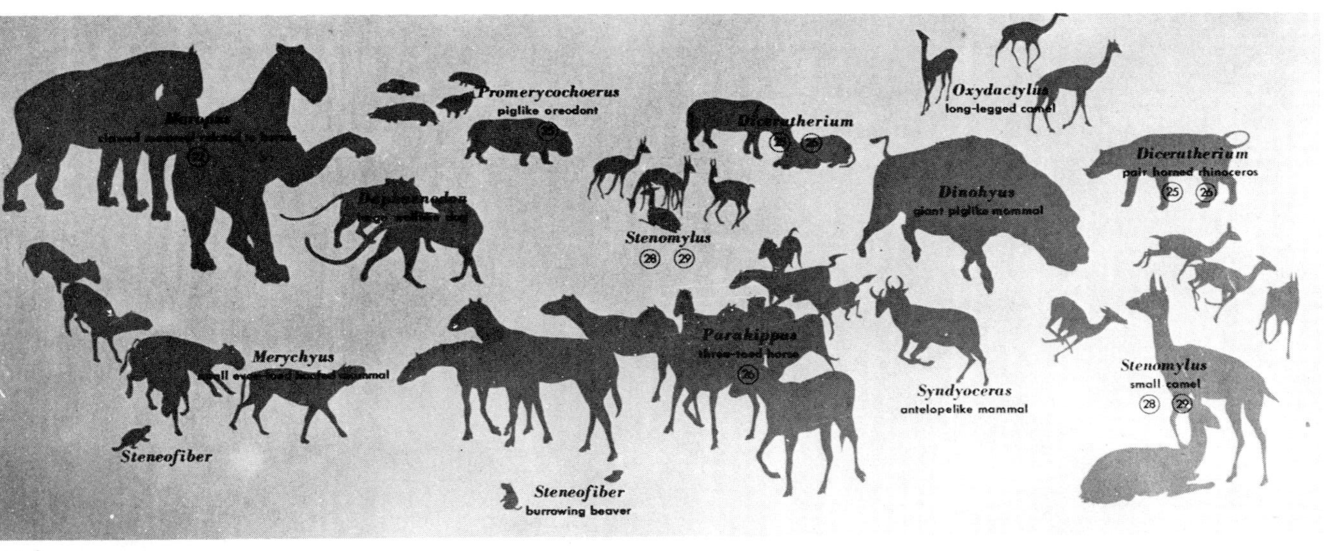

Part 2

FIGURE 10.26
Lower Miocene of the Harrison Formation of western Nebraska. Painting by J. H. Matternes, Smithsonian Institution.

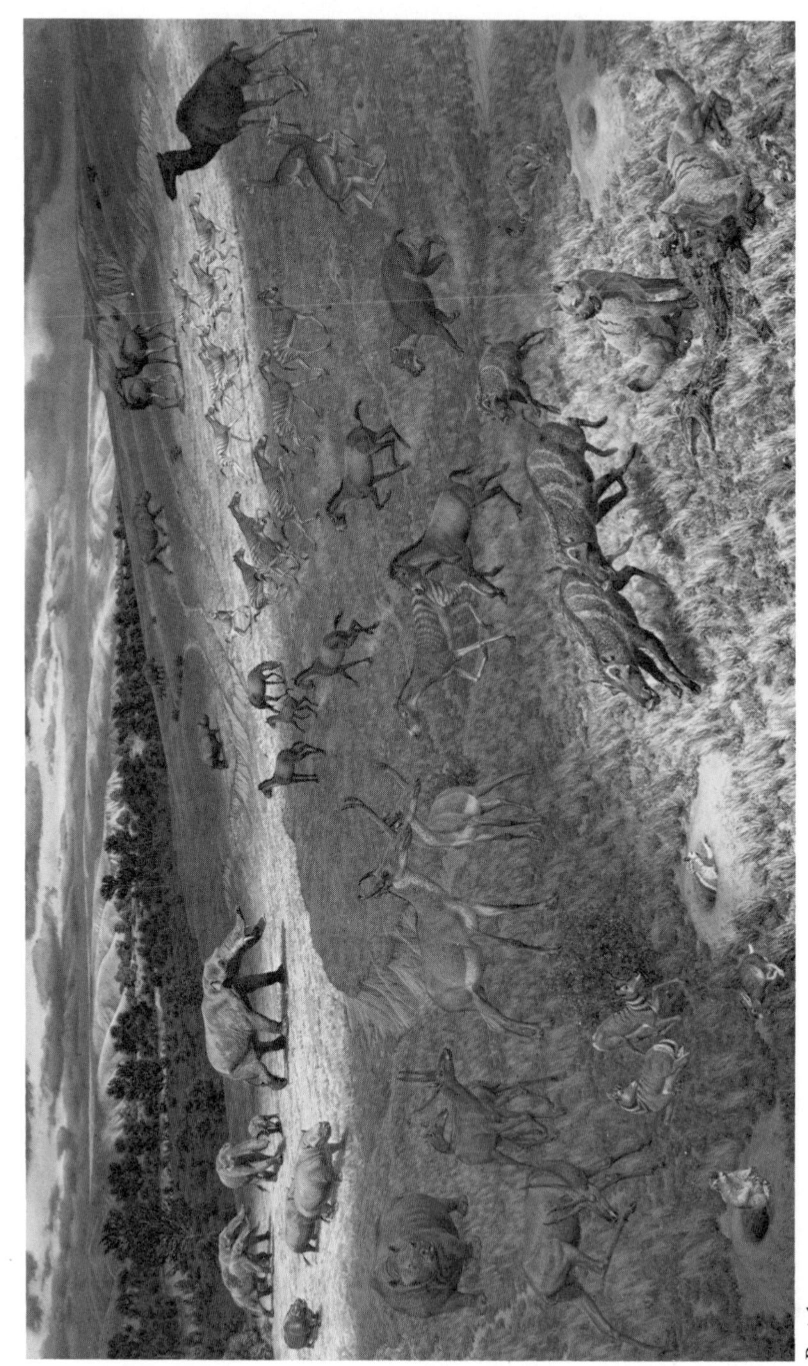

Part 1

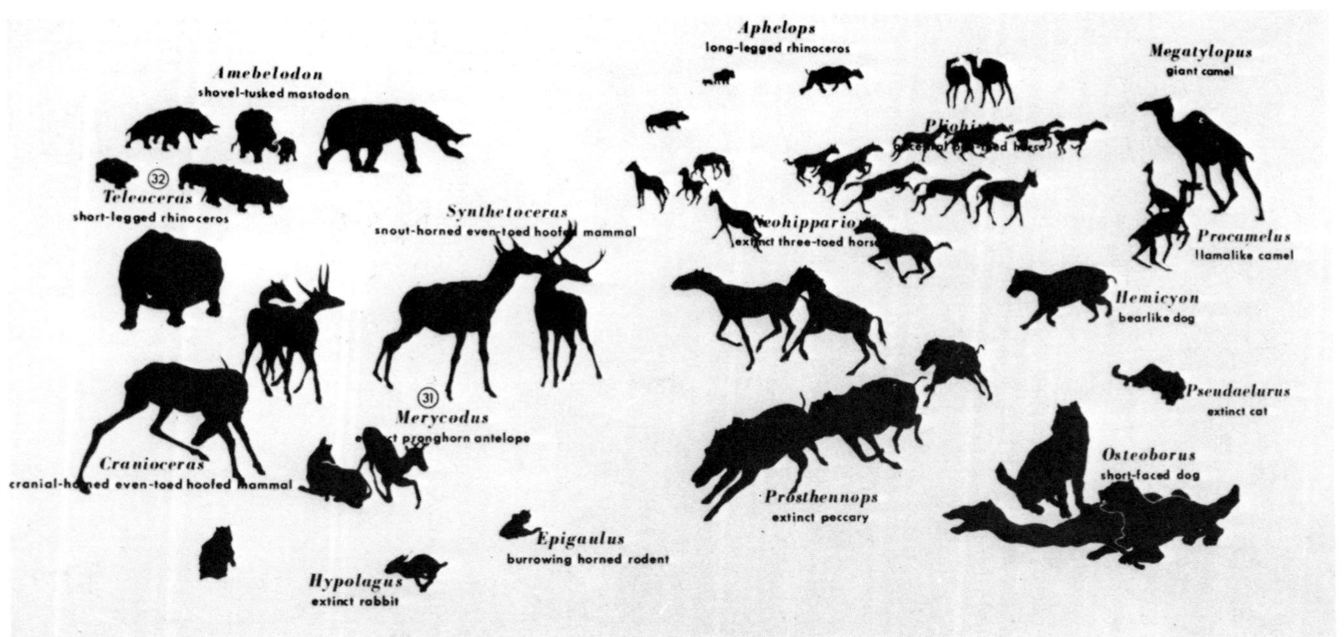

FIGURE 10.27
Early Pliocene of the high plains region. Painting by J. H. Matternes, Smithsonian Institution.

FIGURE 10.28
Pleistocene scene at La Brea tar seeps near Los Angeles, California. Saber-toothed cat, wolf, and vultures have gathered to feed on animals caught in the tar. Painting by C. R. Knight, Field Museum of Natural History. Negative No. CK.73799.

FIGURE 10.29
Pleistocene scene in cool climate. Woolly mammoth and rhinoceros. Painting by C. R. Knight, Field Museum of Natural History. Negative No. CK.71158.

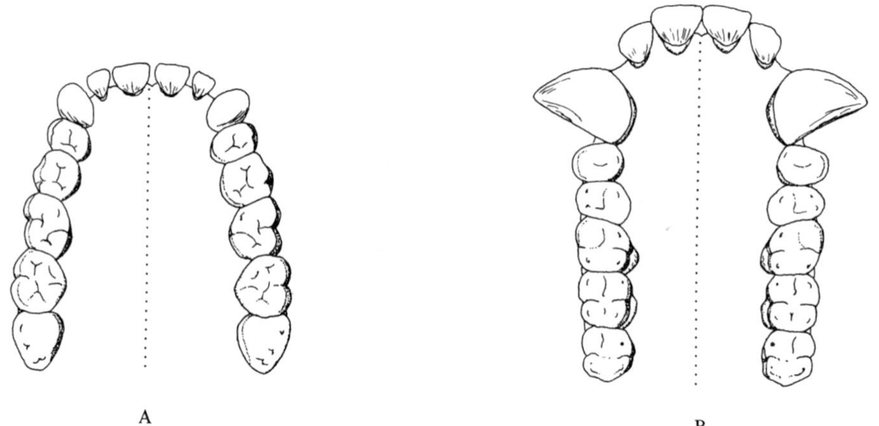

FIGURE 10.30
Upper jaw differences between A. *Australopithecus* and B. gorilla.

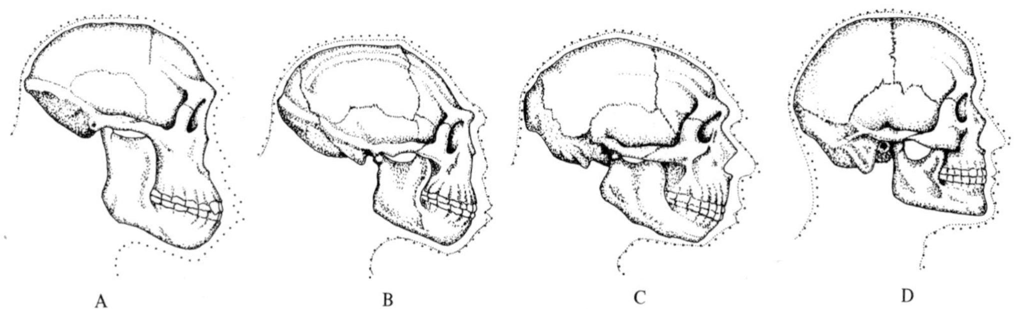

FIGURE 10.31
Comparison of skulls of A. *Australopithecus,* B. *Homo erectus,* C. *Homo sapiens neanderthalensis,* and D. *Homo sapiens.*

where *Homo sapiens* first appeared is now the question, and the answer is ambiguous. Some fossils indicate that *Homo erectus* and *Homo sapiens* may have overlapped considerably in time, or possibly that the progress of evolution of *Homo erectus* varied greatly over its tremendous geographical range, affected locally by climatic conditions and migratory limitations.

Neanderthal fossils occur near the boundary between middle and upper Pleistocene, and they are now classified as *Homo sapiens.* Extensive remains have been found in Europe and around the Mediterranean, dated from nearly 100,000 years ago until about 35,000 years ago. Neanderthal appears to have developed considerable variation. Those inhabiting western Europe, the best and earliest known, are the most different from modern humans in limb and skull shape; as time progressed, they seem to have developed farther from, instead of closer to, modern humans until their disappearance. This seeming "regression" may simply represent an isolated race's adaptation to difficult climatic conditions. Finds in the Middle East indicate great variety and possible transitional forms between those with the extreme western-European characteristics and Crô-Magnon, the oldest modern human in the strict sense. This variety exists even in contemporaneous fossils in single locations.

Modern humans emerged approximately 35,000 years ago, and the older forms disappeared. The most studied and best known is Crô-Magnon, whose remains are nearly indistinguishable from those of modern humans. Crô-Magnon appears suddenly in Europe in beds of about the same age as late Neanderthal, and then Neanderthal disappears. Apparently these people migrated into Europe, probably from the Middle East, considering the possibly transitional forms recently found there. An important problem is whether Crô-Magnon evolved from Neanderthal or simply first intermingled with them in the Middle East and evolved from some other stock. Crô-Magnon is presumed to be the ancestor of the modern European races. Nothing is known of the recent ancestors of the modern races of humans outside Europe.

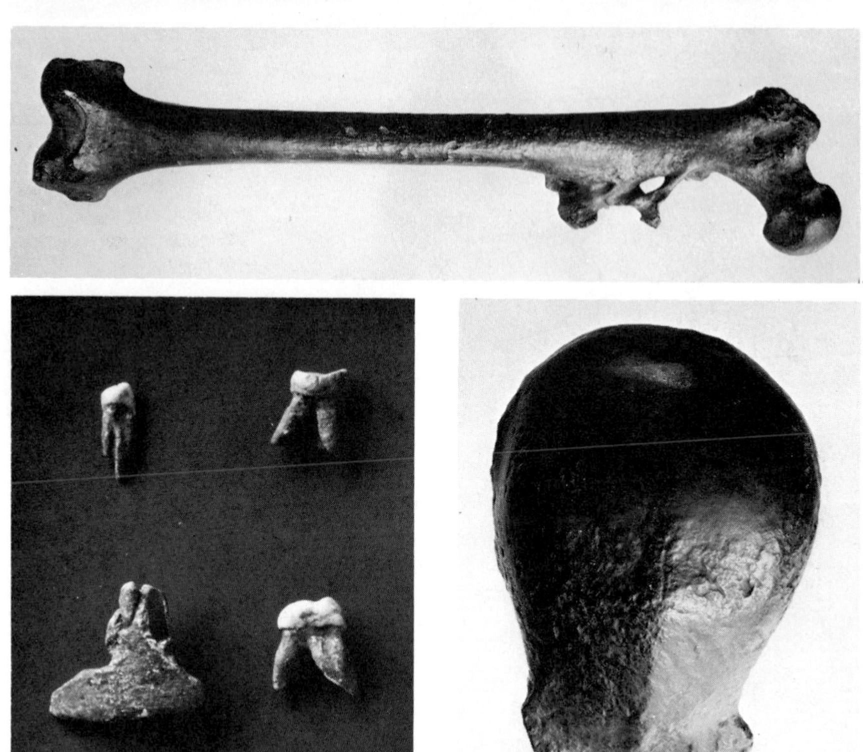

A

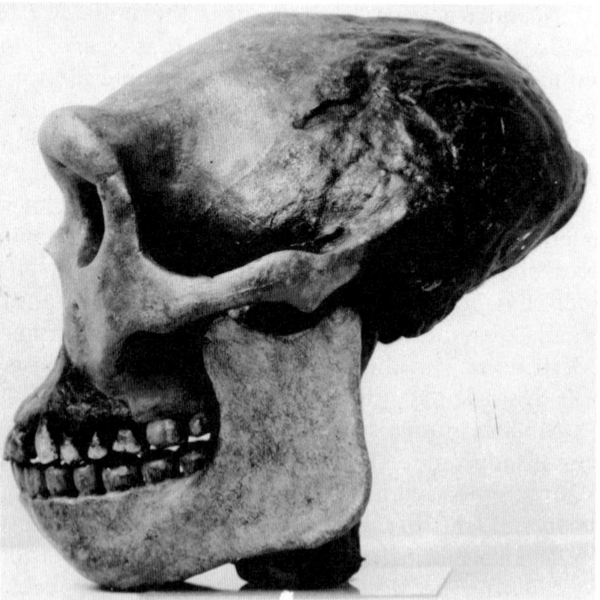

B

FIGURE 10.32
Homo erectus. A. Typical fossils—teeth, femur, and skull cap. B. Restoration of skull by Weidenreich. Note the thick, flat skull with protruding brow ridges and a ridge at the back. The jaw has a receding chin, and the teeth are somewhat primitive. Compare with your skull by running your fingers above and beside your eyes and over the back of your skull and over your chin. Photos from Smithsonian Institution.

FIGURE 10.33
The Heidelberg jaw *(Homo erectus)*. A. A cast of the fossil. B. The restoration by McGregor, based on the jaw. Photos from Smithsonian Institution.

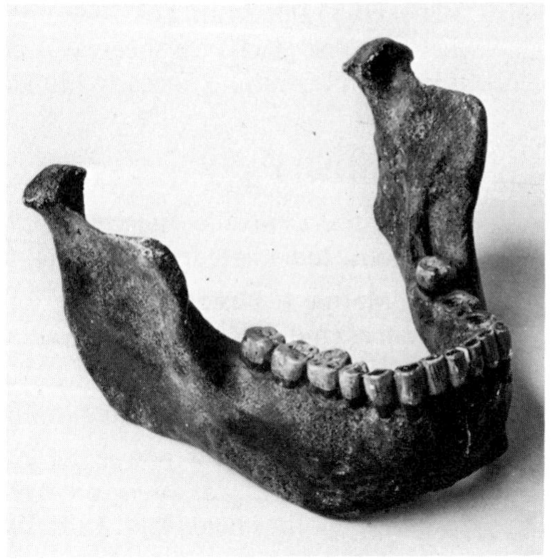

A

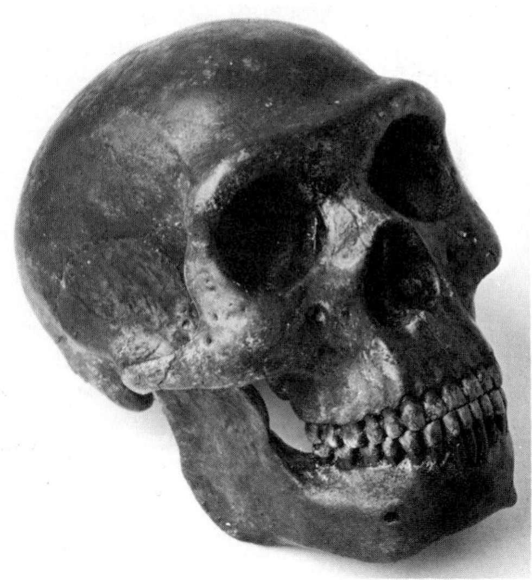

B

The oldest human fossils found in the New World are 11,000 to 13,000 years old on the Palouse River in Washington, and 11,000 to 12,000 years old at Tepexpan, in Mexico. Sites as old as 15,000 years in Alaska and South America, and up to 20,000 years in Mexico, are known from artifacts, but these sites contain no human bones. Many anthropologists believe that human evolution can be described better from artifacts than from bones. Humans must have reached the New World over the Bering land bridge between Siberia and Alaska. They must have crossed the bridge about 35,000 years ago because the bridge was submerged between 35,000 years and 25,000 years ago, and at 25,000 years ago the corridor between the glacial ice sheets was closed by the meeting of the glaciers. Thus the search will continue for early humans in the New World.

SUMMARY

During the Cenozoic, the continents continued their movements to the present locations. India collided with Asia, forming the Himalaya Mountains.

Marine deposition occurred on the Atlantic and Gulf coasts, with repeated advances and retreats of the seas. Up to 15,250 m (50,000 ft) of clastics were deposited on the Gulf Coast.

Limestone was deposited in Florida.

Erosion surfaces reveal the events in the Appalachian Mountains. Coastal plain deposition was on a pre-Cretaceous peneplain. Uplift caused erosion to produce another peneplain. Later uplifts caused dissection.

Orogeny in the Rocky Mountains in late Cretaceous and early Cenozoic time produced mountains and basins. Erosion reduced the ranges and filled the basins by the Oligocene. In Oligocene and Pliocene time, rivers spread thin clastics over the plains area. Uplift or climatic change caused the rivers to downcut, producing the present topography.

During the Eocene, a freshwater lake in which oil shale was deposited covered much of Utah.

The Colorado Plateau was uplifted but otherwise was not greatly deformed during the Cenozoic, even though the surrounding areas were severely deformed.

Faulting in the middle and late Cenozoic produced the Basin and Range topography of the Great Basin.

Northwestern United States is dominated by volcanic rocks. Marine clastics near the coast grade inland with volcanics and nonmarine rocks in the Cascade Mountains. East of the Cascades, about a thousand meters of Columbia River Basalt cover thousands of square kilometers. Pleistocene volcanoes cap the Cascade Range.

The California coast was the site of marine clastic deposition in early Cenozoic that gave way to red beds in the Oligocene. Marine deposition, volcanism, and orogeny all occurred in the Miocene. In the Pleistocene, basins developed.

In the Arctic, some nonmarine clastics were deposited and folded. Subsidence produced the present islands.

Much of the present topography of northern North America was produced by glaciation.

The extent of continental glaciation is approximately marked by the Missouri River and Long Island.

In east-central Washington, the sudden release of an ice-dammed lake of glacial meltwater eroded the Channeled Scablands.

The Great Lakes were formed by continental glaciation.

Clams, oysters, and snails were the main elements in Cenozoic seas.

The plant life of the Cenozoic was essentially modern, with angiosperms abundant.

The Cenozoic is the age of mammals.

Marsupials migrated to Australia and South America in the Late Cretaceous and became isolated. In the Pliocene, the more successful placentals migrated to South America and replaced most of the marsupials.

The Paleocene was warm and temperate in North America, and mammals were archaic and mainly small.

The Eocene was subtropical. Rhinoceroses, deer, pigs, camels, and toothed whales appeared.

The Oligocene was mild, and the dry climate favored faster, long-legged animals. Cats and dogs appeared. *Brontotherium* was the largest mammal ever in North America.

The Miocene was cool and dry, favoring horses, pigs, rhinoceroses, camels, antelopes, deer, and mastodons.

The drying trend continued through the Pliocene.

The Pleistocene was dominated by large animals such as mastodons, mammoths, ground sloths, saber-toothed cats, bears, and giant beavers. Why these animals died out in North America is not known.

The oldest apes are early Oligocene, although other primates are older.

Australopithecus, about 3.5 million years old, used crude stone and bone tools.

Homo erectus, the first human, has been found in beds as old as 1 million years.

Homo sapiens has been found in beds as old as 100,000 years and appears to overlap *Homo erectus.*

Modern humans emerged about 35,000 years ago.

Humans must have come to the New World about 35,000 years ago, but the oldest artifacts are only 20,000 years old.

QUESTIONS

1. When did the four major advances of the sea occur on the Gulf Coast?
2. Describe the erosional history of the Appalachian Mountains.
3. Describe the topography of the Pacific coast from Washington to California.
4. Outline the Cenozoic history of arctic Canada.
5. Outline the development of the Rocky Mountains.
6. When and how did the Basin and Range structure of the Great Basin develop?
7. Where is the thickest section of marine Cenozoic rocks found in North America?
8. How much of North America was covered by ice during the Pleistocene? How many advances occurred?
9. Describe as many of the features of North America caused by glaciers as you can.

10. Describe the Pleistocene events in northwestern United States.
11. How does the marine life of the Cenozoic differ from that of the Mesozoic?
12. What are the two main types of mammals?
13. Outline the development of marsupials and account for their present distribution.
14. Describe the changing climates of the Cenozoic and how these affected the development of mammals.
15. What problems will future geologists have in correlating the present African "big game" fauna with North American fossils?
16. Why is it difficult to determine the development (that is, describe the family tree) of modern humans?
17. When and where did modern humans first appear?

SUGGESTED READINGS

Baldridge, W. S., and K. H. Olsen, "The Rio Grande Rift," *American Scientist* (May-June 1989), Vol. 77, No. 3, pp. 240–247.

Cartmill, Matt, David Pilbeam, and Glynn Isaac, "One Hundred Years of Paleoanthropology," *American Scientist* (July-August 1986), Vol. 74, No. 4, pp. 410–420.

Eckhardt, R. B., "Population Genetics and Human Origins," *Scientific American* (January 1972), Vol. 226, No. 1, pp. 94–103.

Holloway, R. L., "The Casts of Fossil Hominid Brains," *Scientific American* (July 1974), Vol. 231, No. 1, pp. 106–115.

Krantz, G. S., "Human Activities and Megafaunal Extinctions," *American Scientist* (March-April 1970), Vol. 58, No. 2, pp. 164–170.

Marshall, L. G., "Land Mammals and the Great American Interchange," *American Scientist* (July-August 1988), Vol. 76, No. 4, pp. 380–388.

Molnar, Peter, "The Geologic History and Structure of the Himalaya," *American Scientist* (March-April 1986), Vol. 74, No. 2, pp. 144–154.

Raven, P. H., and D. I. Axelrod, "History of the Flora and Fauna of Latin America," *American Scientist* (July-August 1975), Vol. 63, No. 4, pp. 420–429.

Simons, E. L., "Human Origins," *Science* (September 22, 1989), Vol. 245, No. 4924, pp. 1343–1350.

Trimble, D. E., *The Geologic Story of the Great Plains.* U.S. Geological Survey Bulletin 1493, Washington, D.C.: U.S. Government Printing Office, 1980, 55 pp.

Trinkaus, Erik, and W. W. Howells, "The Neanderthals," *Scientific American* (December 1979), Vol. 241, No. 6, pp. 118–133.

Walker, Alan, and R. E. F. Leakey, "The Hominids of East Turkana," *Scientific American* (August 1978), Vol. 239, No. 2, pp. 54–66.

Wernicke, Brian, G. J. Axen, and J. K. Snow, "Basin and Range Extensional Tectonics at the Latitude of Las Vegas, Nevada," *Bulletin, Geological Society of America* (November 1988), Vol. 100, No. 11, pp. 1738–1757.

Wolfe, J. A., "A Paleobotanical Interpretation of Tertiary Climates in the Northern Hemisphere," *American Scientist* (November-December 1978), Vol. 66, No. 6, pp. 694–703.

Appendix A
Rocks

ROCK CYCLE

Rocks are classified into three types according to their origin. Two of these types are formed by processes deep in the earth and, therefore, tell us something about conditions within the crust. They are

1. **igneous rocks,** which solidify from a melt or magma, and
2. **metamorphic rocks,** which are rocks that have been changed—generally by high temperature and pressure within the crust.

The third type, which records the conditions at the surface, is

3. **sedimentary rocks,** which are composed of the weathered material that is deposited in layers near the earth's surface by water, wind, ice, chemical precipitation, or organic activity.

Much of geology is concerned with the interactions among the forces that produce these three rock types. The relationships are quite involved, but can be illustrated by the rock cycle (Fig. A.1).

IGNEOUS ROCKS

Igneous rocks are classified on the bases of composition and texture. Because almost all coarse-grained rocks contain feldspar, the most common mineral family,

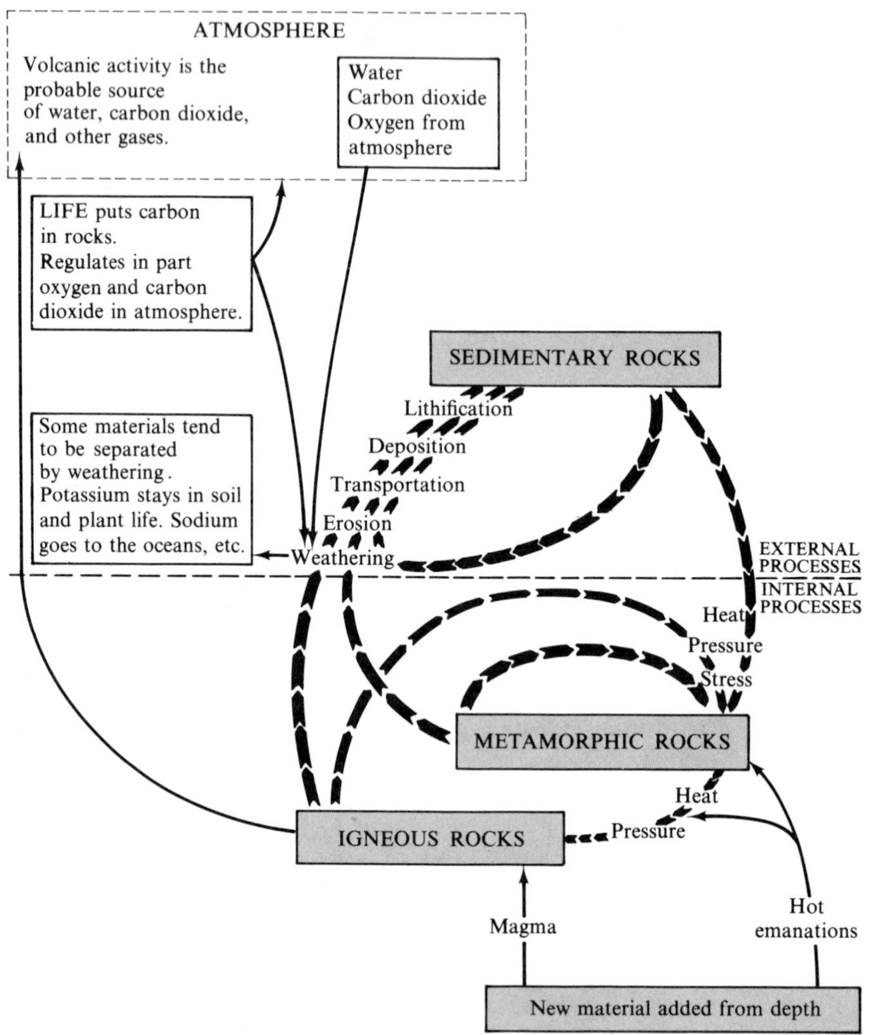

FIGURE A.1
The middle of this diagram, showing the relationships among igneous, sedimentary, and metamorphic rocks, is what is generally considered the rock cycle. The upper and lower parts of the figure show how material is added to and subtracted from the rock cycle.

the type of feldspar is the most important factor in the composition. Of secondary importance is the type of dark mineral that generally accompanies each type of feldspar. See classification diagram, Figure A.3.

In a similar way, the texture tells much about the cooling history of an igneous rock. In igneous rocks the texture refers mainly to the grain size. Rocks that

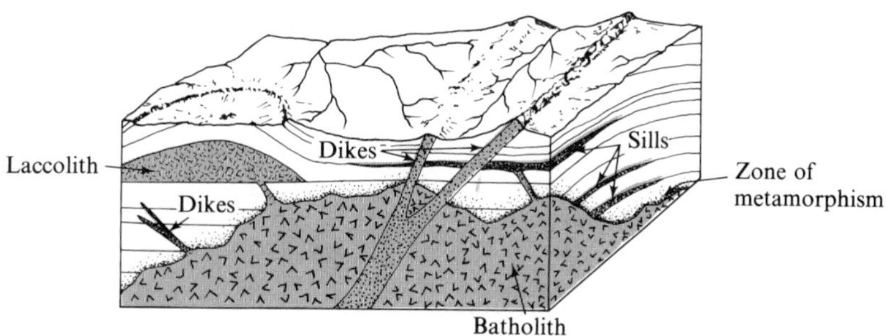

FIGURE A.2
Intrusive igneous rock bodies. The laccolith and sills are concordant with the enclosing sedimentary beds, and the batholith and dikes are discordant. The heat from the batholith metamorphoses the surrounding rocks.

cool slowly are able to grow large crystals; quickly chilled rocks, such as volcanic rocks, are fine grained or glassy.

Igneous rocks occur in two ways, either as intrusive (below the surface) bodies or as extrusive (on the surface) rocks. The ultimate source of igneous magma is probably deep in the crust or in the upper part of the mantle, and the terms intrusive and extrusive refer to the place where the rock solidified. Intrusive igneous rocks can be seen only where erosion has uncovered them. They are described as concordant if the contacts of the intrusive body are more or less parallel to the bedding of the intruded rocks and discordant if the intrusive body cuts across the older rocks. (See Fig. A.2.)

On the chart shown as Figure A.3, the term granite is used for most of the coarse-grained, quartz-bearing rocks. A number of other names are also in use for these rocks, depending on whether they contain one feldspar, two feldspars, or mixed feldspar (perthite). (See Fig. A.4.) Diorite generally has more dark minerals than does granite, and the feldspar is plagioclase. Quartz diorite is distinguished from granite in that it contains only plagioclase feldspar. The latter distinction, however, is not always easy to make.

SEDIMENTARY ROCKS

Clastic Sedimentary Rocks

These rocks are composed of rock fragments or mineral grains broken from any type of preexisting rock. They are subdivided according to fragment size. Commonly, sizes are mixed, requiring intermediate names such as sandy siltstone. They are recognized by their clastic texture. The fragments originate from mechanical weathering.

Composition is not used in the general classification of clastic sedimentary rocks. Composition can be used to subdivide sandstones, as indicated in Table A.1.

The actual formation of shale is somewhat more complex than indicated in

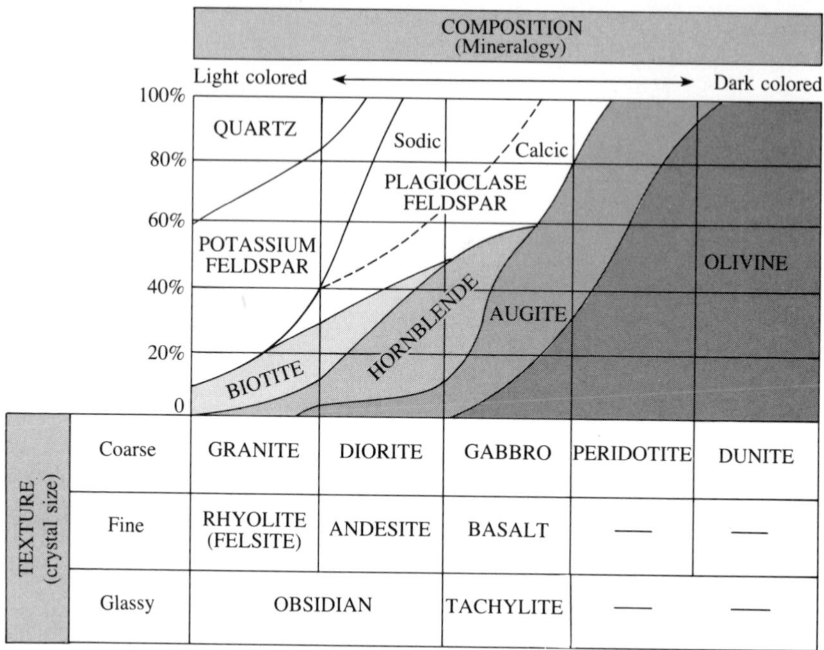

TEXTURE (crystal size)					
Coarse	GRANITE	DIORITE	GABBRO	PERIDOTITE	DUNITE
Fine	RHYOLITE (FELSITE)	ANDESITE	BASALT	—	—
Glassy	OBSIDIAN		TACHYLITE	—	—

FIGURE A.3
Chart showing the classification of the igneous rocks. Composition is indicated horizontally, and texture is indicated vertically. The upper part of the figure shows the range in mineral composition of each rock type.

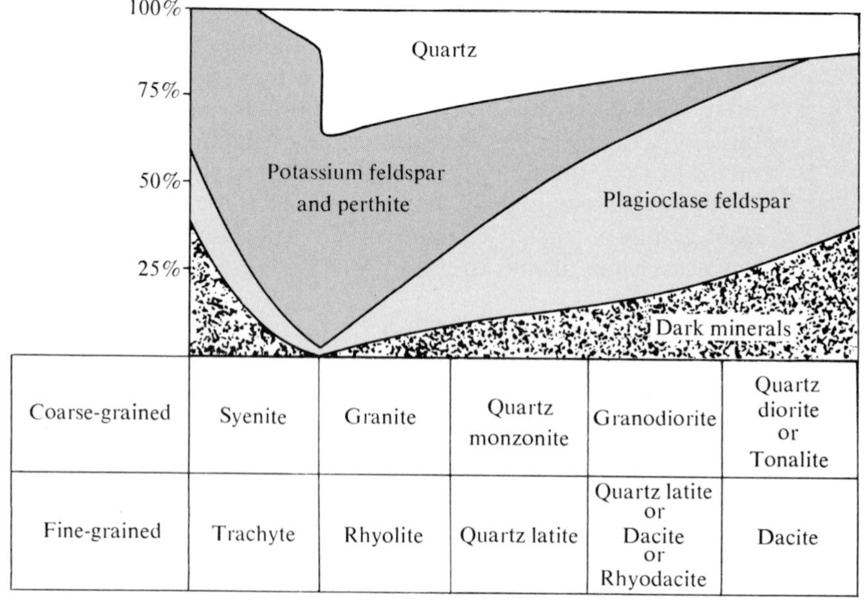

Coarse-grained	Syenite	Granite	Quartz monzonite	Granodiorite	Quartz diorite or Tonalite
Fine-grained	Trachyte	Rhyolite	Quartz latite	Quartz latite or Dacite or Rhyodacite	Dacite

FIGURE A.4
Subdivision showing the names in use for the granitic rocks. The chart is an amplification of the left end of Figure A.3.

TABLE A.1
Clastic sedimentary rocks.

Size	Sediment	Rock
>2 mm	Gravel	*Conglomerate*—generally has a sandy matrix. May be subdivided into *roundstone* and *sharpstone conglomerate* (*breccia* is a synonym) depending on the fragment shape.
$\frac{1}{16}$–2 mm	Sand	*Sandstone*—recognized by gritty feel. Generally designated as coarse, medium, or fine if well sorted. They have also been subdivided on the basis of composition. The more important types are *Quartzose sandstone* (mainly quartz). *Arkosic sandstone* (*arkose*)—over 20 percent feldspar. *Graywacke*—poorly sorted, with clay or chloritic matrix.
$\frac{1}{256}$–$\frac{1}{16}$ mm	Silt	*Siltstone (mudstone)*—may be necessary to rub on teeth to detect grittiness, thus distinguishing it from shale.
$<\frac{1}{256}$	Clay	*Shale*—distinguished from siltstone by its lack of grittiness and its *fissility* (ability to split very easily on bedding planes). Rocks composed of clay that lack fissility are called *claystone.*

Table A.1. Much of the clay that gives a shale its fissility apparently develops or is mechanically reoriented after deposition. This is suggested by the lack of geologically young shale. Young fine-grained clastic rocks are mainly mudstones.

Nonclastic Sedimentary Rocks

These rocks are formed by chemical precipitation, by biologic precipitation, and by accumulation of organic material. These processes extract specific materials from their surroundings, generally seawater, and precipitate these substances, forming rocks. The rocks are classified mainly by composition. (See Table A.2.) As with the clastic rocks, these rocks are commonly mixed, both among themselves and with the clastic rocks.

Limestone is composed of calcite. It is recognized by effervescence with dilute hydrochloric acid. It is generally of biologic origin and may contain fossils. A rock composed mainly of fossils or fossil fragments is called **coquina.**

Oolites are tiny spherical grains formed by layers of calcium carbonate deposited around a small sand grain or shell fragment. Rocks composed of these grains are called **oolitic limestones.**

Chalk is soft, white limestone formed by the accumulation of the shells of microscopic animals. The shells are composed of calcium carbonate, and chalk is recognized by effervescence with acid.

Dolomite is composed mainly of the mineral dolomite. It is recognized by

TABLE A.2
Nonclastic sedimentary rocks.

Rock	Principal Texture	Principal Composition	How Recognized
Limestone	Fine	Calcite $CaCO_3$	Effervesces with dilute hydrochloric acid. Generally of biologic origin and may contain fossils. *Coquina* is limestone composed mainly of fossils or fossil fragments.
Chalk	Fine	Calcite $CaCO_3$	Soft white limestone formed by the accumulation of microscopic shells. Effervesces with dilute hydrochloric acid. Compare with diatomite.
Dolomite	Fine	Dolomite $CaMg(CO_3)_2$	Effervesces with dilute hydrochloric acid after scratching (to produce powder). May have irregular holes.
Chert	Fine	Silica SiO_2	Hardness 6 or 7. Similar to chalcedony. *Flint* is black or dark gray.
Gypsum	Fine to coarse	Gypsum $CaSO_4 \cdot 2H_2O$	Hardness 2. One good cleavage and two poorer cleavages.
Rock salt	Fine to coarse	Halite NaCl	Cubic crystals and cleavage may be visible. Salty taste.
Diatomite	Fine	Silica SiO_2	Soft white rock formed by the accumulation of microscopic shells composed of silica. Distinguished from chalk by lack of effervescence.
Coal	Fine to coarse	Carbon C	Soft, black, burns.

effervescence with dilute hydrochloric acid after scratching (to produce powder); it will also react (without scratching) with concentrated or with warm dilute hydrochloric acid. Dolomite is generally formed by replacement of calcite, presumably while in contact with seawater, soon after burial. Some of the calcium atoms in calcite are replaced by magnesium atoms. The reduction in volume in this replacement may produce irregular voids and generally obliterates fossils.

Chert is microscopically fine-grained silica (SiO_2). It is equivalent to chalcedony. It is unfortunate that many names have been applied to the fine-grained varieties of silica. These names were based on minor differences, such as color and luster, that cannot be substantiated by modern methods, and there is little agreement among mineralogists on their naming. Both chert and chalcedony may contain opal and fine-grained quartz. Dozens of names have been applied to chalcedonic silica; of these, a few may be useful, e.g., agate for banded types and flint for dark-gray or black chert.

Chert originates in several ways. Some may precipitate directly from seawater in areas where volcanism releases abundant silica. Most comes from the accumulation of silica shells of organisms. These organisms remove dissolved silica from seawater to form their shells or skeletons. These silica remains come from diatoms, radiolarians, and sponge spicules, and are composed of opal. Opal is easily recrystallized to form chert. Thus much chert is recrystallized, making the origin difficult to discern. In recrystallization the silica may replace other materials, and fossils replaced by silica are common. If this occurs in limestone, beautifully preserved fossils with delicate features intact can be recovered by dissolving the limestone with acid. Chert may form either beds or nodules.

Gypsum forms from the evaporation of seawater. Rocks formed in this way are termed evaporites; they are discussed below.

Rock salt is composed of halite. It is recognized by its cubic cleavage and its taste. It is deposited when restricted parts of the sea are evaporated.

Diatomite is a soft, white rock composed of the remains of microscopic plants. Because the remains are composed of silica, it is distinguished from chalk by lack of effervescence.

Coal is formed by the accumulation of plant material.

Evaporite deposits are formed by the evaporation of seawater. Gypsum and rock salt are the main rocks formed in this way. When seawater is evaporated at surface temperatures, such as in a restricted basin, the first mineral precipitated is calcite. Dolomite is the next mineral precipitated, but only very small amounts of limestone and dolomite can be formed in this way. Evaporation of a kilometer column of seawater would produce only a few centimeters of limestone and dolomite. After about two-thirds of the water is evaporated, gypsum is precipitated; when nine-tenths of the water is removed, halite forms. During the last stages of precipitation, potassium and magnesium salts form. Thick beds of rock salt imply evaporation of large amounts of seawater.

The rocks just described are the most important, but by no means the only, sedimentary rocks. Any mixture of types is possible and, because any weathering product may form a sedimentary rock, endless variety is possible. Some less abundant types include economically important deposits such as iron oxides, phosphorous rocks, bauxite (aluminum ore), and potash.

METAMORPHIC ROCKS

For classification and identification, metamorphic rocks can be subdivided into two textural groups:

1. Nonfoliated—homogeneous or massive rocks.
2. Foliated—having a directional or layered aspect.

Nonfoliated Metamorphic Rocks

Because the nonfoliated rocks are the simpler, we will consider them first. There are two types of nonfoliated metamorphic rocks. The first type consists of thermal or contact metamorphic rocks called **hornfels** (singular). They generally occur in narrow belts around intrusive bodies and may originate from any type of parent rock. They are generally fine-grained, tough rocks that are difficult to identify without microscopic study unless the field relations are clear. They range from being completely recrystallized, with none of their original features preserved, to being slightly modified rocks, with most of the original features preserved.

The second type of nonfoliated metamorphic rock develops if the newly formed metamorphic minerals are equidimensional and so do not grow in any preferred orientation. The best examples of this are the rocks that result when monomineralic rocks such as limestone, quartz sandstone, and dunite are metamorphosed. In the case of limestone, which is composed of calcite, no new mineral can form; thus, the calcite crystals, which are small in limestone, grow bigger, develop an interlocking texture, and become **marble.** Marble can be distinguished from limestone only by its larger crystals and lack of fossils; being composed of calcite, both effervesce in acid. Marble may have any color; pure white calcite rocks are more apt to be marble than limestone. Impure limestones, as just noted, develop new minerals, particularly garnet. Dolomite marble also occurs.

Because quartz sandstone similarly can form no new minerals, the quartz crystals enlarge and intergrow to form quartzite. **Quartzite** is distinguished from chalcedony, chert, and opal by its peculiar sugary luster.

The olivine of dunite is changed to the mineral serpentine, and the metamorphic rock produced is called **serpentinite.** In other instances, instead of serpentine, talc is the metamorphic mineral that forms. If nonfoliated, the talc rock is called **soapstone;** if foliated, it is called **talc schist.** Slight differences in the composition of the parent rock or of the water solutions that cause this type of metamorphism probably determine which mineral is formed.

Foliated Metamorphic Rocks

Foliated rocks are the more common of the metamorphic rocks and are the product of regional metamorphism. Regional metamorphism generally takes place in areas undergoing deformation, and these very active areas are generally exposed in the cores of mountain ranges. Regional metamorphic rocks are produced by the same stress that makes mountain structures. The forces that fold and fault the shallow rocks exposed on the flanks of mountain ranges can provide the stress fields under which the deep-seated metamorphic rocks, now exposed in the uplifted cores of mountain ranges, were recrystallized.

An example of the rocks produced by increasing metamorphic grade will show most of the foliated metamorphic rock types. These same rock types will form when many other parent rocks are metamorphosed; the differences in original chemical composition will change the relative amounts of the metamorphic minerals only somewhat.

Shale is composed mainly of clay. During lower-grade metamorphism, the clay is transformed to mica, and at higher grades of metamorphism, to feldspar. Which micas and which feldspars form depends on the bulk chemical composition. Because mica is a flat, platy mineral, it tends to grow with its leaves perpendicular to the maximum stress, forming a preferred orientation. Microscopic examination of some foliated rocks shows that the bedding planes are slightly displaced in the plane of the foliation. This suggests that foliation, at least at some places, forms in shear directions. The shear direction is in general at about 45 degrees to the maximum stress in uniform materials. Thus foliation may form perpendicular to the direction of maximum stress or at about 45 degrees to that direction.

The sequence in the metamorphism of a shale is given in Table A.3. The rock names are applied for the overall texture of the rock and not strictly for the mineral transformations, which occur over a range in temperature and pressure. Most **schists** contain feldspar, but the name **gneiss** is reserved for rocks with much more feldspar than mica. Other rocks that have the same general bulk chemical composition, thus producing the same metamorphic rocks, are certain pyroclastic and volcanic sedimentary rocks, many sandstones, arkose, granite, and rhyolite. The coarse-grained rocks in this list are rarely the parents of slate or fine-grained schist but, in general, remain more or less unaffected until medium-grade metamorphism is reached. Occurrence of **slate** in areas where the nonshale rocks

TABLE A.3
Metamorphism of shale.

Sedimentary Rock	Low-grade Metamorphism	Medium-grade Metamorphism	High-grade Metamorphism
Shale →	→ **Slate** →	→ **Schist** →	→ **Gneiss**
Clay	Clay begins to be transformed into mica. Mica crystals are too small to see, but impart a foliation to the rock. May also form by mechanical rearrangement of clay alone.	Mica grains are larger so that the rock has a conspicuous foliation.	Mica has transformed largely to feldspar, giving the rock a banded or layered aspect.

TABLE A.4
Metamorphism of basalt.

Parent Rock	Low-grade Metamorphism	Medium- and High-grade Metamorphism
Basalt ———	→ Greenschist ——— (Greenstone) Fine-grained, foliated Chlorite (green "mica") Green amphibole Quartz	→ Amphibolite (Amphibole schist) Coarse-grained, foliated Plagioclase Dark amphibole

are not metamorphosed suggests that slate may be formed by a mechanical reorientation of clay particles during folding, together with limited recrystallization.

Only one other group of regional metamorphic rocks must be considered among the most common metamorphic rocks. These rocks are formed by the metamorphism of basalt, certain pyroclastic and volcanic sedimentary rocks, gabbro, graywacke, and some calcite-bearing or dolomitic sedimentary rocks. The sequence is given in Table A.4.

Cataclastic metamorphism results when rocks are broken, sheared, and ground near the surface, where the temperature and pressure are too low to cause any significant recrystallization. Thus, these rocks are commonly associated with fault zones, but there are all gradations between these rocks and ordinary schist, especially lower-grade schist formed from coarse-grained rocks. The fine-grained, ground rock in a fault zone is called fault gouge, or fault breccia if it has little cohesion, and similar cohesive rocks are generally called **mylonite.** As the amount of recrystallization of such rocks increases, they become gradational with schist and gneiss. The origin of these rocks is clear when they are encountered in the field, but they are very difficult to identify in hand specimen.

The origin, texture, and mineralogy of the common metamorphic rocks are summarized in Figures A.5 and A.6.

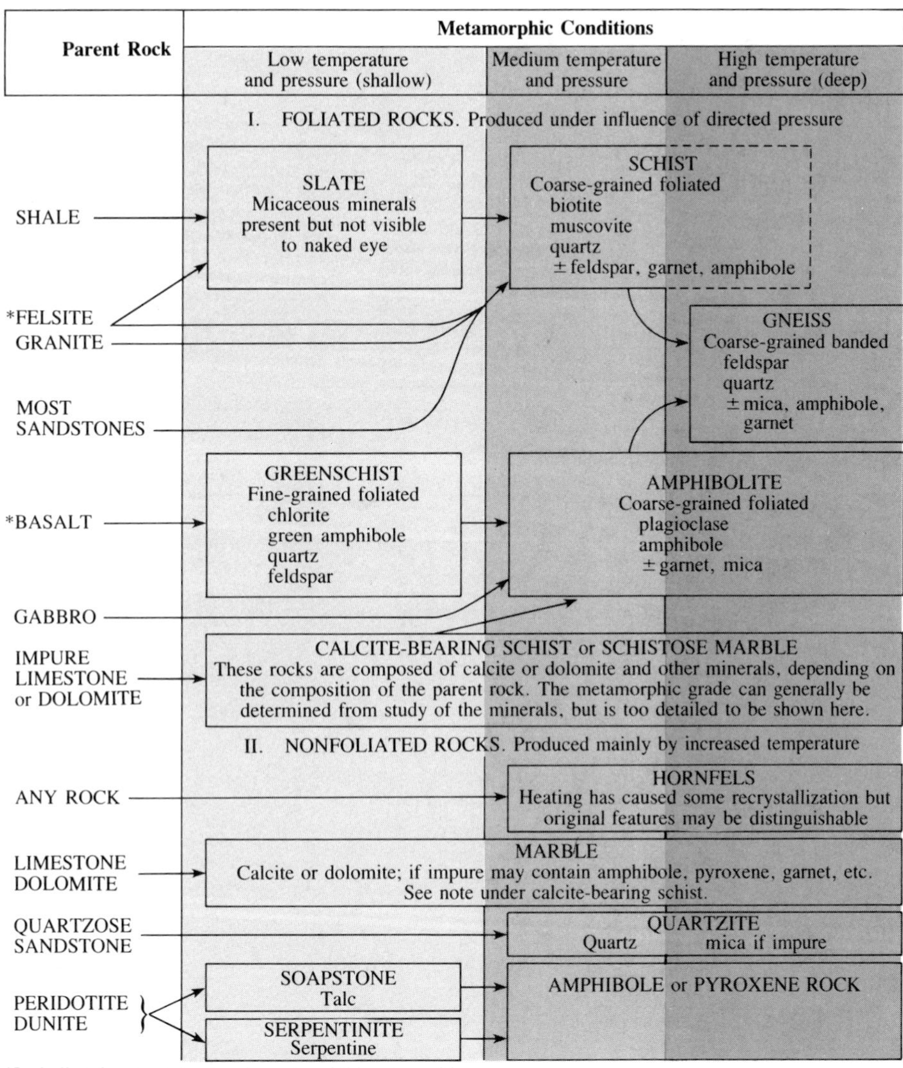

FIGURE A.5
A generalized chart showing the origin of common metamorphic rocks.

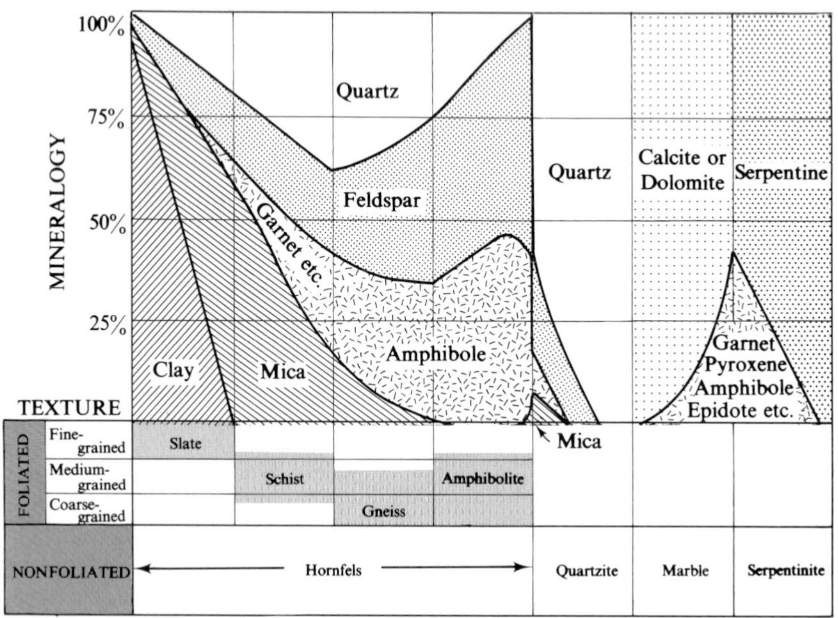

FIGURE A.6
Texture and mineralogy of common metamorphic rocks.

Appendix B Topographic Maps

TOPOGRAPHIC MAPS

Maps are simply scale drawings of a part of the earth's surface. Thus they are similar to a blueprint of an object or to a clothing pattern. Most maps are drawn on sheets of paper and so show only the two horizontal dimensions. Geologists, together with civil engineers and many other users of maps, require that the third dimension, elevation, be shown on maps. Maps that indicate the shape of the land are called *topographic maps.* A number of different ways of showing topography are in current use, such as the familiar color tint that uses shades of green for low elevations and brown for higher elevations. The most accurate method, which will be described here, involves the use of contour lines.

MAP SCALES

Map scales are designated in several ways. (1) A scale can be stated as a specified unit of length on the map corresponding to a specified unit of length on the ground—for example, one inch equals one mile, meaning that one inch on the map represents one mile on the ground. (2) A scale can also be stated as a scale ratio or representative fraction, meaning that one unit (any unit) on the map equals a specified number of the same units on the ground. In the example just used, the scale would be given as 1:63,360, and 1 in. on the map would represent 63,360 in. on the ground (1 mi = 12 × 5280 = 63,360 in.). (On that same map, 1 cm would also represent 63,360 cm on the ground.) (3) Most maps contain a graphic scale, generally in the lower margin, for measuring distances.

CONTOUR LINES

A contour line is an imaginary line, every part of which is at the same elevation. Thus a shore line is an example of a contour. Because a single point can have only one elevation, two contour lines can never cross (except on an overhanging cliff).

By convention, contour lines on each map are a fixed interval apart. This is called the contour interval. Typical contour intervals are 5, 10, 20, 40, 80, or 100 feet or meters. The contour lines are always at an elevation that is a whole number times the contour interval. That is, the contour lines are drawn at 20, 40, 60, 80, and so forth, feet or meters, never at 23, 43, or 63.

A little imagination will reveal that if the contour lines are close together, the surface is steeper than if they are far apart. (See Fig. B.1.)

If we think of an island and remember that the shore line is the zero contour, we can then imagine the successive contours, or shore lines, if the water level rises.

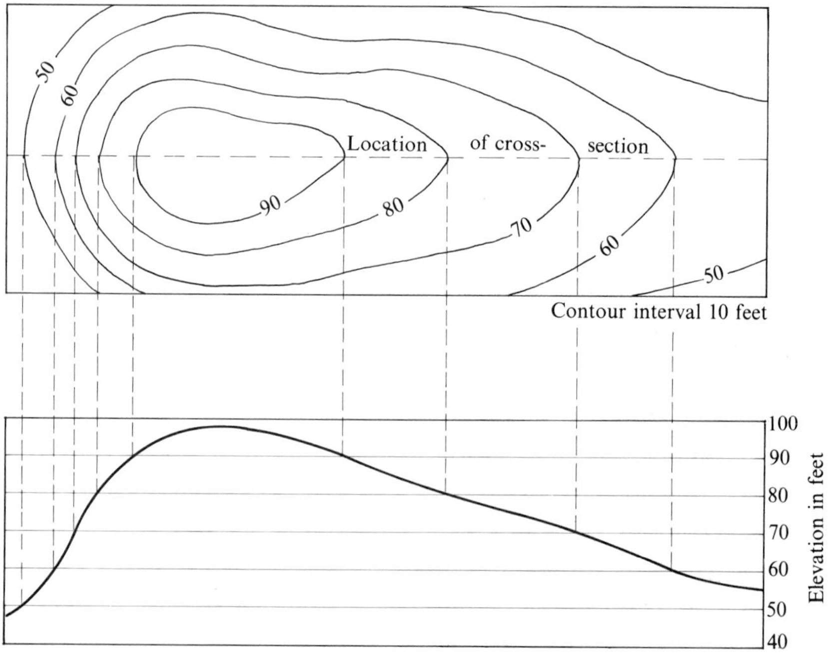

FIGURE B.1
Contour map of a simple hill. The cross-section shows that where the contour lines are closer together the hill is steeper.

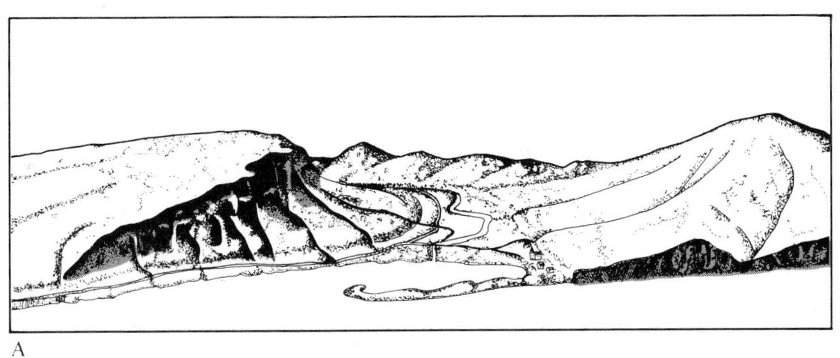

A

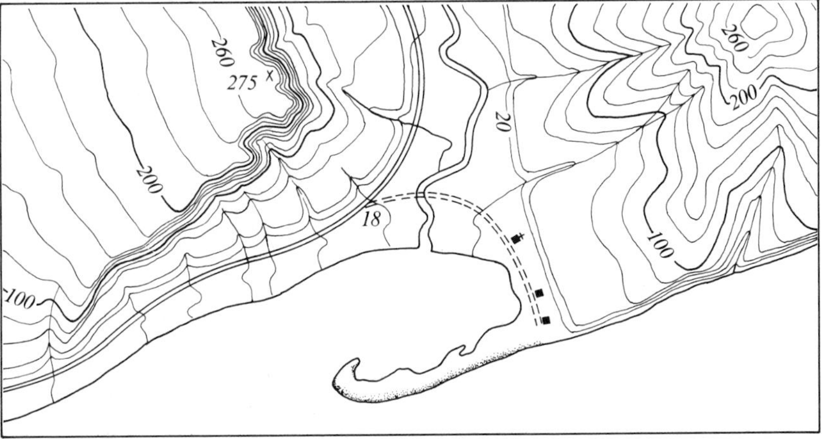

B

THE USE OF SYMBOLS IN MAPPING

These illustrations show how various features are depicted on a topographic map. The upper illustration is a perspective view of a river valley and the adjoining hills. The river flows into a bay which is partly enclosed by a hooked sandbar. On either side of the valley are terraces through which streams have cut gullies. The hill on the right has a smoothly eroded form and gradual slopes, whereas the one on the left rises abruptly in a sharp precipice from which it slopes gently, and forms an inclined tableland traversed by a few shallow gullies. A road provides access to a church and two houses situated across the river from a highway which follows the seacoast and curves up the river valley.

The lower illustration shows the same features represented by symbols on a topographic map. The contour interval (the vertical distance between adjacent contours) is 20 feet.

FIGURE B.2
Perspective view of an area and a contour map of the same area. After U.S. Geological Survey.

From this we learn that a hill is represented on a contour map by a series of concentric closed lines, and, of course, a ridge would be similar, but elongate.

If we think of a valley, we see that the contour lines from Vs pointing upstream. Remember a stream flows downhill at the lowest point in a valley; hence, if we start on one of the valley sides and walk in the upstream direction at a given elevation, we will eventually reach the stream.

A few minutes study of Figure B.2 will reveal much information about reading contour maps.

Contour lines have the inherent limitation that the elevations of points between contours are not designated, although the approximate elevation can be determined by interpolation. In an effort to overcome this limitation, the elevations of hills, road junctions, and other features are commonly given on the maps.

CONVENTIONS USED ON TOPOGRAPHIC MAPS

Because the United States Geological Survey produces most of the topographic maps in use in the United States, these are described here. Until recently these maps were made by actual field surveying, but are now made from aerial photographs with some field surveying to establish horizontal and vertical control points. The maps made since 1942 meet high accuracy standards. Ninety percent of all well-defined features are within one-fiftieth of an inch of their true location on the published map, and the elevations of ninety percent of the features are correct within one-half of the contour interval.

The maps, which are called quadrangles, are bounded by latitude and longitude lines. They are named for features within the map area, and because the names are not duplicated within a state, some are named for less important features. The sizes and scales used are shown in Table B.1. The number of square miles in a map of a given scale varies because of the convergence of longitude lines at the north pole.

COLORS

Water features are in blue.
Human structures—roads, houses, and so forth—are in black.
Contour lines are in brown.

In addition, some maps show important roads, urban areas, and public land subdivision lines in red, and woodlands, orchards, and so forth, in green.

If you would like to obtain topographic maps of any area, write to the addresses below for an index to topographic maps for the state desired. The indexes are free on request and show all of the maps available.

TABLE B.1
Sizes and scales of U.S.G.S. topographic maps.

Map Designation	Scale		Quadrangle Size (Lat.-Long.)	Quadrangle Area (Sq. Miles)	Paper Size (Inches)
	Ratio	One inch equals			
7½ minute	1:24,000	exactly 2000 feet	7½′ × 7½′	49–70	22 × 27
					23 × 27
	1:31,680	exactly ½ mile	7½ ′ × 7½′	49–68	17 × 21
15 minute	1:62,500	approximately 1 mile	15′ × 15′	197–282	17 × 21
					19 × 21
30 minute	1:125,000	approximately 2 miles	30′ × 30′	789–1082	17 × 21
1 degree	1:250,000	approximately 4 miles	1° × 1°	3173–4335	17 × 21
1:250,000	1:250,000	approximately 4 mile	1° × 2°	6349–8669	24 × 34

For states east of the Mississippi River:

Eastern Distribution Branch
U.S. Geological Survey
1200 South Eads Street
Arlington, VA 22202.

For states west of the Mississippi River:

Western Distribution Branch
U.S. Geological Survey
Box 25286, Federal Center
Denver, CO 80225.

Appendix C Geological Maps

Geological mapping is one of the most important methods used in the study and interpretation of the earth and earth history. A geologic map is simply a map on which the various rock types that make up the earth's surface are plotted. Generally, the rock types are plotted on topographic (contour) maps because the topography is commonly related to the rock types, and knowing the topography helps another geologist to interpret the geologic map. To further aid in the interpretation of geologic maps, other symbols are used to indicate such things as the attitude or slope of planar features such as the bedding planes of sedimentary rocks. From the information on the geologic map, which is the surface geology, the underlying structure of the area is interpreted. The main difficulty in geologic mapping is that outcrops of bedrock are relatively rare in most regions because weathered material, soil, and alluvium cover much of the earth's surface.

The preparation of a geologic map begins with the field geologist locating the outcrops of the various rock types on a map or an aerial photograph of the area (Fig. C.1B). Generally, the units that are mapped are **formations.** Formations are rock units that are generally easy to recognize and so reasonably easy to map. The geologic map should show the boundaries between formations, and these boundaries are called **contacts** (Fig. C.1C). If the units shown on the map have directional properties, such as bedding in sedimentary rocks or foliation in metamorphic rocks, the **attitude** of such features should also be noted. The attitude is shown by the T-shaped **strike and dip** symbol shown on the maps. The longer

FIGURE C.1
The stages in making a geologic map.

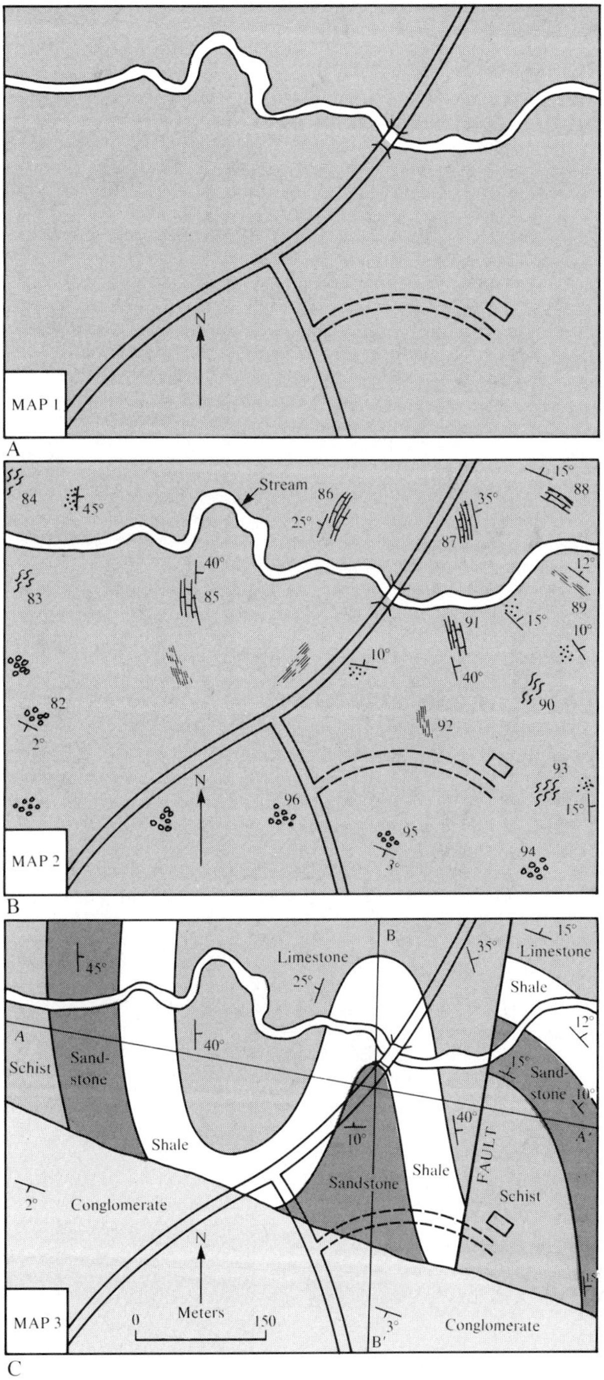

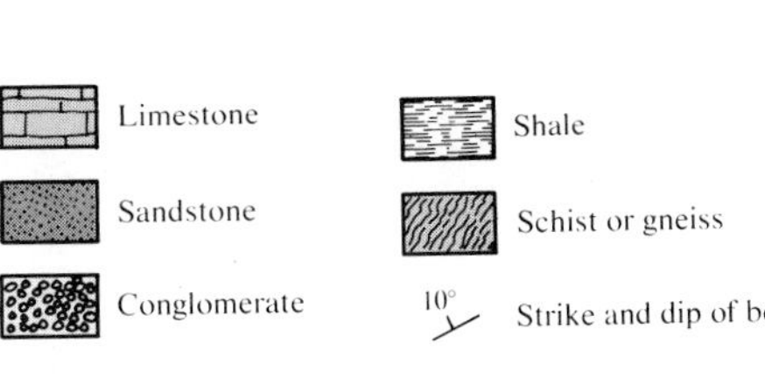

FIGURE C.2
Strike and dip.

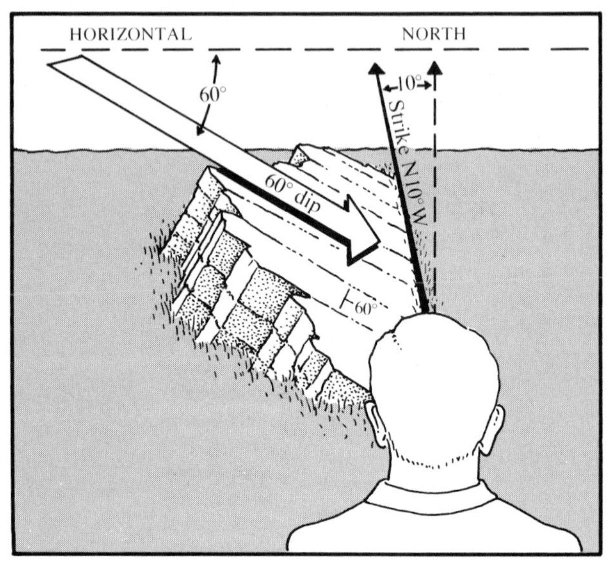

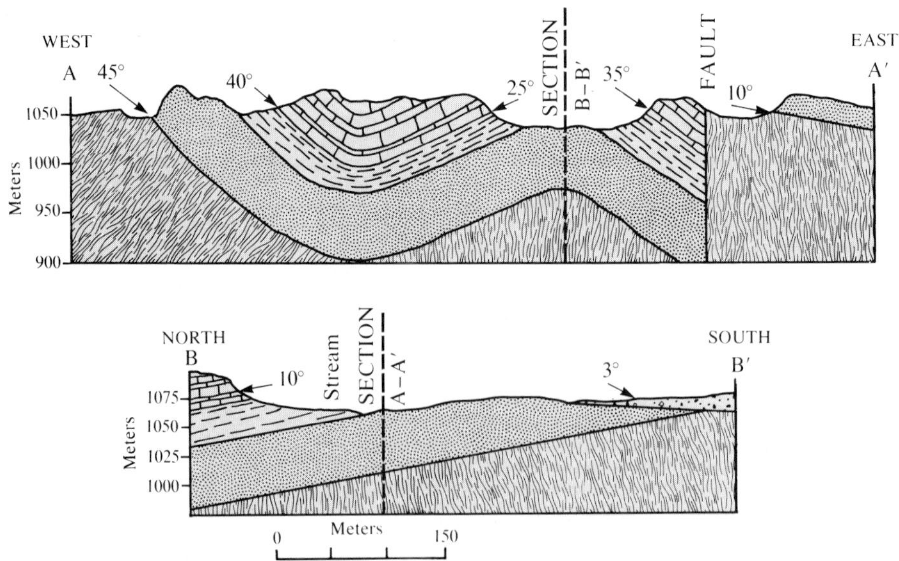

FIGURE C.3
Cross sections A-A′ and B-B′ of the map in Figure C.1C.

line shows the strike, which is the direction of the intersection of the bed and a horizontal plane. The shorter line shows the direction of the dip—the downslope perpendicular to the strike—and the number near it shows the dip angle (Fig. C.2). In addition, a geologic map should show the contacts of such features as intrusive rock bodies. Faults should also be shown.

Typically, different colors are used to show the various rock units on a geologic map. The ages and relationships among the rock units, and the color used for each, are shown in the legend of the map. The legend is generally at the side or bottom of a geologic map. Any other symbols used on the map will also be explained in the legend. Cross-sections may also be included to show the structures (Fig. C.3).

Appendix D
Fossil-Forming Organisms

Only the more important fossil-forming organisms are shown in the following classification. Much more detailed classifications can be found in specialized books. No classification scheme is satisfactory to all users, and, as a result, the classification in every text is different. Even the number of kingdoms is in flux. The four kingdoms used here are convenient for fossils, but only a few years ago three were used, and many biologists now favor five kingdoms.

In the following pages, the organisms in Table D.1 are described.

Kingdom Monera
Division Schizomycophyta

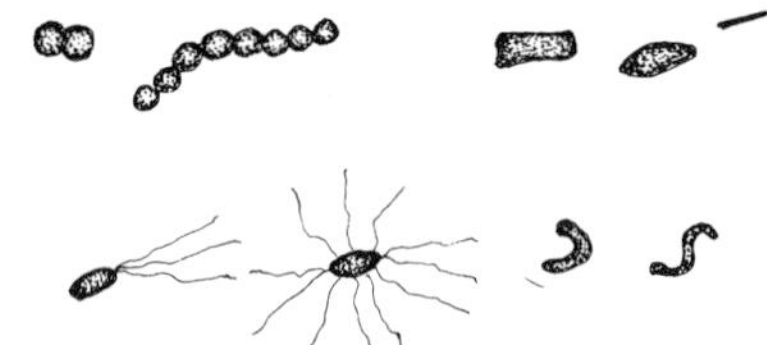

FIGURE D.1
Bacteria are rarely sought as fossils except in ancient Precambrian rocks. Bacteria are shown in Figure 6.5.

Kingdom Protista
Phylum Protozoa
Many of the one-celled organisms in this phylum are important rock formers, and many are used widely to date rocks. Because these small organisms are recovered in drill cores, they are especially useful in correlating strata in oil wells.

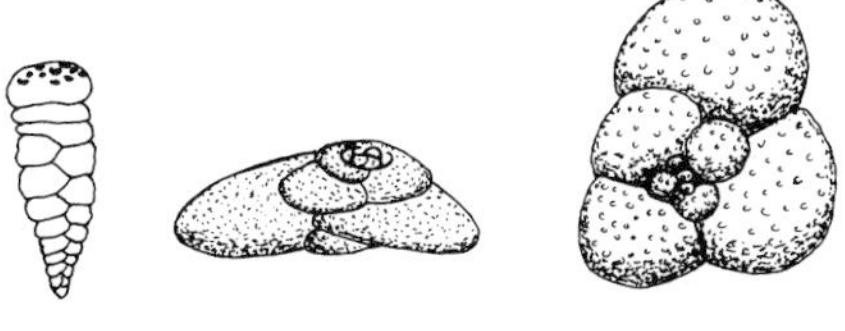

FIGURE D.2
Foraminifers are mainly less than 1 mm (0.04 in.) in diameter, although a few types are much larger. Only a few of the many diverse forms are shown.

FIGURE D.3
Fusulinids are large foraminifers and were abundant in the late Paleozoic. Their size and shape are similar to the size and shape of wheat grains. See Figure 8.25.

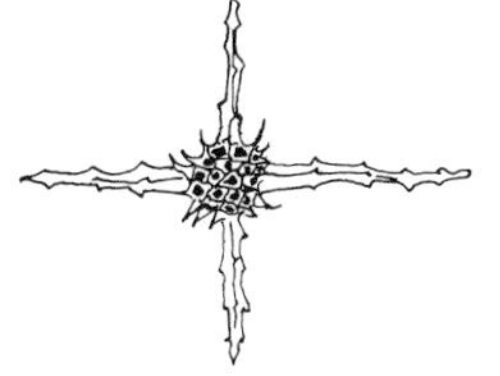
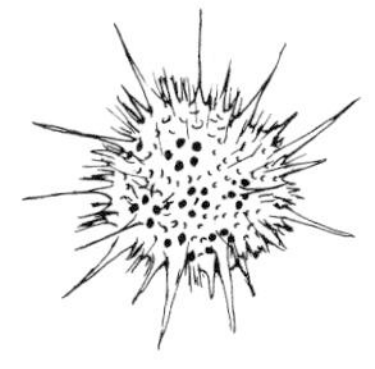

FIGURE D.4
Radiolarians are very small (less than 1 mm) and have shells composed of silica. They are seldom used in geologic dating in spite of their great diversity and numbers.

Kingdom Animalia
Phylum Porifera

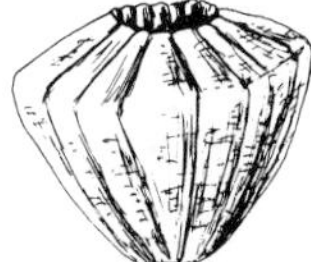

FIGURE D.5
Sponges are simple aquatic animals that feed by drawing water into their bodies through pores and expelling it through a central opening. The most common fossil remains are the spicules that help to maintain the body shape. See Figures 8.22 and 8.23.

Phylum Coelenterata or Cnidaria

This phylum includes jellyfish, hydra, and corals. Of these, only the corals are important fossils. Coelenterates have a simple saclike body with a mouth surrounded by tentacles at one end.

Class Anthozoa (Cnidaria)

Corals are both colonial and solitary. All are marine. Corals are widespread in Paleozoic rocks and are less important in Mesozoic and younger rocks. They

TABLE D.1
Classification of the more important fossil-forming organisms.

Kingdom Monera—single-celled prokaryotes
- Division* Schizomycophyta—bacteria
- Division Cyanophyta-cyanobacteria—blue-green algae

Kingdom Protista—eukaryotes, mainly single-celled
- Phylum Protozoa—foraminifers, radiolarians

Kingdom Animalia
- Phylum Porifera—sponges
- Phylum Coelenterata or Cnidaria—corals
- Phylum Bryozoa
- Phylum Brachiopoda
- Phylum Mollusca
  - Class Bivalvia—clams, oysters
  - Class Gastropoda—snails
  - Class Cephalopoda
    - Subclass Nautiloidea—nautiloids
    - Subclass Ammonoidea—goniatites, ceratites, ammonites
    - Subclass Coleoidea
      - Order Belemnoidea—belemnites
- Phylum Arthropoda
  - Class Arachnoidea—scorpions, spiders, eurypterids
  - Class Trilobita
  - Class Insecta
- Phylum Echinodermata
  - Subphylum Crinozoa—mainly attached forms (sessile)
    - Class Cystoidea
    - Class Blastoidea
    - Class Crinoidea
  - Subphylum Echinozoa—free-moving forms
    - Class Asteroidea—starfish
    - Class Echinoidea—sea urchins
- Phylum Hemichordata
  - Class Graptolithea—graptolites
- Phylum Chordata
  - Class Agnatha—jawless fish
  - Class Placodermii—armored fish
  - Class Chondrichthys—sharks
  - Class Osteichthys—bony fish
  - Class Amphibia
  - Class Reptilia
    - Subclass Anapsida
      - Order Cotylosauria
      - Order Chelonia

*Division is similar to phylum.

TABLE D.1
Classification of the more important fossil-forming organisms. *Continued*

- Subclass Synapsida
  - Order Pelycosauria
  - Order Therapsida
- Subclass Euryapsida
  - Order Sauropterygia
  - Order Ichthyosauria
- Subclass Diapsida—includes lizards and snakes also
  - Order Thecodonta
  - Order Crocodilia
  - Order Saurischia } The dinosaurs may not be reptiles
  - Order Ornithischia }

Class Aves—birds

Class Mammalia
- Subclass Prototheria
  - Order Monotremata
- Subclass Theria
  - Infraclass Metatheria
    - Order Marsupialia
  - Infraclass Eutheria—placentals
    - Order Insectivora
    - Order Chiroptera
    - Order Creodonta
    - Order Carnivora
    - Order Perissodactyla
    - Order Artiodactyla
    - Order Proboscidea
    - Order Edentata
    - Order Cetacea
    - Order Rodentia
    - Order Lagomorpha
    - Order Primates

Kingdom Plantae
- Phylum Bryophyta—liverworts, mosses
- Phylum Psilophyta—early land plants
- Phylum Lycopodophyta—lycopods, club mosses
- Phylum Sphenophyta—horsetails
- Phylum Filicinophyta—ferns
- Phylum Cycadophyta—seed ferns, cycads }
- Phylum Coniferophyta—conifers } Gymnosperms (naked-seed plants)
- Phylum Ginkgophyta—ginkgos }
- Phylum Angiospermophyta—flowering plants

are shown in many figures, and the differences between some Paleozoic and Mesozoic corals are shown in Fig. 9.15.

Phylum Bryozoa

FIGURE D.6
Bryozoans are small colonial animals. Their fossil remains are generally lacy or stony fragments with many holes, or the framework that holds such fronds. See Figures 7.17, 8.18, and 8.19.

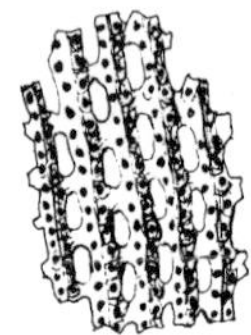

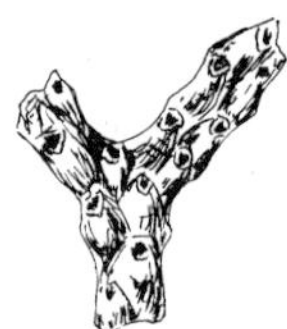

Phylum Brachiopoda

Brachiopods are bivalved animals. They were important in the Paleozoic and have declined in numbers since then. They are illustrated in many figures, especially Figs. 7.14, 8.16, 8.23, and 8.24. Figure 9.24 shows the differences in symmetry between clams and brachiopods.

Phylum Mollusca

This diverse phylum includes clams, oysters, snails, and cephalopods, as well as other forms that are rarely fossilized.

Class Bivalvia

FIGURE D.7
Clams and oysters. Clams have two shells that are mirror images. The forms are diverse. Most are bottom dwellers, but *Pecten*, a scallop, is a swimming type. The shells of oysters and rudistids do not have the bilateral symmetry of clams. See Figures 9.24 and 9.25.

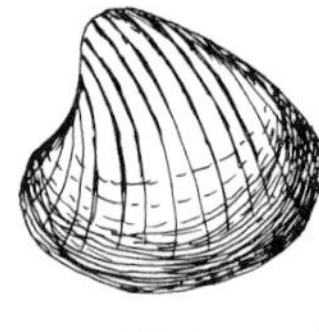

Clam

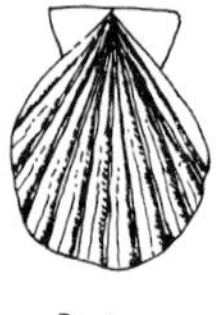

Pecten

Oyster

Rudistid

Class Gastropoda

FIGURE D.8
Snails have a single shell, generally without symmetry. They are the only terrestrial mollusks, although most are aquatic. See Figures 4.5A and B, and 9.16.

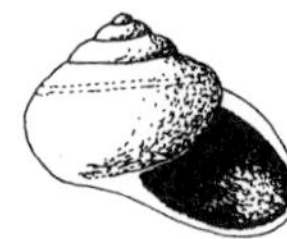

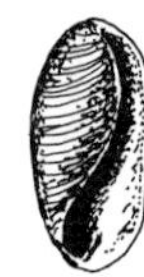

Class Cephalopoda

Subclass Nautiloidea

Nautiloids have simple sutures separating the chambers.

Subclass Ammonoidea

Cephalopods are both straight and coiled. See Figs. 7.15, 7.18, 8.19, 18.22, and 9.17. They can be distinguished from snails by their symmetry, as shown in Fig. 9.16. Ammonites have complex sutures. The distinctions among the ammonoids are shown in Fig. 9.18.

Subclass Coleoidea

Order Belemnoidea

Belemnoids had cigar-shaped shells that were internal. See Figs. 9.21 and 9.22.

Phylum Arthropoda—the joint-footed animals

Class Arachnida

The arachnids include spiders, scorpions, and eurypterids. Eurypterids are shown in Figs. 7.21 and 7.22.

Class Trilobita

Trilobites were important in the Paleozoic, especially in the Cambrian Period. They are illustrated in Figs. 7.12, 7.13, 7.18, and 8.19.

Class Insecta

The insects are the largest group of animals. They are relatively uncommon fossils and so are of little stratigraphic value. See Fig. 4.5D.

Phylum Echinodermata

The echinoderms are all marine and are "five-sided."

Subphylum Crinozoa

Class Cystoidea (See Fig. 7.18.)

Class Blastoidea (See Figs. 8.20 and 8.21.)

Class Crinoidea (See Figs. 8.19, 8.20, and 8.22.)

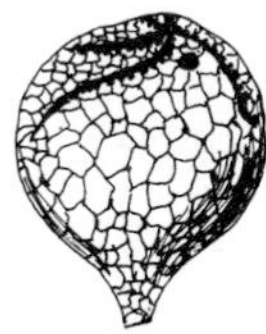

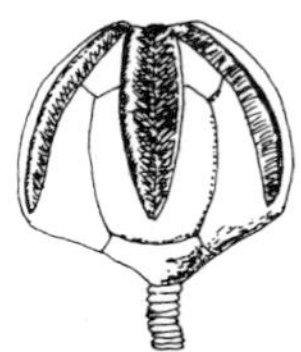

FIGURE D.9
Cystoids are less regular than other crinozoans. Blastoids are budlike. Crinoids are sea lilies.

Subphylum Echinozoa

These echinoids are free moving.

Class Asteroidea

FIGURE D.10
Starfish. See also Figure 8.20.

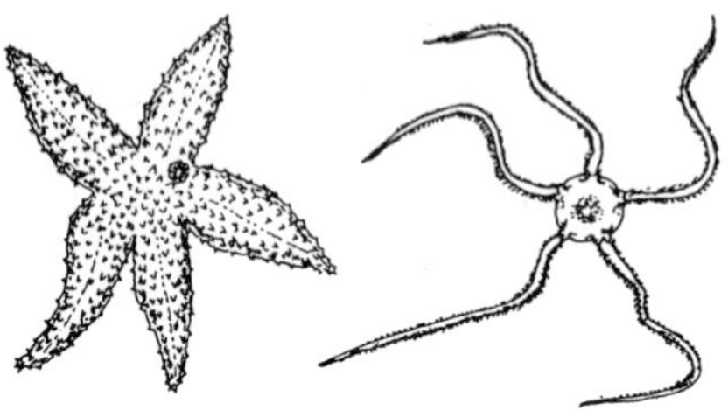

Class Echinoidea

FIGURE D.11
Echinoids differ from starfish in their lack of arms. See Figure 10.19.

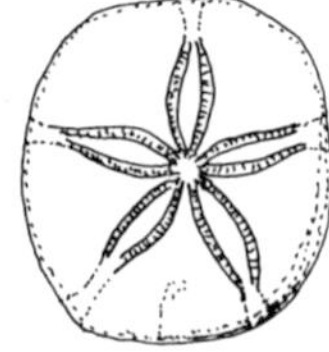

Phylum Hemichordata
Class Graptolithea
The graptolites are extinct marine, colonial animals. They are too poorly known to classify accurately. They are important Ordovician and Silurian fossils.

FIGURE D.12
Graptolites. See also Figure 7.20.

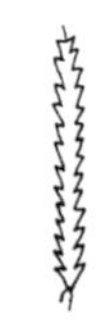

Phylum Chordata
Class Agnatha

FIGURE D.13
The agnathids were primitive jawless fish with bony armor. See also Figure 8.26.

Class Placodermii

FIGURE D.14
The placoderms were early jawed fish. Some were very large. This fish *Dinichthys* reached 9 m (30 ft) long. See also Figure 8.27.

Class Chondrichthys

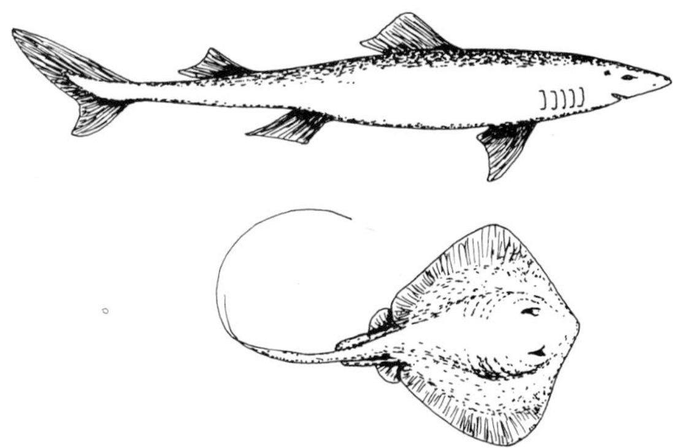

FIGURE D.15
Sharks and rays have been predators since the Devonian.

Class Osteichthys
This class includes all of the bony fish.
Class Amphibia
All amphibians must return to water to lay their eggs. Most live on land most of their lives. Familiar amphibians include toads, frogs, and salamanders. See Figs. 8.33 and 8.37.
Class Reptilia
- Subclass Anapsida
  - Order Cotylosauria
    The stem reptiles include *Seymouria,* shown in Fig. 8.36.
  - Order Chelonia—turtles. See Fig. 10.23.
- Subclass Synapsida—mammal-like reptiles
  - Order Pelycosauria—"sail lizards"
  - Order Therapsida—advanced mammal-like reptiles
- Subclass Euryapsida—swimming reptiles.
  - Order Sauropterygia—plesiosaurs
  - Order Icthyosauria—icthyosaurs. See Fig. 9.29.
- Subclass Diapsida—includes lizards and snakes also
  - Order Thecodonta
    A thecodont is shown in Fig. 9.28.

Order Crocodilia—crocodiles. See Fig. 10.23.
Order Saurischia
Order Ornithischia
} The dinosaurs may not be reptiles. They are shown in Figs. 9.32 to 9.37.
Class Aves—birds. See Figs. 9.30, 9.31, and 10.28.
Class Mammalia
Subclass Prototheria
Order Monotremata
The monotremes are egg-laying mammals. The duck-billed platypus and the spiny anteater are examples. Monotremes are unimportant as fossils.
Subclass Theria
Infraclass Metatheria
Order Marsupialia

FIGURE D.16
Marsupials carry their young in pouches.

Infraclass Eutheria—placental mammals
Order Insectivora

FIGURE D.17
The insectivores include shrews, moles, and hedgehogs. See Figure 10.25.

Order Chiroptera—bats
Order Creodonta
The creodonts were primitive carnivores from which the present carnivores evolved.
Order Carnivora

FIGURE D.18
The carnivores include dogs, cats, bears, weasels, hyenas, and civets. They are swift, smart hunters. See Figures 10.23, 10.25, 10.27, and 10.28.

Order Perissodactyla
These are the "odd-toed" hoofed animals. They are characterized by having the axis of the foot go through the middle toe and by the loss of some of the other toes by straightening out of the foot.

FIGURE D.19
The "odd-toed" animals include horses, rhinoceroses, and tapirs. See Figures 10.23, 10.25, 10.26, and 10.27.

Order Artiodactyla
These are the "even-toed" hoofed animals. They are more abundant than the "odd-toed" mammals. See Figs. 10.23, 10.25, 10.26, and 10.27.

FIGURE D.20
The "even-toed" mammals include camels, deer, and cattle.

Order Proboscidea
This order includes the elephants and mammoths that have trunks. See Figs. 4.5E, 10.27, and 10.29.
Order Edentata

FIGURE D.21
Sloths and armadillos are modern edentates.

Order Cetacea—whales, porpoises. See Fig. 10.24.
Order Rodentia—mice, rats. See Figs. 10.23, 10.25, and 10.27
Order Lagomorpha—hares, rabbits. See Figs. 10.25 and 10.27.
Order Primates—monkeys, apes, humans. See Figs. 10.30 through 10.33.

FIGURE D.22
The prosimians include lemurs, tarsiers, and lorises.

FIGURE D.23
Monkeys, apes, and humans.

Kingdom Plantae
Phylum Bryophyta

FIGURE D.24
The bryophytes include liverworts and mosses.

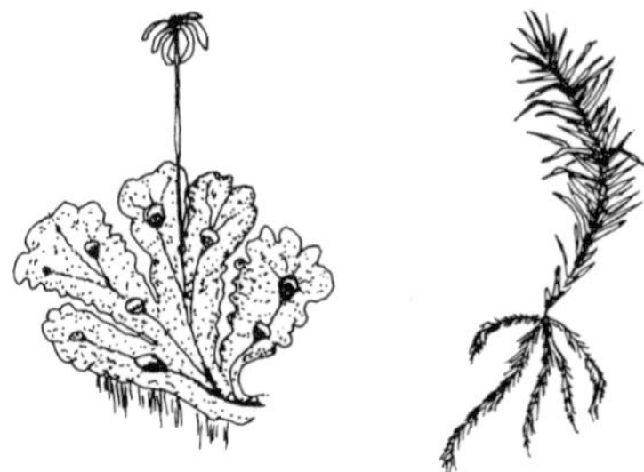

Phylum Psilophyta
These are the earliest vascular plants, and they did not have true roots.

FIGURE D.25
Psilopsids. See Figure 8.30.

Phylum Lycopodophyta

FIGURE D.26
Lycopsids such as *Lepidodendron* and *Sigillaria* were spore-bearing plants with true roots. See Figures 8.31 and 8.32.

Lepidodendron

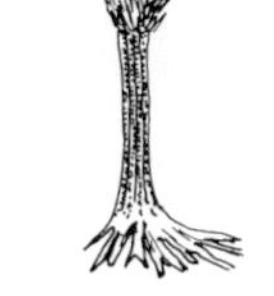
Sigillaria

Phylum Sphenophyta

FIGURE D.27
Equisetum, a "scouring" rush, or horsetail. See Figure 8.30, and *Calamites,* Figure 8.31.

Phylum Filicinophyta—ferns

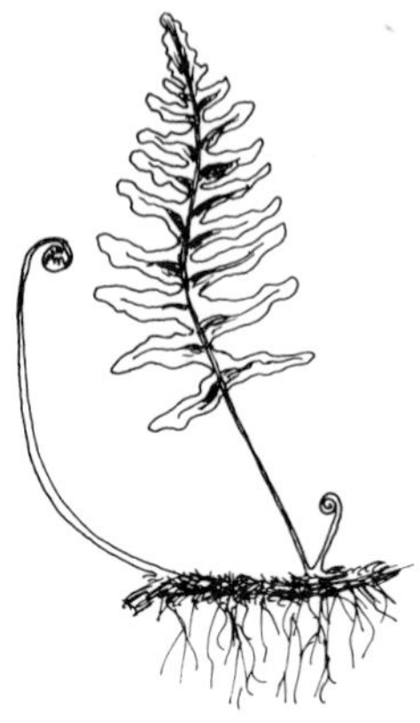

FIGURE D.28
Ferns. See Figure 8.32.

Gymnosperms—plants with naked seeds
Phylum Cycadophyta—seedferns, cycads
Cycads are shown in Fig. 9.30. *Glossopteris* is typical of this phylum.
Phylum Coniferophyta
Pines, firs, cedars, junipers, and redwoods are common members of this order. See Fig. 10.20B. *Cordaites* was abundant in late Paleozoic.
Phylum Ginkgophyta

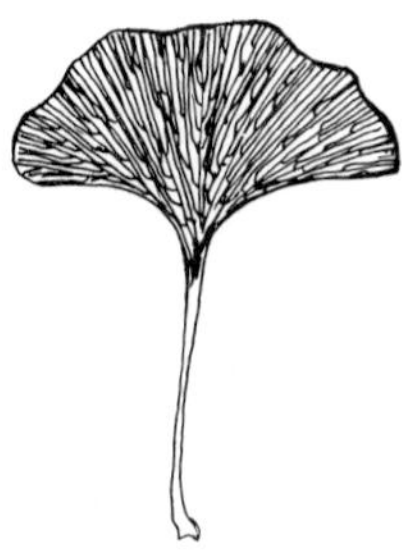

FIGURE D.29
Ginkgo leaf.

Phylum Angiospermophyta—flowering plants
See Figs. 4.5C and 10.20A and C.
The angiosperms are predominant in modern floras.

Glossary

Absolute date Geological date based on the radioactive decay of elements in the rock.

Actualism See uniformitarianism.

Adaptation Change by an organism to take advantage of new or changing conditions.

Age Subdivision of an epoch.

Algae Group of organisms that includes single-celled plants and seaweeds.

Alpine glacier Mountain glacier. Glacier in a mountain valley.

Amino acid One of the nitrogenous organic compounds that are the building blocks of proteins.

Amphibian "Cold-blooded" vertebrate that lives in water; breathes through gills in the early stages of development and through lungs in later stages.

Amphibole Ferromagnesian mineral group. The most important rock-forming amphibole is hornblende.

Amphibolite Metamorphic rock composed mainly of amphibole, such as hornblende, and plagioclase feldspar.

Anaerobic Able to live in the absence of oxygen.

Andesite Fine-grained igneous rock; intermediate in color and composition between felsite and basalt.

Angiosperm Plant with true flowers; bears seeds enclosed in ovaries.

Angular unconformity Unconformity in which the older strata dip or slope at a different angle (generally steeper) than the younger strata.

Anorthosite Coarse-grained rock composed almost entirely of plagioclase feldspar.

Anticline Type of fold in which the strata slope downward in opposite directions away from a central axis, like the roof of a house. The oldest rocks are in the core.

Aphanitic Having a texture in which the minerals are too small to see with the naked eye.

Archean Subdivision of the Precambrian. It includes all time before about 2500 million years ago.

Arkose Sandstone containing much feldspar.

Arthropod Member of the phylum Arthropoda; characterized by segmented bodies and jointed legs.

Asteroid One of a group of small planetlike bodies whose general orbit around the sun is between Mars and Jupiter.

Asthenosphere Layer in the mantle beneath the low-velocity zone in which the rocks appear to flow slowly. See lithosphere.

Atom Smallest unit of an element; composed of protons, electrons, and neutrons.

Atomic fission Reaction in which an atom is split. The atom bomb reaction is an example.

Atomic fusion Reaction in which two or more atoms are combined to form a different element. The hydrogen bomb reaction is an example.

Atomic mass number Total number of protons and neutrons in the nucleus of an atom.

Atomic number Number of protons, or positive charges, in the nucleus of an atom.

Augite Common rock-forming pyroxene mineral; generally black and has two cleavages at right angles. $(Ca,Na)(Mg,Fe^{II},Fe^{III},Al)(Si,Al)_2O_6$.

Backarc basin Area between a volcanic arc and a continent or another volcanic arc.

Banded iron formation Sedimentary rock consisting of alternate layers of iron-rich and iron-poor silica.

Barrier reef Reef between the main ocean and an island or continent.

Basalt Fine-grained volcanic rock; dark in color and composed mainly of plagioclase feldspar and pyroxene. It is one of the most common rock types.

Basement rocks Rocks that underlie the sedimentary cover. The basement is generally a complex of igneous and metamorphic rocks.

Batholith Very large discordant intrusive rock body, generally composed of granitic rocks. Parts of some batholiths may be of metamorphic origin.

Bedding Layers, or laminae, of sedimentary rocks.

Bedrock Unweathered rock; generally below the overburden.

Belemnoid Squidlike animal with an internal shell. A member of the phylum Mollusca, which includes clams, snails, oysters, and ammonites.

Bench, wave-cut Flat surface cut or eroded by wave action.

Biotite Common mineral, a member of the mica family. Recognized by its shiny black or dark color and perfect cleavage. $K(Mg,Fe)_3(AlSi_3O_{10})(OH)_2$.

Blastoid Stalked echinoderm. Also called "sea bud."

Block fault Fault that bounds an uplifted or downdropped segment of the crust.

Blueschist Foliated metamorphic rock containing blue minerals such as glaucophane. Blueschist forms under high pressure and low temperature. See greenschist.

Brachiopod Member of the phylum Brachiopoda. Marine animal with two unequal shells, each of which is bilaterally symmetrical. Also called "lamp shell."

Breccia Clastic sedimentary rock with angular fragments larger than 2 mm in diameter. A volcanic breccia is composed of pyroclastic fragments greater than 2 mm in size.

Brown clay Deep-sea deposit formed by fine material, mainly volcanic dust. The color is probably caused by oxidation. Also red clay.

Bryozoa Phylum of colonial animals that build lacy, branching, or other type structure with many small holes. Also called "moss animals."

Calcic plagioclase Plagioclase feldspar with a high calcium content. Plagioclase is a continuous series between $CaAl_2Si_2O_8$ and $NaAlSi_3O_8$.

Calcite Common rock-forming mineral; composes limestone and marble. Recognized by three directions of cleavage not at right angles and reaction with dilute hydrochloric acid. $CaCO_3$.

Carbon 14 Radioactive isotope of carbon having atomic mass number of 14. Used in radiometric dating.

Carbonate Rock or mineral containing carbonate (CO_3). The main minerals are calcite and dolomite, and the main rocks are limestone and dolomite.

Carboniferous Period Part of the geologic time scale. It is so called because many coal deposits were formed at that time. It lasted from about 360 million to about 286 million years ago. The Mississippian and Pennsylvanian Periods combined.

Carbonization Fossilization process in which the organism becomes a film of carbon, generally on a bedding surface of a sedimentary rock. Many leaf fossils form in this way.

Cast, fossil Fossil formed by the infilling of a cavity produced by the decay of an organism.

Cataclastic Having a texture in which at least some of the minerals have been broken and flattened during dynamic metamorphism.

Cataclastic metamorphism Metamorphism caused by deformation of the rock.

Catastrophism Doctrine that held that most geologic changes were caused by sudden violent events.

Cenozoic Era of the geologic time scale. It has lasted from about 66 million years ago to the present.

Cephalopod Marine invertebrate characterized by tentacles and a straight or coiled calcareous shell with chambers.

Chalcedony Extremely fine-grained variety of quartz. Agate and chert are made of chalcedony.

Chalk Soft, usually white variety of limestone; composed of shells of micro-organisms and very fine calcite.

Chert Rock composed of extremely fine-grained quartz. See chalcedony.

Chlorite Group of platy, micalike minerals. Also, a compound containing chlorine.

Chronozone Rocks deposited during the time span of a fossil.

Cirque Glacial erosion feature. The steep amphitheaterlike head of a glacial valley.

Clam Bivalved animal with two shells that are mirror images. A member of the phylum Mollusca, class Bivalvia.

Clastic Pertaining to rocks or sediments composed of fragments broken from some preexisting rock.

Clastic sedimentary rock Rock formed of fragments or clasts broken from older rocks. Examples are sandstone and conglomerate.

Clay Mineral family. Most clays are fine-grained. Much of the material in soils and shales is clay.

Claystone Rock composed largely of fine clay.

Cleavage In a mineral, a direction of weakness. Breakage along this direction produces a smooth plane.

Coal Sedimentary rock formed of altered plant material that is burnable.

Coarse-grained Having crystals that are easy to see because they are over about ⅙ mm in size.

Collision Convergent plate boundary. A place where plates moving in opposite directions meet.

Colonial animal Animal that lives in a colony. Most of the colonial animals mentioned in this book live in groups or colonies that are organized so that there is a division of labor among the individuals. This term can also mean a type of animal that lives in close association with other similar individuals.

Columnar jointing Jointing that breaks the rock into rough columns, generally six-sided. Common in basalt; probably caused by shrinking due to cooling.

Comet Solar system body believed to be composed of a mass of frozen gas with some small particles.

Conglomerate Coarse-grained clastic sedimentary rock. The fragments are larger than 2 mm.

Contact metamorphic rock Thermally metamorphosed rock near the boundary of contact of an intrusive body such as a batholith.

Continental drift Movement of continents or plates on the earth's surface.

Continental glacier Large glacier caused by the buildup of an ice sheet on a continent. Greenland is an example.

Continental rise Transition between the continental slope and the deep ocean floor.

Continental shelf Shallow ocean area between the shoreline and the continental slope that extends to about a 180-m (600-ft) depth surrounding a continent. The underlying crust is generally continental.

Continental slope Relatively steep slope between the continental shelf and the ocean floor.

Contour line Line on a map connecting points of equal elevation.

Convection current (cell) Movement in a fluid caused by differences in density. The density differences are generally caused by heating.

Convergence In evolution, the formation of similar structures in different types of organisms.

Convergent plate boundary Place where two plates are moving toward each other, as in a collision or at a volcanic arc.

Coprolite Fecal pellet or casting. Fossilized excrement or feces.

Coquina Type of limestone composed largely of shells or shell fragments.

Coral Bottom-dwelling marine animal. Corals are both solitary and colonial and secrete external calcium carbonate skeletons.

Core Innermost layer of the earth; consists of an inner core believed to be solid and an outer liquid core.

Correlation Determination of the equivalence of two exposures of a rock unit.

Crater, impact Depression caused by impact of a falling object. Meteorites cause most large impact craters.

Crater, volcanic Volcanic depression typically developed at the top of a volcano.

Cretaceous Period in the geologic time scale. It lasted from about 144 million years ago to about 66 million years ago.

Crinoid Type of echinoderm attached to the sea bottom by a stalk.

Cross-beds, cross-strata Beds or laminations at an oblique angle to the main bedding.

Cross-cutting feature Fault or discordant intrusive body.

Crust Outermost layer of the earth. Above the Moho.

Crystal Geometrical solid bounded by smooth planes that result from an orderly internal atomic arrangement.

Crystalline Having crystal structure, that is, an orderly internal structure.

Curie point Temperature above which a substance loses its magnetism.

Cut-and-fill feature Channel eroded (cut) in sediments and later filled (deposited) with sediment. All this occurs during deposition.

Cycle of erosion Sequence of changes that are believed to occur during erosion of a landscape.

Cyclothem Repeated cycles of marine and nonmarine rocks, generally including coal beds.

Daughter element Element formed from another element in radioactive decay.

Decay, radioactive Spontaneous change of one element into another.

Deep-focus earthquake Earthquake with focus below 290 km (180 mi).

Density Denseness or compactness. Mass of a material in grams per cubic centimeter.

Deuterium Isotope of hydrogen.

Devonian Period in the geologic time scale. It lasted from about 408 to 360 million years ago.

Diatomite Rock composed largely of the siliceous shells of diatoms.

Dike Tabular discordant intrusive body.

Dike swarm A number of dikes either more or less parallel or radial from a single source.

Dinosaur One of a group of extinct Mesozoic reptiles characterized by the structures of skulls and hips.

Dipole magnetic field Type of magnetic field produced by a bar magnet.

Disconformity Type of unconformity in which the beds above and below the unconformity are parallel.

Discontinuity, Mohorovičić (M-discontinuity, Moho) Level at which a distinct change in seismic velocity occurs. It separates the crust from the mantle.

Discordant intrusive Intrusive rock body that cuts across the bedding of the enclosing sedimentary rocks.

Divergence (radiation) Process of spreading out into new, different habitats.

Divergent plate boundary Mid-ocean ridge or spreading center. Place where tectonic plates are moving apart.

Dolomite Mineral, $CaMg(CO_3)_2$, or a rock composed largely of dolomite.

Dominant character Character inherited from one parent that appears in the offspring to the exclusion of the corresponding recessive character from the other parent.

Drift See continental drift.

Drift, glacial General term for glacier-transported and -deposited material.

Dune Accumulation of windblown sand.

Dunite Coarse-grained igneous rock composed almost entirely of olivine.

Earthquake Vibrations of the earth caused by sudden internal movements in the earth.

Echinoderm Member of the phylum Echinodermata. Marine invertebrate animals, most of which have five-sided symmetry and many of which have spines.

Eclogite Rock that forms under very high pressure; composed of garnet and pyroxene. Its chemical composition is similar to that of basalt.

Electron Fundamental subatomic particle. It has a very small mass and a negative charge.

Element Material that cannot be changed into another element by ordinary chemical means. Just over 100 elements are known. All atoms of an element have the same atomic number (number of protons).

Eocene Subdivision of the Cenozoic Era in the geologic time scale. It lasted from about 58 to 37 million years ago.

Eon Time unit covering more than one era.

Epoch Subdivision of a period in the geologic time scale.

Era Major subdivision of the geologic time scale.

Erathem Rocks deposited during an era.

Erosion Wearing away and removal of material on the earth's surface.

Estuary Tidal inlet along a sea coast. Many form at river mouths.

Eukaryote An organism with cells that have an organized nucleus. See also prokaryote.

Eurypterid Extinct arthropod.

Evaporite Sedimentary rock formed by the evaporation of water, leaving the dissolved material behind.

Extrusive igneous rock Igneous rock that erupted on the earth's surface. Volcanic rock.

Fault Break in the earth's crust along which movement has occurred.

Fault breccia Rock composed of angular fragments that were formed by fault movements.

Fault gouge Material ground up by fault movements.

Fauna All the animals of a given area.

Feldspar Most abundant mineral family. Consists of plagioclase, $CaAl_2Si_2O_8$ and $NaAlSi_3O_8$, and potassium feldspar, $KAlSi_3O_8$ (orthoclase and sanidine). Mixtures of potassium and sodium feldspar are termed perthite.

Felsite Light-colored, fine-grained igneous rock; most felsites have chemical compositions similar to that of rhyolite.

Ferromagnesian mineral Silicate mineral containing iron and magnesium; generally dark-colored.

Fine-grained Having crystals that are difficult to see because they are smaller than about ¼ mm in size.

Fissility Ability of a rock to split easily along parallel planes.

Flood basalt Basalt that forms thick extensive flows that appear to have flowed rapidly (flooded).

Flood plain Flat part of a river valley subject to floods at times.

Flora All of the plants of a given area.

Focus, earthquake Point where an earthquake originates.

Folding Bending of strata.

Foliation (foliated) Directional property of metamorphic rocks caused by the layering of minerals. Generally caused by crystallization under pressure. Also, the flat arrangement of features in any rock type.

Foraminifer One-celled animal. Their generally microscopic shells are useful in dating some rocks.

Forearc basin Area between a submarine trench and a volcanic arc.

Formation Rock unit that can be mapped.

Fossil Any evidence of once-living organisms.

Fracture Break in a mineral that is not along a cleavage direction.

Fracture zone Long linear area of apparent faulting and irregular topography on the ocean floor. See transform fault.

Fusulinid Type of foraminifer shaped like a wheat grain. Important fossil in the Pennsylvanian and Permian systems.

Gabbro Coarse-grained igneous rock composed mainly of plagioclase feldspar and augite.

Garnet Mineral family.

Gene Unit of inheritance, transmitted in the sex cells of the parents, that controls the characteristics of the offspring.

Genus Group of species believed to have descended from a single ancestor.

Geosyncline (geocline) Large elongate downsinking area in which many thousands of feet of sedimentary rocks are deposited. It is later deformed to form a mountain range.

Glacier Mass of moving ice.

Glassy Having no crystals.

***Glossopteris* flora** Fossil assemblage characterized by the distinctive *Glossopteris* leaves. The flora occurs in rocks in the southern hemisphere and is an important line of evidence for continental drift.

Gneiss Coarse-grained foliated metamorphic rock; contains feldspar and is generally banded.

Graded bedding Bedding that has a gradation from coarse at the bottom to fine nearer the top.

Grade, metamorphic Measure of the intensity of metamorphism.

Gradualism Slow evolution by small steps.

Granite Coarse-grained rock containing quartz and feldspar.

Graphite Soft mineral composed of carbon.

Graptolite Extinct colonial animal. Found in early Paleozoic rocks.

Gravel Clastic sediment with fragments larger than 2 mm.

Gravity anomaly Departure from expected gravitational attraction; area where the attraction of gravity is larger or smaller than expected.

Graywacke Type of sandstone, generally containing rock fragments and poorly sorted, with a clay or chloritic matrix.

Greenschist Schist with much chlorite and other green minerals; generally formed by the metamorphism of basalt.

Greenstone Low-grade metamorphic rock. See greenschist.

Group Number of formations.

Half-life Time required for one-half of the atoms of a radioactive element to decay.

Hardness Mineral's resistance to scratching.

Heat Form of energy.

Hematite Ore of iron. Fe_2O_3.

Holocene Youngest subdivision of the geologic time scale. The Recent, about the last 10,000 years.

Hornblende Common rock-forming mineral. An amphibole. $Ca_2Na(Mg,Fe^{II})_4(Al,Fe^{III},Ti)(Al,Si)_8O_{22}(O,OH)_2$.

Hornfels Metamorphic rock formed by thermal or contact metamorphism of any rock type.

Ice-sheet (icecap) glacier See continental glacier.

Ichthyosaur Extinct fishlike reptile.

Igneous rock Rock that formed by the solidification of a magma. Rock that was once melted or partially melted.

Index fossil Fossil that can be used to date the enclosing rocks.

Insect Arthropod of the class Insecta; has a three-part body, three pairs of legs attached to the middle body part, one pair of antennae, and usually wings.

Interlocking texture Texture, mainly in igneous and metamorphic rocks, in which the minerals grew together, filling all the available space.

Intrusive igneous rock Rock that solidified below the surface.

Invertebrate Animal without a backbone.

Ion Atom that has gained or lost electrons and so has an electrical charge.

Iron meteorite Meteorite composed of iron, nickel, and other metals. Also called metallic meteorite.

Island arc Curving group of volcanic islands associated with a subduction zone.

Isotope Variety of an element having a different number of neutrons in its nucleus.

Jurassic Period in the geologic time scale. It lasted from about 208 to 144 million years ago.

Lacustrine Pertaining to lakes.

Lagoon Shallow body of water with restricted connection with the ocean.

Lava Molten volcanic rock on the surface; also, the rock that solidified from it.

Limestone Sedimentary rock composed mainly of calcite.

Lithification Process whereby a sediment is turned into a sedimentary rock.

Lithosphere Crust and upper part of the mantle, above the low-velocity zone, in which the rocks behave as solids. See asthenosphere.

Lithospheric plate Tectonic plate composed of lithosphere.

Loess Wind-deposited silt.

Low-velocity zone Layer in the upper mantle where the rocks are near their melting temperature and so seismic waves are slowed.

Magma Natural hot melt composed of a mutual solution of rock-forming materials (mainly silicates) and some volatiles (mainly steam) that are held in solution by pressure. It may or may not contain suspended solids.

Magmatic (volcanic) arc Area of volcanic and intrusive rocks associated with a descending tectonic plate at a subduction zone.

Magnetic stripe One of the bands of alternating high and low magnetic intensity that parallel mid-ocean ridges.

Magnetite Mineral, an ore of iron. Fe_3O_4.

Magnitude Measure of the amount of energy released by an earthquake.

Mammal Warm-blooded vertebrate that bears live young and produces milk to feed them.

Mantle Layer of the earth between the crust and the core.

Marble Metamorphic rock; most commonly composed mainly of calcite. Formed by the metamorphism of limestone. There is also dolomite marble.

Mare Large, low, dark, smooth area on the moon (plural: maria).

Marsupial One of a subclass of mammals in which the mother's nipples are located inside a pouch, or marsupium. The relatively undeveloped young are carried in the pouch for several months after birth.

Mélange Broken and faulted rocks at a subduction zone.

Member Subdivision of a formation.

Mesozoic Era in the geologic time scale. It lasted from about 245 to 66 million years ago.

Metamorphic grade Measure of the intensity of metamorphism.

Metamorphic rock Rock that has undergone change in texture, mineralogy, or composition as a result of conditions below the depth affected by weathering processes.

Meteor "Meteorite" burning because of friction in the atmosphere. Shooting star.

Meteorite Matter that has fallen on the earth from outer space.

Methane Marsh gas. CH_4.

Mica Mineral family. Biotite and muscovite are common rock-forming micas.

Microfossil Small fossil seen only by using a microscope.

Micrometeorite Tiny meteorite that must be viewed under a microscope. See meteorite.

Mid-ocean ridge (spreading center, divergent plate boundary) Submarine ridge that is active both seismically and volcanically. New ocean floor is created at such a ridge.

Mineral Naturally occurring, crystalline, inorganic substance with a definite small range in chemical composition and physical properties.

Miocene Subdivision of the Cenozoic Era of the geologic time scale. It lasted from about 24 to 5 million years ago.

Mississippian Period in the geologic time scale. It lasted from about 360 to 320 million years ago.

Mohorovičić discontinuity (M-discontinuity, Moho) Level at which a distinct change in seismic velocity occurs. It separates the crust from the mantle.

Mold, fossil Impression left in the surrounding rock by the decay of organic material.

Monotreme One of a primitive subclass of mammals that lay eggs and that secrete milk through a number of nonunited glands rather than through true nipples. The duck-billed platypus and spiny anteater are monotremes.

Moraine Distinctive landform built by deposition of unsorted material (till) by glaciers.

Mud cracks Irregular polygonal cracks that form because of the shrinkage of mud when it dries.

Mudstone Fine-grained sedimentary rock, without fissility, that formed from the consolidation of mud.

Multiple working hypotheses Thought process requiring that as many hypotheses as possible are devised and then are tested in an attempt to arrive at the best reasoned theory.

Mummification Fossilization process that involves drying out and preservation of the soft parts.

Muscovite White mica. Recognized by its color and perfect cleavage. An important rock-forming mineral. $KAl_2(AlSi_3O_{10})(OH)_2$.

Mutation Spontaneous, inheritable change in an organism.

Mylonite Fine-grained metamorphic rock formed by the milling of rocks on fault surfaces.

Nappe Large plate of rocks moved from its place of origin by thrust faulting or recumbent folding.

Nautiloid Cephalopod in which the septa separating the chambers are either straight or have simple curves.

Nebula Cloud of gas or dust found in space.

Neutron Fundamental particle with no electrical charge; found in the nuclei of atoms.

Nonclastic sedimentary rock Chemically or biologically precipitated sedimentary rock.

Nonconformity Type of unconformity in which younger layered rocks overlie older intrusive or metamorphic rocks.

Nucleus Center of an atom; composed of protons and neutrons.

Obsidian Volcanic glass.

Ocean basin floor Uniform, generally flat, ocean floor away from plate boundaries and continents.

Oceanic crust That part of the earth above the Moho and under the oceans; composed largely of basaltic rocks.

Oligocene Subdivision of the geologic time scale. It lasted from about 37 to 24 million years ago.

Olivine Common rock-forming mineral. $(Mg,Fe)_2SiO_4$.

Omphacite Dense, green pyroxene found in the rock eclogite.

Oolite Small spherical or ellipsoidal body, generally of calcite. Also a rock composed of oolites cemented together.

Ooze Fine-grained sediment found on deep ocean floors; composed of more than 30 percent organic material.

Ophiolite Part of ocean lithosphere thrust onto a continent by a plate collision.

Ordovician Period in the geologic time scale. It lasted from about 505 to 438 million years ago.

Original horizontality Principle stating that sedimentary beds are deposited, in most cases, in horizontal layers.

Ornithiscian Dinosaur with pelvic bones similar in construction to those of birds.

Orogeny Process of making the internal structures of mountains; especially folding, faulting, and igneous and metamorphic processes.

Orthoclase Common rock-forming mineral. Potassium feldspar. $KAlSi_3O_8$.

Overturned fold Fold at least one limb of which has been rotated more than 90 degrees so that the beds are overturned.

Oxidation Process of combining with oxygen.

Paleocene Subdivision of the geologic time scale. It lasted from about 66 to 58 million years ago.

Paleoenvironment Environment of some time in the past.

Paleogeography Geography of some time in the past.

Pegmatite Very coarse-grained igneous or metamorphic rock. The crystals are larger than 1 cm in size.

Peneplain Smooth, almost flat erosion surface that develops very late in the cycle of erosion.

Pennsylvanian Period in the geologic time scale. It lasted from about 320 to 286 million years ago.

Peridotite Coarse-grained igneous rock composed mainly of olivine and pyroxene with little or no feldspar.

Period, geologic time Main subdivision of the geologic time scale.

Permian Period in the geologic time scale. It lasted from about 286 to 245 million years ago.

Permineralization Process of fossilization in which material is deposited in pore spaces.

Petrifaction Process of changing organic material into stone.

Phosphate rock Sedimentary rock containing calcium phosphate.

Photosynthesis Production of carbohydrates from water, carbon dioxide, and solar energy by chlorophyll in plants.

Phylum One of the major divisions of the plant and animal kingdoms. A group of closely related classes.

Pitchblende Ore of uranium.

Placental mammal Mammal whose young develop within the mother's body. The embryo's nourishment and waste disposal take place through means of an organ called the placenta, which is a highly selective filter between the mother's bloodstream and that of the offspring.

Placoderm Extinct primitive jawed fish.

Plagioclase One of the feldspar minerals. A continuous series from $CaAl_2Si_2O_8$ to $NaAlSi_3O_8$. Recognized by its hardness, two cleavages, and striations on one of the cleavages.

Plate One of a number of divisions of the lithosphere that move across the earth's surface.

Plate tectonics Concept that the earth's surface is composed of a number of thin slabs or plates that move across the surface.

Pleistocene Subdivision of the geologic time scale. The last glacial age.

Pleochroic halo Colored zone or halo surrounding a tiny radioactive mineral grain. The color of the halo changes when the grain is rotated under a polarizing microscope. Also radiation halo.

Pliocene Subdivision of the geologic time scale. It lasted from about 5 to 2 million years ago.

Pluton Any body of igneous rock that formed below the surface.

Polar wandering Movement of the geographic poles (as opposed to continental drift).

Precambrian Oldest subdivision of the geologic time scale. It includes all the time from the origin of the earth to about 570 million years ago.

Precession of the equinoxes Slow change from year to year of the apparent position of the stars at the time of the equinox.

Preservation Process in which some portion of once-living material is preserved in essentially unaltered condition.

Primate One of an order of mammals that includes apes, monkeys, lemurs, and humans.

Prokaryote A single-celled organism with nonnucleated cells. See also eukaryote.

Proterozoic Subdivision of the Precambrian. It lasted from about 2500 million years ago to the beginning of the Cambrian.

Proton Fundamental subatomic particle with a positive charge; found in the nucleus of an atom.

Punctuated equilibrium Theory that at times evolution is rapid, but most of the time very little change occurs.

Pyroclastic rock Rock formed of volcanic ejecta.

Pyroxene Mineral family. Augite is the common rock-forming member.

Quadrangle Topographic map produced by the United States Geological Survey.

Quartz Common rock-forming mineral. Recognized by its hardness, 7, and its lack of cleavage. SiO_2.

Quartz diorite Coarse-grained igneous rock; composed largely of quartz, plagioclase, and dark minerals.

Quartzite Metamorphic rock composed mainly of quartz. It forms from the metamorphism of sandstone or chert.

Quartz (quartzose) sandstone Sandstone composed largely of quartz.

Radiation halo Colored zone or halo surrounding a tiny radioactive mineral grain. The color of the halo changes when the grain is rotated under a polarizing microscope. See pleochroic halo.

Radioactivity Property of some elements that enables them to change spontaneously into other elements by the emission of particles from the nucleus.

Recent Most recent subdivision of the geologic time scale. The Holocene.

Red beds Sedimentary rocks colored red because of the presence of iron oxide.

Red clay Deep-sea deposit formed by fine material, mainly volcanic dust. The color is probably caused by oxidation. Also brown clay.

Reef (biologic) Rock built of organic remains, especially of coral.

Relative date Date based on superposition of beds and cross-cutting features.

Remanent magnetism Permanent magnetism induced in a rock by a magnetic field.

Reptile Vertebrate that has a dry, hardened, usually scaled skin, breathes with lungs, lays shelled eggs, and whose temperature is dependent on environment.

Reversed magnetism Magnetism with polarity opposite to that of the present field.

Rhyolite Fine-grained igneous rock similar in composition to granite.

Rifting In plate tectonics, a place where a plate is broken apart.

Rift valley Valley formed along a rift or lateral fault or at the crest of a mid-ocean ridge.

Ripple-mark Undulating surface on a sediment; caused by water movement.

Rock Natural aggregate of one or more minerals, or any essential part of the earth (or any other part of the solar system).

Rock cycle Sequence through which rocks may pass when subjected to geological processes.

Salt dome Structure caused by the upward movement of rock salt through overlying sedimentary rocks.

Sand Clastic sediment with fragments that range in size from $1/16$ to 2 mm.

Sandstone Clastic sedimentary rock with fragments between $1/16$ and 2 mm in size.

Saurischian Dinosaur with reptilelike pelvic bones.

Scarp Steep slope or cliff.

Schist Foliated metamorphic rock, generally containing conspicuous mica.

Sea-floor spreading Concept that new ocean floor is created at mid-ocean ridges (spreading centers) and moves slowly toward the volcanic arcs, where it is consumed.

Sedimentary rock Rock formed near the earth's surface, generally in layers.

Sedimentation phase Time of accumulation of sedimentary rocks in the history of a geosyncline.

Series Subdivision of a system.

Serpentine Mineral family. $Mg_3Si_2O_5(OH)_4$.

Serpentinite Rock composed largely of serpentine.

Shale Fine-grained sedimentary rock composed largely of clay and silt and characterized by its fissility.

Sharpstone conglomerate Conglomerate composed of angular pebbles. See also breccia.

Shelf, continental See continental shelf.

Shield Continental area that has been relatively stable for a long time. Shields are composed mainly of ancient Precambrian rocks.

Sill Concordant igneous intrusive body.

Silt Clastic sediment with particles that range in size between $1/256$ and $1/16$ mm.

Siltstone Clastic sedimentary rock composed largely of silt.

Silurian Period in the geologic time scale. It lasted from about 438 to 408 million years ago.

Sink Depression out of which no surface stream flows.

Slate Fine-grained metamorphic rock characterized by well-developed foliation.

Slope, continental See continental slope.

Slump features (sedimentary rocks) Features such as small folds that form by downslope movement of soft sediments.

Soapstone Metamorphic rock composed of talc.

Sodic plagioclase Sodium-rich end member of the plagioclase feldspar series. $NaAlSi_3O_8$.

Solar system The sun and planets that orbit it, including the satellites of the planets, comets, meteors, and asteroids.

Sorting Measure of the range in size of the fragments in a clastic sediment or sedimentary rock. If the fragments are all similar in size, the sediment is well sorted.

Species Group of similar organisms that can interbreed to produce fertile offspring.

Spreading center (mid-ocean ridge, divergent plate boundary) Submarine ridge that is active both seismically and volcanically. New ocean floor is created at such a place.

Stage Subdivision of a series.

Stony-iron meteorite Meteorite composed of both silicate minerals and metals such as iron and nickel.

Stony meteorite Meteorite composed of silicate minerals.

Stratigraphic correlation See correlation.

Subduction Descent of a plate, generally an oceanic plate, into the asthenosphere.

Supernova Very luminous exploding star.

Superposition Principle stating that in undisturbed sedimentary beds the oldest is on the bottom and the youngest is on the top.

Syncline Fold in which the beds slope inward toward a common axis. The youngest rocks are in the core.

System Rocks deposited during a geologic period.

Tachylite Volcanic glass of basaltic composition.

Tectonic Pertaining to deformation of the earth's crust, especially the rock structure and surface forms that result.

Terrane A portion of the crust with distinctive rock types and/or structures.

Tertiary Period in the geologic time scale. It lasted from about 66 to 2 million years ago.

Texture Size and arrangement of grains in a rock.

Thecodont One of an extinct group of reptiles. Ancestor of the dinosaurs, as well as of several other groups of reptiles.

Thrust fault Reverse fault with a fairly flat fault plane.

Tide Periodic rise and fall of the ocean; caused by the gravitational attraction of the moon and, to a lesser extent, the sun.

Time line Horizon in strata that formed at the same time over a wide area; for example, a layer of volcanic ash.

Titanothere One of an extinct group of hoofed Cenozoic animals. Some of the later titanotheres were large.

Transform fault Fault caused where two tectonic plates move past each other. It offsets a mid-ocean ridge but the actual movement is in the opposite direction to the apparent offset of the ridge.

Trench Very deep, elongate area of the ocean; associated with volcanic arcs.

Triassic Period of the geologic time scale. It lasted from about 245 to 208 million years ago.

Trilobite Extinct arthropod that lived in the Paleozoic and was most abundant in the Early Paleozoic. Characterized by segmented bodies with longitudinal grooves that divide the body into three segments.

Tuff Pyroclastic volcanic rock with fragments smaller than 2 mm. Consolidated ash.

Turbidite Sedimentary rocks deposited by turbidity currents.

Turbidity current Current or flow caused by the movement downslope of water of high density because of its suspended material (muddy water).

Type area Place at which a formation or system is well exposed and is defined.

Type specimen Typical specimen or specimens used to define a fossil species.

Unconformity Surface of erosion and/or nondeposition. A gap in the local geologic record.

Uniformitarianism Concept that ancient rocks can be understood in terms of the processes presently occurring on the earth. Also called actualism.

Varved clay Fine, thinly bedded glacial sediments deposited in still water. Each pair of dark- and light-colored beds is assumed to be a yearly deposit.

Vertebrate Animal with a backbone.

Volcanic (magmatic) arc Area of volcanic and intrusive igneous rocks associated with a descending tectonic plate at a subduction zone.

Volcanic island arc Volcanic arc where the volcanoes form islands.

Welded tuff Tuff that was erupted when hot and whose fragments have fused together as a result of the action of heat and the gases present.

Index